Syntactic Wordclass Tagging

Text, Speech and Language Technology

VOLUME 9

Syntactic Wordclass Tagging

edited by

Hans van Halteren
University of Nijmegen

KLUWER ACADEMIC PUBLISHERS
DORDRECHT / BOSTON / LONDON

A C.I.P. Catalogue record for this book is available from the Library of Congress.

ISBN 0-7923-5896-1

Published by Kluwer Academic Publishers,
P.O. Box 17, 3300 AA Dordrecht, The Netherlands.

Sold and distributed in North, Central and South America
by Kluwer Academic Publishers,
101 Philip Drive, Norwell, MA 02061, U.S.A.

In all other countries, sold and distributed
by Kluwer Academic Publishers,
P.O. Box 322, 3300 AH Dordrecht, The Netherlands.

Printed on acid-free paper

Printed in the Netherlands

Contents

PREFACE

In both the linguistic and the language engineering community, the creation and use of annotated text collections (or *annotated corpora*) is currently a hot topic. Annotated texts are of interest for research as well as for the development of *natural language processing* (NLP) applications. Unfortunately, the annotation of text material, especially more interesting linguistic annotation, is as yet a difficult task and can entail a substantial amount of human involvement. All over the world, work is being done to replace as much as possible of this human effort by computer processing. At the frontier of what can already be done (mostly) automatically we find *syntactic wordclass tagging*, the annotation of the individual words in a text with an indication of their *morphosyntactic* classification. This book describes the state of the art in syntactic wordclass tagging.

As an attempt to give an overall view of the field, this book is of interest to (at least) two, possibly very different, types of reader. The first type consists of those people who are using, or are planning to use, tagged material and taggers. They will want to know what the possibilities and impossibilities of tagging are, but are not necessarily interested in the internal working of automatic taggers. This, on the other hand, is the main interest of our second type of reader, the builders of automatic taggers and other natural language processing software.

As a consequence of having two types of reader, the book has been split into two parts: Part I examines the field from the point of view of the user and Part II from that of the implementer. In this way each type of reader knows what they have to read to fulfil their needs. It would, of course, have been possible to produce two books, one for each type of reader, but we have decided against this. We think that each reader should be aware of what is going on on the other side of the fence and hence we would like to encourage everyone to read the (to him or her) less interesting part as well. The users will then learn why automatic taggers behave as they do and as a result may possibly be able to make better use of tagged material. The software builders will get a better view of how their products are used and as a result may get ideas leading to better taggers.

To conclude this preface, I want to draw your attention to the authors. As you will see, the book is made up out of chapters and sections on subfields, each written by an expert on that specific subfield. However, the book as a whole is meant to be more than just a collection of papers. All contributors have done their best to write an integral part of the book rather than an individual paper.

Furthermore, the contributors have sometimes worked together, giving each other suggestions for changes and extensions. There have even been a number of instances of one or more paragraphs from one chapter being moved to another chapter, by another contributor. As a result, attribution of chapters and sections can never be provided fully accurately. I have chosen to list the main author for each chapter/section, the one who has been primarily responsible, who has done the lion's share of the work and who consequently deserves most of the credit, but it should be noted that in reality credit should be spread out much more evenly than is suggested. In addition, credit is due to the anonymous reviewers for their many useful suggestions.

HANS VAN HALTEREN

CONTRIBUTING AUTHORS

Eric Brill
Department of Computer Science, John Hopkins University, Baltimore MD.
brill@cs.jhu.edu

Nicoletta Calzolari
ILC-CNR, Pisa.
glottolo@ilc.pi.cnr.it

Jan Cloeren
TOSCA Research Group, University of Nijmegen.
Now at Ernst Klett Verlag GmbH, Stuttgart.
j.cloeren@klett-mail.de

Walter Daelemans
ILK, Tilburg University.
Walter.Daelemans@kub.nl

Marc El-Beze
Laboratoire d'Informatique, Avignon.
marc.elbeze@lia.univ-avignon.fr

Gregory Grefenstette
Xerox Research Centre Europe, Grenoble.
Gregory.Grefenstette@xerox.fr

Hans van Halteren
TOSCA Research Group, University of Nijmegen.
hvh@let.kun.nl

Lauri Karttunen
Xerox Research Centre Europe, Grenoble.
Lauri.Karttunen@xrce.xerox.com

Geoffrey Leech
UCREL, Lancaster University.
G.Leech@lancaster.ac.uk

Bernard Merialdo
Institut EURECOM, Valbonne.
merialdo@eurecom.fr

Monica Monachini
ILC-CNR, Pisa.
monica@ilc.pi.cnr.it

Kemal Oflazer
Dept. of Computer Engineering, Bilkent University.
ko@cs.bilkent.edu.tr

Anne Schiller
Xerox Research Centre Europe, Grenoble.
Anne.Schiller@xrce.xerox.com

Nicholas Smith
UCREL, Lancaster University.
n.i.smith@lancaster.ac.uk

Atro Voutilainen
Department of General Linguistics, University of Helsinki.
Atro.Voutilainen@ling.helsinki.fi

Andrew Wilson
Linguistics Department, Lancaster University.
Now at Englische Sprachwissenschaft, Technische Universität Chemnitz.
Andrew.Wilson@phil.tu-chemnitz.de

I The User's View

1 ORIENTATION

Atro Voutilainen

1.1 MORPHOSYNTACTIC TAGS

Tags are descriptive symbols that are assigned to (simple or complex) words in text, by hand or by machine. Often these tags are called *part-of-speech* tags, since they typically indicate the part of speech (noun, verb, preposition, etc.) of the word. Autonomous linguistics traditionally recognizes five main levels of analysis: *phonology*, *morphology*, *syntax*, *semantics* and *pragmatics*. The status of parts-of-speech is not obvious in this classification: does parts-of-speech constitute a level of its own; is it included in one or more of the other levels; it there even need to postulate such a linguistic phenomenon as parts-of-speech?

In terms of past linguistic descriptions, parts-of-speech have certainly had a long life in every major linguistic tradition. In the Indian tradition, Pāṇini (c. 350 B.C.) stipulated that "a word is that which has nominal or verbal endings"; for simplicity, indeclinable words were decreed to have nominal endings which were automatically deleted afterwards in the course of derivations. Beginning with Sībawaihi (d. 793), the Arab tradition distinguished between three wordclasses, i.e. nouns (including adjectives, pronouns and active or passive participles), verbs and particles. In the oldest extant morphological analysis of the Western tradition, Dionysios Thrax (c. 100 B.C.) distinguished between eight wordclasses, using mainly formal criteria: noun, verb, participle, article (including the relative pronoun), pronoun, preposition, adverb, conjunction. Marcus Terentius Varro (116-27 B.C.) established a four-way classification

H. van Halteren (ed.), Syntactic Wordclass Tagging, 3-7.

of the wordclasses based on the presence vs. absence of case and/or tense: noun = +c, -t; verb = -c, +t; participle = +c, +t; adverb = -c, -t.

According to Schachter (1985), every natural language seems to have some kind of part-of-speech system, though there is certainly considerable variation in different languages. The best criteria for parts-of-speech seem to be grammatical rather than semantic (see e.g. the discussion in Fries 1952). Schachter (op.cit.) gives three main grammatical criteria for identifying parts-of-speech in different languages: (i) syntactic distribution, (ii) syntactic function and (iii) the morphological and syntactic classes that different parts-of-speech can be assigned to. These criteria suggest that parts-of-speech mainly belong to the domain of syntax (cf. the distributional and functional criteria) though in the case of languages with a morphology, the lower level also has a role (criterion (iii)).

Whatever the linguistic status of parts-of-speech, large amounts of part-of-speech tagged electronic text are presently available. For instance, consider the following sample, taken from the beginning of a well-known tagged text corpus, the tagged version of the Brown University Corpus of Present-Day American English (Kučera and Francis 1967):

```
the/AT Fulton/NP County/NP Grand/NP Jury/NP
said/VBD Friday/NR an/AT investigation/NN
of/IN Atlanta's/NP$ recent/JJ primary/NN
election/NN produced/VBD no/AT evidence/NN
that/CS any/DTI irregularities/NNS took/VBD
place/NN ./.
```

Here the words in the sentence are followed by a tag; the slash "/" separates the word from the tag. The tag AT indicates that the word belongs to the grammatical category of articles; NP is reserved for proper nouns; VBD for verbs (in the past tense); IN for prepositions; and so on.[1]

If we look at such samples of annotated English corpora, we will find that the *tagsets* (also known as *tag palettes*) in fact contain much more tags than the ten-odd that could be expected on the basis of traditional part-of-speech descriptions of English (cf. Chapter 4). For instance, consider annotations for various kinds of nouns as given in the Lancaster-Oslo/Bergen (LOB) corpus (Garside *et al.* (eds.) 1987; Johansson 1986). We find e.g. the following tags: NN, NNS, NN$, NN'$, NP, NPS, NR. The tag NN represents only common nouns in the nominative singular, NNS represents the nominative plural forms of common nouns, NP represents proper nouns in the nominative singular, NR represents a small group of nouns whose distribution and function overlaps with that of adverbs, and so on.

[1] For a full list of Brown Corpus tags, see Appendix 17.6.

Obviously, the tags used in annotated corpora contain not only information about parts-of-speech, but also about *inflectional* and even *lexico-semantic* properties. The tags used in annotated corpora like the Brown University Corpus (Francis and Kučera 1982), LOB (Garside *et al.* (eds.) 1986), British National Corpus (Leech *et al.* 1994) and University of Pennsylvania Treebank (Marcus *et al.* 1993) are designed to represent several kinds of information at once and their number ranges from some dozens to over a hundred. Tagsets have also been designed for other languages and occasionally the number of tags is rather large, even a few hundreds in languages with a rich morphology. Clearly, calling these tags 'part-of-speech tags' is short of the mark; a more adequate name would be *morphosyntactic tag*.

In most annotated corpora, each word is given only a single information-rich tag. However, it is also possible to represent this information more transparently by using several tags that together convey the same information, e.g. the information represented by the tag NN could also be given as a list of three tags, N NOM SG (noun in the nominative singular), or as a feature structure, such as [part-of-speech=noun, case=nominative,number=singular]. This latter representation is probably an economical alternative when describing languages where the combinatorics in the inflectional system would in the unitag system require hundreds or thousands of tags instead of tag strings based on an inventory of a few dozen property-specific tags (see also 4.4).

Generally, the choice of the appropriate tag for a word depends not only on the word itself; the context is important as well. This means that a word by itself is often *ambiguous*: without a linguistic context, there is no way of knowing which of the alternative tags should be assigned. For instance, the English word "round" could be a preposition, adverb, noun, verb or adjective. Usually it is only in a linguistic context that "round" can be unambiguously analysed, e.g. in the sentence "It came round the corner", the word "round" is analysed as a preposition. What makes designing accurate taggers a difficult problem is mainly the question of how to model the linguistic context of homographs like "round" so fully and accurately that the contextually correct analysis (or, in the case of genuine grammatical ambiguity, analyses) can be predicted automatically.

Presently there is considerable interest in tags and tagging (roughly: automatic assignment of contextually appropriate tags to words in running text) and there are many different tagsets available or under construction. There has also been work on advocating standards for tagsets; witness recent *EAGLES* (Expert Advice Group for Language Engineering Standards) work on morphosyntactic tagsets (cf. Chapter 5). Increasing the commensurability and translatability between tagsets will probably make taggers and tagged texts more useful still, which seems to be motivation enough for promoting standards.

1.2 AUTOMATIC TAGGING

Tagged text is useful for many purposes, e.g. linguistic studies of text corpora, speech processing applications, as a first step in proper computer parsing of text, automatic recognition of terms and translation units and so on (cf. Chapter 3). Texts can be annotated by hand, but ensuring a high annotation quality for large amounts of text in this way is expensive (cf. Chapter 7). There is an obvious need for reliable automatic taggers.

The design of reliable automatic taggers is not an easy problem. A considerable amount of effort has been spent recently on making taggers, especially during the 1980s and 1990s. Presently there are at least a few dozen taggers for English alone. These systems typically share the following characteristics:

- A *tokenizer* prepares the text for further analysis, e.g. by identifying words, punctuation marks and other utterance boundaries (cf. Chapter 9).
- A *lookup module* assigns all possible morphosyntactic analyses for each word in the input utterance as alternatives (cf. Chapters 10 to 13). The lookup module typically uses (i) a *lexicon* representing word stems and affixes and (ii) a *guessing module* for analysing those words not represented in the lexicon. The lookup module considers one word at a time, i.e. at this stage, linguistic context has no role. Either integrated into or separate from the lookup module, there may be a module for analysing multi-word expressions, especially those idioms that, from the point of view of grammar, act as single words.
- Often many words get more than one analysis as alternatives, i.e. ambiguities arise (e.g. the English word "look" can be used as a noun and as a verb). For choosing the correct alternative(s), there is a *disambiguation module* that uses contextual, and often also statistical, information as the basis of its predictions (cf. Chapters 14 to 17).

Optimally, these three modules produce correctly disambiguated text that no longer contains superfluous analyses. Realistically, one must accept that a certain percentage of the text is either not tagged correctly or has been left partially ambiguous (cf. Chapter 6).

Tagging has been a hot research topic since the early 1980's. Why is this the case? As we already noted, there are practical needs for large amounts of tagged text. Other reasons may be the following: (i) making a tagger is easier (i.e. less frustrating) than another eagerly-studied problem in Computational Linguistics, namely designing a full-fledged syntactic parser, (ii) the necessary computing resources are easily available and (iii) both tagged and untagged electronic text is available in large quantities, thus enabling the creation and testing of the knowledge bases of the taggers.

Furthermore, tagging in itself has turned out to be a methodologically interesting issue. Analysis of words that are not in the lexicon is a non-trivial problem, as is the implementation problem of making the tagger fast and space-effective. However, the most difficult problem in the design of taggers is the question about the kind of information needed for correct disambiguation (choosing the correct alternative tag). This, then, will also fill most of Part II of this book and we will see topics addressed like:

- the kind of information that is is needed, e.g. statistics or rules;
- the representation of this information, e.g. in the form of rules, collocational matrices, hidden Markov models or neural network;
- the acquisition of the necessary information, e.g. on the basis of observations by linguists, automatic learning algorithms or combinations thereof.

2 A SHORT HISTORY OF TAGGING

Atro Voutilainen

2.1 APPROACHES TO WORDCLASS TAGGING

Outlining the development of even such a young field as the design of automatic wordclass taggers is not an easy task.[1] Different systems and approaches could be categorized in several ways, i.e. the classification given below is by no means the only known one, probably not even the most popular one. In this chapter, we will make a main distinction between the *linguistic approach*, based on hand-written rules, and the automatic *data-driven approach.*

Characteristic of the linguistic approach is that thc knowledge base used by the tagger is written by an expert grammarian in the form of rules. The rules are based on generalizations about the language that may be based on the grammarian's knowledge of the language as well as on observations of large text samples, descriptive grammars and dictionaries.

In the data-driven approach, the relevant language model is derived from automatically conducted statistical studies of large text samples. Typically the *training corpora* have been manually annotated using tags defined by linguists. The statistically derived language model may be represented in different ways, e.g. as collocational matrices, Markov models, local rules or neural networks. Usually the data-driven language

[1] The author wishes to thank Fred Karlsson, Pasi Tapanainen and especially Hans van Halteren for useful comments on earlier drafts of this chapter.

H. van Halteren (ed.), Syntactic Wordclass Tagging, 9-21.

model concerns very short word sequences and their frequencies. On the basis of the corpus-based local statistical language model, the tagger selects from the alternatives the one with the highest probability.[2]

Though we make a binary division of tagging approaches into the data-driven and the linguistic, it should also be noted that usually different systems are not methodologically as 'pure' as the dichotomy might suggest. The data-driven approach presupposes considerable linguistic expertise, e.g. in the design of tagsets (cf. Cutting 1994), lexicons, biases (hand-coded preferences; cf. Chanod and Tapanainen 1995a) and idiom lists (cf. Garside *et al.* (eds.) 1987). Another linguistic resource needed by the best data-driven taggers is an annotated corpus, the creation of which requires considerable linguistic effort. On the other hand, linguistic systems generally need not only rules that are believed to be 'linguistic' in the sense that the linguist believes them to be absolutely true of the language; a considerable part of disambiguation is usually also carried out with heuristic rules that are empirically known to be effective disambiguators, but not quite true. A part of the grammarian's workmanship is the ability to make generalizations about tendencies rather than about 'absolute' facts alone and these tendencies (expressed as heuristic rules) may have a natural counterpart in corpus probabilities.

The following exposition will often refer to various performance figures based on metrics like 'accuracy', 'recall' and 'precision'. One should note that these figures only convey a rough idea about the performance of the systems; they are not strictly comparable e.g. because of differences in languages, tagsets, principles of applying them and test corpora (cf. Chapter 6).

2.2 PIONEERING WORK

The first automatic taggers from the late 1950s and 1960s were based on hand-written disambiguation rules. These systems used small lexicons that gave all possible analyses to some of the input words. For the analysis of words not represented in the lexicons, heuristic rules were used that relied on affix-like letter sequences at word-boundaries, capitalization and other graphemic clues about word category. Those words not analysed by the pattern rules were given several open-class analyses as alternatives (noun, verb, adjective readings).

Ambiguity resolution was carried out with reductionistic linguistic rules that eliminate alternative analyses on the basis of the local context (e.g. two words to the left and to the right of the ambiguous word). For instance, a rule might discard a verb reading from an ambiguous word if the preceding word is an unambiguous article. These rules were written by linguists and evaluated against text samples. The tags that remained intact after the application of the rules were optimally the correct analyses of

[2] Optionally, a tagger can also select all analyses above a given probability threshold; probabilistic taggers of this type are generally called *n-best taggers*.

the words. In practice, some words retained more than one tag; the output was further disambiguated and corrected by human posteditors.

The first system the author is aware of is a finite-state parser created at the University of Pennsylvania in the late 1950s (cf. Joshi and Hopely 1997)[3]. One of the components in this partial parser was a tagging part where lexical ambiguities were resolved with a small hand-coded finite-state disambiguator using a grammar of 14 ordered context rules. Another early system was created by Klein and Simmons (1963). It used 30 tags and it was reported to reach an accuracy of 90% (i.e. with unambiguously tagged output, 90% of all words received the correct tag). Unfortunately, the tagger was tested only against small amounts of text from very few genres, so the performance of the system in a more realistic situation remains unknown.

A large-scale system, modelled on the Klein and Simmons tagger, was implemented in connection with the linguistic annotation of the Brown University Corpus that contains some 1.1 million words of American English representing 15 text genres (Kučera and Francis 1967). The tagger, called TAGGIT (Greene and Rubin 1971), used a larger tagset (71 word tags and separate tags for punctuation marks) than its predecessor, as well as a somewhat larger lexicon. The TAGGIT disambiguation grammar contained 3,300 rules that, depending on rule type, selected or discarded a tag on the basis of the local context (a five-word window). Together, these rules made about 77% of all words in the Brown corpus unambiguous; the remaining 23% of words remained to be resolved by human posteditors. The error rate of TAGGIT (the portion of words without a correct tag) was not precisely reported. According to Francis and Kučera (1982), the tagged Brown corpus needed several rounds of proofreading for correcting tagging errors.

2.3 THE BREAKTHROUGH OF DATA-DRIVEN METHODS

The next major advance in tagging took place in the late 1970's. The first effort in this new 'wave' of taggers was carried out as part of the annotation of the British English counterpart of the Brown corpus, the Lancaster-Oslo/Bergen Corpus (Johansson 1986). For annotating this corpus, a system called CLAWS1 (Constituent-Likelihood Automatic Word Tagging System, version 1) was developed by the UCREL team at the University of Lancaster (Marshall 1983; Garside *et al.* (eds.) 1987) and by Norwegian collaborators from the universities of Oslo and Bergen.[4]

[3] Thanks to Lauri Karttunen for pointing out this reference.

[4] By now, the CLAWS system has reached version 4; cf. Garside and Smith (1997).

2.3.1 N-gram taggers

As its name suggests, CLAWS1 was based on radically different design principles than its linguistic predecessors: disambiguation in CLAWS is based on choosing the correct word category on the basis of statistical corpus evidence. Use of linguistic rules was mostly abandoned (with the exception of the hand-coded word pattern rules in the idiom component).

CLAWS1 consists of five main modules. The first module is a semiautomatic pre-editor that prepares the text for later stages, e.g. by identifying words and punctuation marks. The next phase is called word-tagging. This means assigning each wordform with all possible tags as alternatives using a 7,000 word lexicon and a heuristic guesser. Already during lexical lookup, the alternative tags are ranked with regard to likelihood; the least likely alternatives are marked with a rarity symbol that can be used by the disambiguation phase. The lexicon represents most closed-class words, common abbreviations, names and some other open-class words. The tagset was a redevelopment of the Brown corpus tagset and totalled 139 tags.

Before the probabilistic disambiguator is applied, a hand-crafted idiom recognition module is used for identifying certain multi-word expressions behaving as a single grammatical unit (e.g. complex prepositions). This module is useful especially in those cases where the probabilistic component would mispredict.

The fourth main component is called CHAINPROBS. It chooses the most likely tag (or tags) of an ambiguous word on the basis of the likelihood of occurrence in a given local context. The information basis of CHAINPROBS is a matrix of collocational probabilities derived from some 200,000 words of text from the Brown corpus annotated according to the CLAWS1 tagset. The matrix represents bigrams, i.e. sequences of two words or tags. From a matrix of this kind, two kinds of probabilities can be calculated: lexical (the likelihood of word X representing category A) and contextual (the likelihood of category B occurring directly adjacent to category A) A probabilistic disambiguator should determine an optimal balance between lexical and contextual probabilities. The formula used in the final version of CLAWS1 is (Marshall 1987):

$$\frac{Frequency\ of\ the\ tag\ sequence\ A - B}{Frequency\ of\ tag\ A * Frequency\ of\ tag\ B}$$

The last module, postediting, represents a phase where the remaining disambiguation as well as error correction is carried out, mainly by human annotators.

Compared to its predecessors, CLAWS1 is surprisingly accurate: 96-97% of all input words get the correct tag when the tagger is forced to produce only unambiguous analyses (Garside *et al.* (eds.) (1987), Johansson (1986) and Voutilainen *et al.* (1992), report similar figures for CLAWS1). CLAWS1 can also be used as an n-best tagger: if 14% of words are allowed to remain ambiguous, 99% of all words retain the correct tag as one of the surviving analyses (Johansson 1986).

In the wake of CLAWS1, many other N-gram taggers have been developed. Well-known systems close to CLAWS in design principles are Parts of Speech by Church (1988) and Volsunga by DeRose (1988). These systems use larger lexicons (witness parallel advances in computer technology) and they are also much faster than CLAWS1, but not more accurate: Parts of Speech reaches 95-97% (Marcus *et al.* 1993), while Volsunga's accuracy is about 96% (DeRose 1988). Further bigram or trigram taggers have been proposed by Weischedel *et al.* for English (1993), by Åström for Swedish (1995) and by Nagata for Japanese (1994).

2.3.2 Data-driven local rules

Another succesful data-driven approach to tagging in the late 1980's is the derivation of local disambiguation rules automatically from tagged texts. Hindle's algorithm (1989) seems to be the first of its kind successfully applied for large-scale analysis. The rules in Hindle's grammar are local and ordered with regard to reliability and they can refer to words and tags; in this sense they resemble rules in TAGGIT. However, there are also default rules that apply to any pending ambiguities; as a consequence, the system produces fully disambiguated text. To illustrate, the rule

```
[PREP+TNS]=TNS[N+V]
```

applies to words ambiguous due to a preposition and tense marker reading (actually the word "to"). This rule selects the tense marker reading whenever the word is followed by a noun–verb ambiguous word. Obviously this rule does not always produce correct analyses; however, some of the mispredictions can be avoided if another rule with a higher ranking has been applied earlier.

In Hindle's model, corpus-based grammar learning is based on the detection of a disambiguation error. At the outset there may be a number of hand-written or even arbitrary rules. Whenever the system using these rules makes a prediction that disagrees with the manually annotated training corpus, the internal state of the underlying Fidditch syntactic parser as well as the current input configuration is recorded and two things can happen: (i) the mispredicting rule is ranked lower in terms of precedence in the grammar, or (ii) a new more specific rule is learned that analyses the ambiguous word in question as analysed in the corpus whenever the context is the same as the word/tag pattern in the corpus. The new rules gain or lose in application precedence as they are applied to further text in the training corpus. If a new rule makes too many errors, it is deactivated altogether.

After several iterations over 90% of the tagged Brown corpus, Fidditch increased its rule base from the initial hand-crafted 350 rules to 35,000 ordered rules. The high number of rules is partly due to the fact that Hindle's rules can refer to not only tags but also to some 300 common wordforms in order to employ lexical information. When tested against a fresh sample of the Brown corpus (10% of the whole corpus), the Fidditch tagger reached an accuracy of 97%.[5]

2.4 RECENT WORK IN THE DATA-DRIVEN APPROACH

2.4.1 Hidden Markov Models

The CLAWS-type probabilistic method implements a kind of 'open' Markov model (cf. Cutting *et al.* 1992) where the transitions between grammatical categories and their frequencies are given explicitly. Another popular probabilistic tagging method is based on Hidden Markov Models (HMM; cf. Chapter 16), 'hidden' because here state transitions are unobservable (Cutting *et al.* 1992). An advantage of HMM taggers is that only a lexicon and some untagged text is needed for training a tagger, i.e. no laborious tagging of large corpora would be needed.

The goal is to automatically make as simple a language model as possible on the basis of language observations. The basic idea of learning from ambiguous data is actually quite straightforward. Consider the following ambiguous sample from English:

```
the        ART
man        N V
is         V
an         ART
agent      N
.
```

In this sentence there is one ambiguous word: "man" is ambiguous due to N and V. Because we have no lexical or contextual probabilities in this ambiguous corpus, choosing between the alternative analyses of "man" may seem problematic. However, on examining a larger sample of morphologically analysed ambiguous text it becomes evident that in certain configurations verb tags (V) occur only rarely, while noun (N) tags are found quite often. For instance, consider "agent" in the same sample sentence: the lexicon gives only a noun reading for "agent", which, like "man", happens to follow a word labelled as an article (ART). In this corpus, accepting a noun analysis next to an article analysis will not add to the complexity of the model because a noun occurs after an article in any case (e.g. in "the man"). In contrast, proposing a verb reading after an article will make the Markov model more complex than the data actually warrants. The simpler model is therefore preferred, so the noun analysis of "man" will be selected as the more likely one.

[5] Hindle notes in passing that these automatically acquired rules are about 50% more reliable than earlier manually written disambiguation rules used by Fidditch.

As a consequence of one round of training an HMM on an untagged corpus, this 'training corpus' itself becomes less ambiguous, which enables making further generalizations. In this way, the model becomes more accurate, up to a certain number of learning iterations.

In practice, some tag transition probabilities are assigned to the initial HMM. The training cycles refine this initial setting using the Baum-Welch re-estimation algorithm (Baum 1972). An interesting property of HMM taggers is that they operate on full sentences rather than on fixed-length word sequences. In principle, HMM taggers can employ long-distance information. In practice, however, the size of the contextual 'window' is often limited to two or three words. Another attractive feature of HMM taggers is that linguistic information can be incorporated to some extent in the tagger, coded as biases (by manipulating the lexicon, the tagset and the initial tag probabilities).

A problem common to all data-driven taggers is the so-called *sparse data problem*: most wordforms in any corpus of natural text occur with such a low frequency that adequate statistics cannot be calculated for them individually. A solution for this problem was proposed by Kupiec (1989) who introduced the use of equivalence (or ambiguity) classes in HMM tagging: pool middle- or low-frequency words into classes that are ambiguous due to the same categories (e.g. the class of N/V ambiguous words); count statistics for the class (rather than for the words separately); then apply these statistics in the analysis of the class members.

HMM taggers have been implemented for many languages, e.g. English (Cutting *et al.* 1992), Swedish (Cutting 1994), Chinese (Chang and Chen 1993), Spanish, French (Chanod and Tapanainen 1995a) and German (Feldweg 1995). If these taggers are trained on tagged corpora, their accuracy is very much the same as that already reached by the first N-gram taggers: around 95-97%. Unsupervised training tends to decrease the accuracy somewhat. HMM taggers can be implemented very efficiently: some recent implementations analyse over 10,000 words per second on a Unix workstation (Gregory Grefenstette, personal communication).

Overall, the attraction of HMM taggers seems to be that implementing one for any language requires minimal human effort. However, the initial optimism about dispensing with tagged training corpora has been questioned recently: e.g. empirical experiments by Merialdo (1994) and Elworthy (1994) suggest that even a small amount of tagged training corpus is a better source for training an HMM than a large amount of untagged text. Furthermore, work on a French HMM tagger by Chanod and Tapanainen (1995a) suggests that using biases effectively in an HMM tagger actually requires a considerable degree of expertise and experimentation, thus making linguistic craftmanship more essential than was initially obvious.

2.4.2 Recent work on data-driven local rules

Brill (1992; see also Chapter 15) has carried on the data-driven rule-based approach to tagging, originated by Hindle. Unlike Hindle's algorithm, Brill's system uses no higher-level syntax in rule acquisition. The number of rules is much smaller than Hindle's (ranging in different versions from some 70 to over 250 rules, the accuracy ranging from some 95% to 97%; cf. Brill 1994). As in Hindle's system, Brill's rules are local and ordered and they can refer to words and tags.

Here also, rule learning is error-driven. The initial-state tagger produces single lexical analyses for words, based on lexical probabilities derived from tagged corpora. The tagger's predictions are compared to the tagged training corpus and error statistics are generated. A number of predefined local rule templates are used for creating and testing new rules for rewriting the initial-state tagger's analyses into correct ones and the template with the best correction rate is added to the ordered grammar.

Brill (1994) has experimented with learning rules from ambiguous corpora (i.e. where only the lexical analyser but no disambiguation has been applied). The basic idea is much like that used in the construction of an HMM from ambiguous corpora: the most common tags are preferred in unambiguous contexts. For example, following the unambiguous article "the", a noun analysis is preferred as the more common alternative over a verb analysis. Brill (1995) reports that the best overall accuracies (about 97%) seem achievable using both ambiguous and unambiguous learning corpora as the training corpus.

Local rule-based taggers can be implemented very efficiently. Roche and Schabes (1995) report a finite-state implementation that analyses over 10,000 words per second on a Unix workstation; this is comparable to the speed of recent HMM taggers.

2.4.3 Neural taggers

Wordclass tagging has also been investigated in the connectionist paradigm (cf. Chapter 17). Some interesting research has been carried out by Schütze (e.g. 1993) and Schmid (1994a). As an illustration, we outline Schmid's method.

Schmid (1994a) presents a neural network tagger based on multilayer perceptron networks. An artificial neural network consist of simple units (each associated with an activation value) and directed links for passing the values between the units. A multilayer network consists of an input layer and an output layer (and optionally of hidden intermediary layers). Activation values are propagated from input to output layers. At each unit, the input activation values are summed and a bias parameter is added. The network learns by adapting the weights of the connections between units until the correct output is produced.

Schmid's tagger uses a lexicon which gives each word all possible tags as alternatives, each tag associated with a lexical probability (the probability of tag A occurring with word X in a tagged corpus). In the output layer, all units have a value of zero except the correct unit (tag), which gets the value of one. Training on a tagged corpus means adjusting the connections between the input and output units so that the connections to those output units most closely corresponding to the correct tags in the training corpus propagate the highest values.

In a system of this kind, tagging a word means (i) copying the tag probabilities of the word and its neighbours into the input units and (ii) propagating the activations to the output units. The tag corresponding to the output unit with the highest value is then associated to the word. Schmid compares his Net-Tagger to a trigram tagger and an HMM tagger: both Net-Tagger and the trigram tagger reach an accuracy of slightly above 96%; the HMM tagger's accuracy was a little above 94%.

2.4.4 Case-based taggers

Case-based reasoning, a brand of similarity-based reasoning, is a method mainly known for its applications to e.g. grapheme-to-speech conversion and lexical acquisition. Recently, there has been work on tagging within this framework (see Chapter 17). According to Daelemans *et al.* (1996), a case-based tagger employs a data-driven learning technique that uses local context (ambiguous or unambiguous) and the characteristics of the word to be disambiguated. The learning component builds a case-base and the closest-matching cases are then applied for resolving new ambiguities. Case-based taggers have been made for Dutch and English and their accuracy is similar to that of state-of-the-art HMM taggers.

2.4.5 Combined data-driven taggers

Van Halteren, Zavrel and Daelemans (1998) experiment with the idea that idiosyncratic errors produced by a single algorithm can be cancelled by combining the outputs of different data-driven taggers on the basis of various voting strategies and second-stage classifiers. Their experiments show that such combination taggers perform better than their best component tagger. Furthermore, the best combination tagger's error rate was almost 20% lower than that of the best component tagger.

2.5 RECENT WORK IN THE LINGUISTIC APPROACH

Since the introduction of CLAWS1, the data-driven statistical approach has been predominant in tagging. Relatively little published work on the linguistic approach seems to have been done on tagging during the 1980's. One of the few was Milne (1986) who reports on small-scale experiments with deterministic rule-based tagging. In the early 1990's, however, more work on the linguistic approach to tagging has been reported. Next we outline four linguistic rule-based systems.

2.5.1 English Constraint Grammar

The earliest published large-scale work on the rule-based approach in the early 1990's seems to have been done in a framework called Constraint Grammar (cf. Chapter 14; Karlsson 1990; Voutilainen *et al.* 1992; Karlsson *et al.* (eds.) 1995). In overall architecture, Constraint Grammar (CG) resembles TAGGIT: both use a tokenizer, morphological analyser and a rule-based disambiguator. As in TAGGIT, CG disambiguation rules, depending on rule type, select a correct reading or reject an illegitimate reading, on the basis of relevant words or tags in the left- or right-hand context. However, CG rules can refer up to sentence boundaries (rather than to the local context alone). Word positions and simple structural patterns can be identified in context conditions. Another characteristic of CG is that the grammar is partially ordered: the constraints are grouped into subgrammars that are applied in a predefined order. However, no particular application order of constraints in the same subgrammar is presupposed.

A typical constraint states a linear order generalization in the form of a partial, often negative pattern–action statement. For instance, a constraint could remove a subjunctive reading if there is no instance of the subordinating conjunction "that" or "lest" anywhere in the left-hand context. Another, more local rule could remove a finite verb reading if the first word to the left is an unambiguous article. If all context-conditions of a constraint are satisfied, the action (*REMOVE* or *SELECT*) is carried out; otherwise the ambiguity is left pending. What 'survives' all constraints in the grammar is the analysis of the sentence. Optimally the analysis is correct and unambiguous, but often some of the hardest ambiguities remain unresolved.

The first published description is of English; the system is known as EngCG, English Constraint Grammar. Voutilainen *et al.* (1992) report a performance test of the EngCG morphological analyser and disambiguator on a 2,167-word corpus that contains four texts from different genres, all new to the system. In the evaluation, EngCG failed to give the correct analysis only to 5 words of the total 2,167; this means a recall of 99.8%. On the other hand, 4.5% of the words remained (mostly two-ways) ambiguous; i.e. the precision was 95.5%. Several later evaluations using larger texts and more objective evaluation methods have given essentially similar figures (cf. e.g. Voutilainen and Heikkilä 1994; Voutilainen 1995a). EngCG is also reasonably fast: a recent version (EngCG-2 with 4,000 rules) analyses about 3,000 words per second.

2.5.2 A rule-based tagger of Turkish

Oflazer and Kuruöz (1994) document a linguistic rule-based tagger for Turkish.

As the lexical component, the system uses a large two-level morphological analyser of Turkish (cf. Chapter 12). The rule formalism resembles Constraint Grammar, except for two main differences. Firstly, in addition to the selection and rejection rules, there is also a third operation, *COMPOSE*, that is used for recognizing multi-word patterns, e.g. idioms. The COMPOSE rules give their targets new word analyses, i.e. the lexical analyses can be overridden with COMPOSE rules. Secondly, the rules are carefully ordered according to their reliability.

The Turkish grammar contains about 250 rules and it is reported to fully disambiguate 97-98% of all words, with an error rate of 1-2%; i.e. 98-99% of all words retain the correct tag.

2.5.3 A finite-state tagger of French

Chanod and Tapanainen (1995a) report a finite-state tagger of French that uses a large two-level morphological description of French as its basis.

The constraint-based tagger uses a rule formalism related to phonological rewrite rules (Kaplan and Kay 1994); some resemblances can also be found with finite-state parsers developed in Helsinki (Koskenniemi 1990). The grammar consists of three ordered subgrammars: one reliable and two heuristic grammars. The heuristic grammars resolve all remaining ambiguities. Overall, the grammar contains only 75 rules that are represented as 11 transducers. In the one-month (sic!) development phase, the rules were tested and corrected against an annotated corpus of no more than 1,000 words.

The tagger produced unambiguous output, with an accuracy of about 97-99%, when tested against new texts comprising 17,000 words. Before applying the tagger to these texts, a manually analysed benchmark version was prepared; an error was counted when the tagger's analysis differed from the benchmark corpus.

2.5.4 A syntax-based tagger of English

Voutilainen (1995b) documents a new English tagger that uses two main components: the EngCG morphological tagger followed by a surface-syntactic finite-state parser. The purpose of the finite-state parser in this experiment was to resolve those morphological ambiguities left pending by the EngCG morphological tagger. Its main task was to assign a unique surface-syntactic structure to its input by using reductionistic finite-state rule automata as introduced in Koskenniemi (1990); as a side effect of this syntactic parsing, also the remaining morphological ambiguities were resolved.

When tested against new texts of 38,000 words (that were first analysed by two linguists, thus preparing the 'benchmark' corpus against which the tagger's output was automatically compared), the system reached an accuracy of over 99% at the level of morphosyntactic tags (i.e. the tags for actual syntactic functions – subject, object, etc. – were not evaluated).

2.6 THE CURRENT SITUATION

In the 1980's, research on tagging was almost entirely devoted to data-driven statistical methods. In the past few years, the scene has become more varied. The data-driven approach is certainly still the most popular one in tagging. The merits and problems of data-driven methods are reasonably well known, e.g. they are known to suffice for making a tagger with 95-97% accuracy with a modest human effort for several languages. It is also reasonably clear by now that improving beyond the 97% accuracy is very difficult with the present data-driven methods. Recently there has been renewed interest in the alternative linguistic approach to tagging. For example, Constraint Grammars are being written for at least Finnish, Swedish, Danish, Basque, Swahili, German and Portuguese at various institutions and some of these grammars are already quite extensive, though still unpublished. Recent work on local finite-state grammars is also reported, e.g. in (Roche 1992; Silberzstein 1994).

The reports on recent linguistic methods, especially Constraint Grammar, have fueled a discussion about the relative merits of the linguistic and the data-driven approach. Proponents of the data-driven approach sometimes claimed that the only reason EngCG appeared to do much better than other taggers was the different manner of measuring, precision and recall instead of accuracy. This claim was shown to be incorrect in Voutilainen and Heikkilä (1994), by comparing precision and recall of EngCG with that of n-best versions of data-driven taggers.

Proponents of the linguistic approach have in turn made direct comparisons between taggers using identical training and test material. Chanod and Tapanainen (1995a) report on a comparison between a statistical (HMM) tagger and a constraint-based tagger. Their claim of better quality with comparable development time for the constraint-based grammar, however, loses in importance because the HMM tagger was trained using unsupervised training only. A more recent comparison was reported by Samuelsson and Voutilainen (1997). A state-of-the-art statistical tagger was trained on a corpus of over 300,000 words manually analysed (and proofread several times) according to the EngCG grammatical representation. The trained statistical tagger and a recent version of the EngCG tagger (EngCG-2) were run on a test corpus of 55,000 words from three genres. Before the comparison, the corpus was manually disambiguated by two expert linguists that followed the double-blind method (op.cit.) to guarantee that the corpus is analysed as correctly as possible. Both taggers were applied to this corpus in a mode where they left equal amounts of ambiguity unresolved: in these circumstances, the error rate of EngCG-2 was an order of magnitude smaller than that of the statistical tagger.

Apart from demonstrating that comparison of taggers is a difficult matter (cf. Chapter 6), the discussion about relative merits of the different approaches did not turn out to be a very fruitful one. Since then, most proponents of both camps have recognized that the 'enemy' is making a useful contribution to the development of tagging and attempts are being made to combine the merits of the two approaches.

Leech *et al.* (1994) emphasise the role of linguistic knowledge in tagging; they report on a recent version of CLAWS (CLAWS4) that uses an extended hand-coded idiom component to prevent the statistical component from making certain misanalyses. Tapanainen and Voutilainen (1994) report a combination of EngCG and the Xerox

HMM tagger: EngCG is used first for resolving some of the ambiguity with a very small error rate; the probabilistic Xerox tagger resolves the remaining ambiguities with a higher error rate. Both of these new hybrids are reported to reach an accuracy of about 98.5%, i.e. better than has been achieved with purely data-driven methods. Tzoukermann *et al.* (1995) report on ongoing work on French that uses rules along with statistical techniques. A third type of hybrid is writing linguistic rules to correct errors made by a data-driven tagger, the rationale being that rules need to be written only for those phenomena problematic for the data-driven tagger (rather than for all constructions in the language); this approach has been investigated by the TOSCA team in Nijmegen (Hans van Halteren, personal communication) and the UCREL team at Lancaster (Fligelstone *et al.* 1997).

Further advances can perhaps be made in the data-driven rule-based framework by taking the form of the rules written by linguists as a model for data-driven learning algorithms, instead of continuing work only with sequences of two or three words. Promising work in this direction has already been done in probabilistic syntactic parsing (cf. e.g. Black *et al.* (eds.) 1993; Schabes *et al.* 1993; Magerman 1994). One emerging approach to tagging in this vein is *probabilistic Constraint Grammar* (Samuelsson 1995; Samuelsson *et al.* 1996). In probabilistic Constraint Grammar, the expressive power of the probabilistic system is designed to be equal to that in Constraint Grammar and, in addition, this probabilistic variant can use the merits of the data-driven approach.

There are also some areas in the linguistic approach where data-driven techniques might be useful. For instance, hand-written structural heuristic rules might be more predictable if statistical techniques, instead of the rule of thumb, were used for evaluating their performance on large amounts of text. Data-driven techniques could also be useful in making generalizations about sequences of words (rather than tags); lexical information, e.g. in the form of automatically extracted modifier–head sequences, might be useful in resolving some ambiguities that feature-oriented rules alone cannot cope with.

3 THE USE OF TAGGING

Geoffrey Leech and Nicholas Smith

3.1 INTRODUCTION

Wordclass tagging is the most popular and well-established kind of linguistic annotation to be applied to text, for two reasons.[1] Firstly, automatic taggers are a well researched and well understood type of software (cf. Chapter 2) and it is possible to tag large amounts of text to a reasonably high degree of accuracy. Secondly, wordclass tagging is generally considered to be a useful first stage annotation applied to text as a preliminary to other activities in *natural language processing* (NLP).

As a result, tagging is a high-profile activity in many text-related research and application areas. However, at the moment, its popularity may well be due to its being at the intersection of what is desired and what is feasible. In retrospect, the prominence given to word-tagging in the present state of NLP will no doubt seem excessive to succeeding generations of researchers. Current obsession with tagging is a sign that we are still on the nursery slopes of serious NLP applied to unrestricted text data. Yet the wide variety of applications for tagging shows its potential as an *open sesame* to almost all areas of progress in NLP as things are today. For good or for ill, it is currently believed that most of the improvement in NLP will come from learning from annotated data.

[1] The authors would like to thank Eric Atwell and Chris Tribble for useful comments on parts of this chapter.

H. van Halteren (ed.), Syntactic Wordclass Tagging, 23-36.

In this chapter we will look at the various uses to which tagging can be put. We will start by examining the uses of tagging in the field where it was conceived and largely developed: *corpus linguistics*, the study of language with the aid of (large) text collections. Then we will examine how tagging has found its way from this scholarly use into more practical applications.

3.2 TAGGING IN CORPUS LINGUISTICS

It is our contention that tagged corpora are a useful resource for many purposes. However, since even an untagged corpus is already a useful resource, it may be better to think of the uses of tagging in terms of the 'added value' which a corpus gains, as a result of tagging, for further research and development applications.

It may be initially helpful to distinguish between two types of corpus processing: (i) the input of information to a corpus and (ii) the output (or extraction) of information from a corpus. *Annotation* of a corpus – of which wordclass tagging is the best-known case – is a primary example of how information can be input to a corpus, by marking or labelling linguistic segments, categories and structures identified within the corpus itself.[2] This informational input is not simply a matter of making explicit the linguistic phenomena which are implicit in the text. Rather, because there are so many different (and often conflicting) systems of annotation that might be applied to a text, it may be regarded as an interpretative process: the set of symbols used to annotate the text represents an analytic scheme which, in however limited a way, reflects some theoretical perspective. Nevertheless, the sets of categories, features, etc. used in annotating a corpus are generally chosen to be as theoretically neutral as possible. They are intended to be uncontroversial or *consensual* classes of information which will be useful to as many analysts as possible.

Using a *concordance program*, on the other hand, is the primary example of outputting information from a corpus (see especially Sinclair 1991). To output a concor-

[2] As a matter of principle, we distinguish between the annotation of the text (adding extra information) and the representation of the text itself. For electronic storage and processing purposes, text representation typically not only involves encoding the text orthographically, in terms of conventional spelling and punctuation, but also encoding explicitly other visible features of the text, such as special characters, special features of layout, etc. This kind of mark-up should be distinguished from annotation. The same distinction (between text representation and annotation) also applies to transcriptions of spoken discourse. However, in this case, it is difficult to draw a line between the two: to decide, for example, whether the encoding of features of stress and intonation constitutes *prosodic mark-up* or *prosodic annotation*. More relevant here is the observation that grammatical tagging can interact with mark-up, whether of written or of spoken language. (For example, the English word "that" is typically stressed when it is a demonstrative pronoun, but unstressed when it is a conjunction. Only the unstressed form of "that" can have its vowel reduced to schwa). Hence, a sufficiently detailed transcription can provide information to help automatic tagging and, conversely, a wordclass-tagged text can be used to add predictable prosodic and phonetic information to a transcription. Because of this interaction, the differentiation between text representation and mark-up is a technicality which need not concern us further in this section.

```
pposition of Mrs Thatcher , to |hold| a special conference next year to
ellip; the music room can only |hold| about a hundred , comfortably . M
u wind **10;10335;l And leap , |hold| after hold , **10;10346;l A golde
ds were enough **10;11601;l To |hold| an avalanche off . **9;11612;pb *
e emotion of fear . Keep tight |hold| and continue while there 's time
 fire has gained a substantial |hold| and is being retarded in its grow
ead the folded end through the |hold| and tuck the two ends through the
he prince , trying manfully to |hold| back his tears , was led away . A
rillas at one point claimed to |hold| captive . Officially , Britain do
is quiet much longer , we will |hold| classes in Fulham &equo; they pro
96;div2 18 TIE BANDS Tie bands |hold| curtains away from a window or do
o per cent of the population ) |hold| either Certificates of Identity (
. 6 is my favourite ) but also |hold| enough thinned colour to facilita
quare by 5cm ( 2in ) deep , to |hold| fabric firmly in position and pre
onday . Mr Urbanec , who is to |hold| further talks with Mr Havel today
t &mdash; &equo; &bquo; Let 's |hold| hands , &equo; said Jim . He put
```

Figure 3.1 A KWIC Concordance of the word "hold" sorted alphabetically on the first following word. Output extracted using Xkwic (Christ 1994).

dance listing, the program searches the corpus for examples of a particular phenomenon (say a word), then lists them in a form which is helpful to the user, especially by sorting them in a particular order and by providing the user with a context and a corpus 'address' for each occurrence listed. Figure 3.1 gives an example of an extract from concordance output. In this case, we see a screen dump of a *KWIC* concordance (Key Word in Context), in which the word sought is located in the middle of the screen, while its linguistic context ranges to the left and to the right of the *key word.*

A concordance program is an example of a more general type of software, which can be labelled a *corpus search and retrieval tool.* Such tools can normally provide not only concordance output, but also other kinds of output information from a corpus, particularly quantitative information such as word frequency lists, ordered either by frequency rank order or alphabetically, word clusters, type-token ratios, etc. A search and retrieval tool is a user's interface with the contents of the corpus.

The provision of automated input and automated output facilities for a corpus can be enormously elaborated beyond what has been briefly described above. In exploring some of the elaborations, we will also explore the functions of tagging as an important preliminary to other processes. Furthermore, the purposes of tagging may affect the kinds of tagsets (cf. Chapter 4) we develop: specifying some word categories (say, the distinction between proper nouns and common nouns) may be important for meaning-related uses. On the other hand, specifying other categories (say, various syntactic categories of conjunction) may be important for a subsequent syntactic analysis, or *parsing*, of the corpus. The differing tagsets devised for different purposes need not be inconsistent with one another: although tagset A may ignore some distinctions made by tagset B and, vice versa, the differences may be simply differences of *granularity*, or degree of detail.

In considering the 'added value' which tagging provides for the corpus linguist, we may notice the ways tagging enhances the usability of a corpus for the two general types of processing (inputting and outputting information) already mentioned. In the remainder of this section, then, we will explore the usefulness of tagging from these two points of view:

- Input of information, i.e. adding further annotations: 3.2.1.
- Extraction of information, i.e. enhancing text search and retrieval processes: 3.2.2.

3.2.1 Adding further annotations

There are many types of corpus annotation, apart from wordclass tagging. They include syntactic annotation (i.e. parsing or partial parsing), semantic annotation (typically the semantic tagging of words) and discourse annotation (e.g. the annotation of anaphoric relations in a text). Compared to wordclass tagging, all these types of annotation are relatively difficult to achieve and relatively undeveloped. The best developed of them, however, is syntactic annotation.

Where wordclass tagging is a preliminary to other levels of annotation, its primary use is to resolve the homograph ambiguities retained in the orthographic form of a text. For example, the word spelled "cut" in English is multiply ambiguous: it can be (a) a finite present tense verb, (b) a finite past tense verb, (c) an infinitive, (d) a past participle, (e) a noun, or (less commonly) (f) an adjective (as in "cut flowers"). Depending on the granularity of the tagset adopted, we may also need to distinguish "cut" as a finite indicative verb from "cut" in the (g) imperative or (h) subjunctive mood. E.g.:

(a) If they rebel, they *cut* themselves off from the rule of law.

(b) I *cut* myself shaving this morning.

(c) We will have to *cut* several minutes out of the script.

(d) They were *cut* off for weeks last winter.

(e) There was a deep *cut* on his forearm.

(f) My mother always prefers plant pots to *cut* flowers.

(g) *Cut* the stem of the plant down close to the ground.

(h) It was proposed that the bank *cut* interest rates.

On the other hand, the need for distinguishing homographs is more obvious where there is a clear semantic, as well as grammatical difference between them, as in the case of "leaves" as a plural noun and "leaves" as a present tense verb.

Syntactic Annotation. Tagging can be thought of as the first stage of parsing: marking the categories of the *leaves* on the syntactic parse tree. In fact, in the longer term, it can be strongly argued that, as corpus parsers become more and more effective, the need for tagging as a special stage will disappear. The two processes will simply be merged in a single grammar and parser. This trend can be seen in the constraint grammar software (e.g. EngCG) of the Helsinki group (Chapter 14; also Karlsson *et al.* (eds.) 1995), where the morphosyntactic tagging at word level is combined with a partial dependency parse.

Up to the present state of the art, however, it has been frequently found useful to separate grammatical annotation (e.g. Svartvik and Eeg-Olofsson 1982, Garside *et al.* (eds.) 1987 and Oostdijk 1991) into the two stages of tagging and parsing. The algorithm for wordclass tagging is generally much simpler than the algorithm for overall parsing and may therefore be carried out on large amounts of corpus data with comparative ease (e.g. this has been done for the 100 million words of the British National Corpus). Parsing, on the other hand, is likely to be more intensive in the use of computer time and human resources. Currently, in corpus parsing, ambiguity, inaccuracy and incompleteness are prevalent in the output from the parser.

Once a sequence of sentences has been tagged and the error rate and ambiguity rate are negligibly low, we can use the tagged corpus as an enhanced input to parsing (enhanced, that is, as compared with the raw corpus). Clearly, the tagging will have removed many of the ambiguities of parsing. Thus, if "Frank thought it necessary to stock up on iced tea" is correctly tagged, each word will be disambiguated and a very large number of potential interpretations (in some grammars, as many as several thousand) will be reduced to one.

Whereas tagging is the pre-requisite to many types of corpus annotation or, more generally, NLP work, it can be argued that *the* central NLP task is parsing – the task that provides the key to the relation between form and meaning that is the essence of more long-term goals of NLP. (For example, machine translation, information extraction and even speech recognition and synthesis cannot do without semantic encoding/decoding, which in turn cannot easily do without parsing). So, in this respect, tagging is providing an entry to the most central area of corpus processing.

Semantic annotation. In its turn, ideally, parsing is a pre-requisite to semantic annotation: the extraction of the logical form of a sentence, and its consequent semantic interpretation, cannot be easily accomplished without parsing information. For some purposes, however, a rough-and-ready analysis of the content of a text is all that is required and for this we do not necessarily need a full-scale parse, but can move directly from wordclass tagging to semantic tagging, i.e. the tagging of words according to their sense in context.[3]

Here, as in the case of syntactic annotation, we can think of tagging as a step on the way to a more abstract level of analysis. Thus a word such as "spring" has a number of senses (season of the year, mechanical device, issue of water from the ground, take a jump). Some of these senses apply to the noun, some to the verb and others to both. Therefore grammatical tagging will undertake part of the work of disambiguation needed for semantic tagging. Other disambiguations will need to be done at a purely semantic level. Note that sometimes the wordclass tagging will unnecessarily distinguish wordforms which have similar meanings: e.g. "love" (noun) and "love" (verb); but, in general, wordclass tagging is an undeniably useful, if not indispensable, preliminary to word-sense resolution.[4]

Discoursal annotation. For discourse annotation, we take as an example the annotation of text for anaphoric relations. In such cases, tagging is again a useful first step. Anaphor-antecedent links are in general identified by a combination of syntactic clues with semantic and pragmatic clues. For example, anaphors are very frequently pronouns and antecedents are very frequently noun phrases. The most common anaphors, personal pronouns, may have other grammatical features, such as number and gender, to aid the recognition process. Thus it is important, for the identification of antecedents of German "er", "sie" and "es", for instance, to be able to recognize features such as singular/plural and masculine/feminine/neuter, which are frequently encoded through wordclass tagging.

3.2.2 Information extraction

Section 3.2.1 above has taken the usefulness of various layers of corpus annotation largely for granted. But let us now give more practical consideration to how an annotated corpus – whether at the wordtagging level or at other levels – proves itself a more useful resource than a corpus in its 'raw' state. It is easy enough to show that tagging, and *a fortiori* syntactic, semantic and discoursal annotation, are useful for the other tasks to be dealt with in this chapter. For example, the argument in 3.2.1, that tagging is valuable for higher levels of annotation, feeds into the argument that tagging is useful for information extraction. Information extraction from textual data needs as many indicators as possible of what the text means. At the syntactic level, among other things, this means being able to retrieve meaning from the text via the subcategorization frames for verbs, e.g. in French, "repartir" has one meaning as a transitive

[3] The term *semantic tagging* can also apply to the marking of semantic roles, such as Agent and Patient, but this is a less common research activity than word-sense resolution.

[4] This approach has been adopted by, for example, Wilson and Rayson (1993) and Janssen (1990).

verb – "to retort" – and a totally different meaning as an intransitive verb – "to leave again". At the semantic level, it means being able to retrieve the appropriate sense of a word, even if morphologically and syntactically, the senses cannot be differentiated. At the discourse level it means being able to identify the textual coreferent of words like "they", whose interpretation depends crucially on context.

In general, it may be argued that the more linguistic information is encoded in a corpus, by means of annotation, the more useful the corpus is for information extraction. Here it is useful to note a distinction between two kinds of information – linguistic and non-linguistic. In this section, focussing on corpus linguistics, we will concentrate on the linguistic information. We will come back to the extraction of non-linguistic information below.

The simplest kind of linguistic information to extract from a corpus is information about the location, frequency and distribution of particular sequences of characters. At the lexical level, this can easily be found by search/retrieval tools such as Tact, OCP, WordCruncher, Icecup, Sara, WordSmith etc. (see Kirk 1994:26–35 for a survey of different tools). Fairly well-known types of output from a corpus, using these tools, are concordance listings and rank-order or alphabetical-order frequency lists. If one uses a tagged corpus for these purposes, the list(ing)s can be more targeted towards higher-level linguistic categories and therefore more informative and amenable to making relevant generalizations about the text. Compare, for example, the frequency lists in Kučera and Francis (1967), based on the raw Brown corpus, and those in Francis and Kučera (1982), based on the tagged (and lemmatized) Brown corpus (figure 3.2).

Other types of lexical information are collocations (measured by statistics such as mutual information (Church and Hanks 1990) and other kinds of habitual word combination (e.g. idioms, fixed phraseological units). Johansson and Hofland (1989) is an example of a frequency lexicon of idioms and collocations based on a tagged corpus (the LOB tagged corpus).

Wordclass tagging is a first step towards lemmatization, which is the (automated) classification of inflectional wordforms under their appropriate lexeme or lemma, e.g. "eat", "eats", "ate", "eating", "eaten" are all forms of the lemma "EAT". Lemmatization, in turn, is usually a necessary step towards the incorporation of corpus-derived information into a computational lexicon, where the lemma, or headword, is the main unit for the storage of information about words.

The lexical information extracted from a corpus can be used to enrich lexicons of various kinds. Frequency information is one sort of information which conventional (paper) dictionaries lack and which an annotated corpus provides the only feasible means to acquire. Another kind of information derivable from a suitable corpus is the way occurrences of a lexical form or lexeme are distributed through different registers or text types.

word	tokens	genres	texts
cut	192	15	111
cuts	30	8	24
cutting	66	8	24

lemma	form		tag	tokens	genres	texts
cut			noun	35	8	21
	cut		NN	17	7	12
		cut	NN-HL	2	1	2
	cuts		NNS	16	5	10
cut			verb	245	15	117
	cut		VB	87	15	56
		cut	VB-HL	1	1	1
	cuts		VBZ	14	8	14
	cut		VBD	25	11	21
	cut		VBN	60	14	47
	cutting		VBG	56	7	21
		cutting	VBG-HL	1	1	1
		Cutting	VBG-TL	1	1	1

Figure 3.2 Frequencies and distributions of "cut" and its inflections. Top: Kučera and Francis (1967). Bottom: Francis and Kučera (1982). Figures show: raw frequency in Brown corpus; number of genres in which form appears; number of texts in which form appears. In the lower list, the tags distinguish the noun and verb lemmas and various other features (e.g. number, person, finite/non-finite). For details of the Brown corpus tagset, see App. 17.6.

Unlike conventional dictionaries, computational lexicons have typically been closely associated with corpora, including annotated corpora. There is often a synergy between computational lexicon and corpus: on the one hand, information for enriching a lexicon or lexical database can be extracted from the corpus. On the other hand, lexicons are needed for accurate annotation, including grammatical tagging.

Until recently, search and retrieval tools have typically lacked *annotation awareness*, i.e. they have been incapable of searching on annotations as a distinct type of information encoded in a text. Now, however, there are some tools (e.g. Cosmas (al Wadi 1994), Corpus Workbench (Christ 1994), WordSmith (Scott 1996), Sara (Aston and Burnard 1998)) which have the ability to search on wordclass tags as well as words and characters.[5] As more advanced tools become available, it will be possible to search

[5] There are, hoewever, sometimes restrictions on the possible search patterns, e.g. Sara can only search for tags in combination with words.

on other kinds of linguistic annotation, too. Hence, the extraction of linguistic information from corpora will become more powerful as corpora with more levels of annotation become available and as the tools are developed which can extract information at these levels.

3.3 PRACTICAL APPLICATIONS

In the previous section we have shown that tagging is useful for almost any activity in corpus linguistics. Since corpora consist of text, it is obvious to expect that tagging can be useful for other text-related activities as well. This expectation has proved to be correct and tagging has found its way into many practical applications. At the same time, it has to be admitted that at present many of these practical applications are potential rather than actual: much of the research using tagging and tagged text has been undertaken with practical applications in mind, but as yet (in 1998), few products making use of the results of this research have been developed far enough to reach the marketplace.

In principle, we can distinguish between two types of practical uses of tagging. In one type, we see the tagging process itself used directly, embedded into a larger NLP system. In the second type, what is used is the output of the tagging process: tagged (and possibly manually corrected) text. Like most categorizations, the distinction is for some cases a rather artificial one. When tagging is used in information retrieval, the tagged text is probably stored somewhere and hence one could say this was a use of the second type. On the other hand, there is probably no manual intervention and the tagger is likely to be embedded in the indexing software, so one could also argue for the first type. In the sectioning below, we have made our decision as to the type on the basis of the criterion of *reusability*: we speak about the use of tagged text if the text is created once with a number of applications in mind.

3.3.1 Uses of tagging software

Information retrieval. The browsing tools described in 3.2.2 can be used as easily for the extraction of non-linguistic information. The difference lies in the purpose. *Message Understanding*, *Information Extraction (IE)* and *Information Retrieval (IR)* are general terms used for the kind of activity where the function of the software is to get at what a text is about (the objects, events, domains of knowledge, etc. to which it refers), rather than what are the linguistic characteristics of the text in itself. For example, the need may be to extract information from a set of documents about their content for a variety of IR purposes. Thus, from an unorganized text stream, we may wish to filter out documents which have no relevance to our current interest, leaving a residue of documents which are worth further scrutiny. We may wish to select automatically a set of documents which have certain content characteristics in common. We may wish to assign incoming documents automatically to one or more text categories (of domain or genre). We may wish to extract a set of documents with a common topic from an existing database of documents or abstracts. All these needs have a common pre-requisite – the requirement of extracting meanings from the text – and, as we have seen, wordclass tagging can be an important first step towards this goal. One desideratum, for example, may be to extract names, including the names of organizations, which may be recognized by tell-tale tag sequences, especially sequences of proper nouns or common nouns: here again, tagging can help.

Word processing. Most word processors today attempt to provide checks not only on spelling, but also on grammaticality. For both types a (possibly very shallow) grammatical analysis is necessary in order to reach even minimal usefulness for the user. As we have already seen, wordclass tagging is an integral part of parsing and hence of spelling- and grammar-checks. Knowing the category of a misspelled word also helps in cutting down the number of proposed corrections, which should belong to the appropriate category. For example, if "witer" is typed in error for "whiter", it would be helpful to be offered the comparative adjectives "whiter" and (possibly) "wider" as corrections, in preference to "winter" and "wither". Furthermore, the tagged sentences can serve as an explanation to the user for the suggestions the word processor makes. Suggestions, especially the unavoidable incorrect ones, may not be clear to the user. Presenting a complete syntactic analysis will not be very helpful, since the user is unlikely to be linguistically knowledgeable. Wordclass tags may just be sufficient to demonstrate how the word processor has understood the sentence and why it makes its suggestions.

Speech processing (synthesis and recognition). To give just a brief indication of the relevance of grammatical tagging to the development of speech processing software, we may note that for speech synthesis, it is important to be able to distinguish homographs (words which are written alike but are pronounced differently). Wordclass tagging, for example, enables the synthesizer to identify the difference of stress placement between "separate" (adjective) and "separate" (verb), or between "object" (noun) and "object" (verb), and so produce the correct pronunciation in the correct context.

Speech recognition, on the other hand, has to cope with the opposite problem: the occurrence of homophones (words which are pronounced alike, but which have different written forms). Here again, tagging enables the correct discrimination to be made: e.g. when a user is dictating a message to a speech recognition device, the software is able to distinguish between "there" and "their" and to choose the appropriate spelling.

Handwriting recognition. Conceptually very similar to speech recognition is the task of automatically recognizing and interpreting handwriting (Atwell 1996). In both cases, the problem is to decode into a digital, character based form an analog signal which may be degenerate and received across a noisy channel. For wide-vocabulary, open-ended recognition tasks, the automatic decoder therefore needs linguistic knowledge: in particular, knowledge of what kinds of words are likely to follow what other kinds of words in the message sequence. The same kinds of language models (e.g. hidden Markov models) are equally applicable to message recognition in both media. Handwriting recognition is therefore another application where wordclass tagging may produce better results in the longer term. At present, however, in both speech and handwriting recognition, wordclass tagging is a research technique, rather than a technique already built into marketable devices.[6]

Machine-aided translation. For fully-fledged machine translation, grammatical tagging must be regarded as a low-level task indeed. However, with the coming of corpus-based translation methods (e.g. Brown *et al.* 1988) attention has shifted from rule-based to statistical identification of translation equivalence. A strong focus of present-day research is on multi-lingual corpora and particularly on parallel corpora (or corpora of texts in different languages, each of which is matched by a translation into the other language). Here, again, a tagged version of each corpus gives added value in facilitating the automatic identification of translation equivalents, particularly on the word and phrase levels (as in the EUROTRA project – Gaussier and Langé 1994).

3.3.2 Uses of tagged text

Dictionaries and grammars. Over the past twenty years, there has been a large investment in corpus-based dictionary development, especially in the international English language dictionary market. Well-known examples of corpus-based dictionaries are the *Collins COBUILD English Language Dictionary* (1987; 2nd edn. 1995) and the *Longman Dictionary of Contemporary English (LDOCE)* (1978; 3rd edn. 1995).

Tagging is of great benefit to lexicographical work in allowing more discriminating analyses of homographs and collocations. In *LDOCE3*, for example, homographs are broken down according to their part-of-speech, the most common 3000 wordforms being assigned a marker to indicate their frequency across a range of large corpora (Summers 1996). Such information can be helpful to users of the dictionary, such as

[6] Speech recognition researchers at IBM have reported some success in using language models derived from tagged corpora (Jelinek *et al.* 1990); their wordclass-based algorithms have been carried over from English to other languages, e.g. Slovenian (Gros *et al.* 1993). Nevertheless such methods have been slow to come through into commercial products and in particular speech recognizers designed for small-scale, limited vocabulary tasks are almost exclusively trained on text containing no linguistic annotations at all. A similar situation prevails in handwriting recognition, despite some promising work, e.g. by Keenan (1993) and Hanlon (1994), using tag-based modelling.

```
ments . ( Licks his finger and |holds it up| &mdash; facing audience .
sick . ( GUIL licks a finger , |holds it up| experimentally . ) GUIL :
 it 's what is expected . ( He |holds his hand out| for the dagger . GU
turday break when two prefects |held each new boy up| in turn to kiss t
a rotten night for us . I must |hold my hand up| and say that their goa
;9525;l Lawmakers an media try |holding rappers back| **10;9536;l An it
 car , between me and it , and |held one hand up| to Verity . She dance
tle paper Lifeboat flag . Rory |held his collar out| so that he could l
he brown hill . One animal was |holding the rest back| ; limping heavil
bacco tin from the jacket . He |held the tin up| . &bquo; As long as yo
 Fergus rolled over to him and |held the shotgun out| to him . &bquo; H
lf instead that I 'd manage to |hold it in| until I was back in the fla
s . Ash followed my gaze . She |held the bottle up| in front of my eyes
he cone over in his hands . He |held it out| to Janice . She took it .
ransmit energy **10;7978;l And |hold it back| undissipated too . **10;7
ds were enough **10;11601;l To |hold an avalanche off| . **9;11612;pb *
ad , which acts as a weight to |hold them down| . The lead weight shoul
lead strip acts as a weight to |hold the bunch down| . It is important
```

Figure 3.3 Concordance showing inflections of the verb "hold", followed by an adverbial particle within a 4-word window to the right; the lines are sorted on tag immediately following key word. Output extracted using Xkwic (Christ 1994).

language learners and teachers, in acquiring or identifying a core vocabulary.[7] For the analysis of homographs, concordancers now provide the facility to search on a *key tag* as well as a *key word* in context. Thus one can refine the concordance results in figure 3.1 by filtering the output to only the verb instances of "hold" (including its inflections, if the corpus has been lemmatized). One might add a further constraint, stipulating that (for instance) an adverbial particle occur within a few words to the right of "hold", to obtain a set a set of examples of phrasal verb usage, as in figure 3.3.

In this way we can see how the task of the lexicographer or the grammarian is made simpler, by the facility to retrieve illustrative examples of various lexico-grammatical phenomena, more quickly and with fewer unwanted examples than could be done using a 'raw' text corpus. These benefits should lead to a better understanding of language in use and hence better descriptions in dictionaries and grammars.

It is worth mentioning briefly that there are a growing number of tools capable of identifying collocations in corpora automatically (Smadja 1990; Church and Hanks 1990). Many of these tools use tagging as a first step in finding syntactic collocations in text and one of their principle benefits is that they identify some types which the human expert may have overlooked.

[7] Nevertheless, as Summers (1996:262) remarks, it would be still more useful to use the frequency of individual *senses* of a homograph as the basis for dictionary ordering.

Language learning and language awareness raising. The use of corpora as just described need not be confined to experts. Foreign language learners too are beginning to reap the benefits of personal interaction with corpora, using learning approaches that are often freer and more exploratory than has traditionally been the case (Tribble and Jones 1990; Johns 1994). One of the problems learners commonly face is lack of awareness of the kinds of linguistic structures typically used by native speakers in different writing styles. With a suitable tagged corpus and guidance in use of retrieval software, students can investigate how native speakers employ a grammatical structure, such as postmodification of nouns using participles, and see how this is realized in different ways depending on the genre or style. The key advantage of this approach is that students are looking at 'real' data: in many cases they are likely to find – and add to their own repertoire – examples of structures which neither the teacher nor the writer of the textbook had considered.

Another obstacle faced by many language learners is lack of familiarity with grammatical terminology, which is so often taken for granted in textbooks and reference works. However a tagged corpus can be used as the basis for the acquisition of at least a basic knowledge of grammar. As a simple example, the Cy-tor software (McEnery and Wilson 1994) applies the technique of masking wordclass annotations that have been already added to the text, prompting learners to fill in the gaps; their analyses can be checked against the hidden tags and if illustrative examples of the tags are required they can be retrieved from the corpus via a KWIC display. The text may be selected according to level of difficulty, topic or type of grammatical content.

Development of NLP software. The importance of taggers as a first stage in many NLP systems has already been amply described. However, tagging can also have a more indirect influence on NLP systems: it can be useful in their development. A special case here is the development of a new tagger, either to tag with a different tagset or to tag more reliably with the same tagset.

Usually, the resource used in the development of new software is a tagged and (preferably) corrected corpus. In building a new tagger, the value of such a corpus is that it can act as a *training corpus*. That is, the information regarding tag sequences, used to predict the tags of a new text, can be derived from the tagged corpus. Similarly, the tagged corpus can be used as a *test corpus*. This means it can be an instrument for automatically evaluating the success of a new tagger, by comparing the result of automatically tagging a given text or corpus with a previously tagged and corrected version of the same text or corpus. The new tagger may be applied again to another corpus, so that more accurately tagged corpus material results. This process can be repeated several times, leading to a cycle of feedback and improvement.

A tagged or parsed corpus can also be used as a testbed for parser development and parser evaluation. For full evaluation of a parser, the most appropriate testbed corpus is likely to be a *treebank*, i.e. a (partially) parsed corpus. But it is likely (as in

Black *et al.* (eds.) 1993) that this treebank will have been grammatically tagged, as a preliminary stage of syntactic annotation. Furthermore, even a corpus which has only been wordclass tagged (but which must have been corrected) can be used for parser evaluation, as a correct parse will have to contain the correct wordclass selection for each word.

Finally, the contribution of tagging to the development of NLP software need not necessarily go by way of a tagged corpus. The research and development efforts in the area of tagging lead to an increased insight into the processing of natural language in general. This shows itself most clearly in the rule-based approaches, but is certainly present as well in the data-driven approaches where the new insights can lead to better modelling of the data.

4 TAGSETS

Jan Cloeren

4.1 INTRODUCTION

In this chapter we focus on the various aspects of the annotation labels used in tagging and of the design of the *tagsets* they belong to. We start with the determination of the possible linguistic contents of a tagset, that is to say: which level of linguistic description it operates on (morphology, syntax, semantics, discourse, etc.) and what is described on each level. We then discuss problems related to the treatment of multi-token units, ambiguities and underspecification in wordclass tagging. Finally, we deal with the more formal aspects of the tagset, such as the structuring of the information and the representation of the labels. This includes terminological questions (label names) as well as decisions on the manner of encoding. In addition, we discuss a proposal by the Text Encoding Initiative (TEI) for the encoding of linguistic analyses in corpora using Standard Generalized Markup Language (SGML).

4.2 INFORMATION CONTENTS OF THE TAGS IN THE TAGSET

Within the current practice of corpus annotation we can distinguish between various levels of annotation, starting from the addition of *lemma* information to each word in the corpus. A next step is the *morphosyntactic* level of annotation, which concerns the labelling of the grammatical nature of each word in the corpus (wordclass and features). Subsequent steps are *syntactic* annotation (the labelling of the syntactic functions of the tokens or the structures of the utterances with a syntactic parser), *semantic* annotation

H. van Halteren (ed.), Syntactic Wordclass Tagging, 37-54.

(the annotation of the semantic categories of the tokens or their senses as defined in a dictionary) and *discourse* tagging (the annotation of discourse relations in the texts and also within the utterances).[1]

4.2.1 Morphosyntactic tags

At present, most corpora are raw or have been annotated morphosyntactically only. A reason for the focus on morphosyntax is its relative feasibility when compared to the added value for lexicography and grammar development (cf. Chapter 3). We therefore start our examination of tagsets with the morphosyntactic level.

When we look at various prominent tagsets which have been employed in the tagging of wordclass information in major European and American corpora, we can observe a number of characteristics that hold for morphosyntactic tagsets in general (cf. Cloeren 1994). In the first place, all tagsets account for major wordclass information. This means that each token in the text receives a label which, for example, indicates that it is a verb, a noun or a preposition. Furthermore, it is obvious that all tagsets coincide largely on the wordclasses that are distinguished. Generally the major wordclasses are the following:

adjective	pre/postposition	adverb
article	conjunction	interjection
noun	numeral	pronoun/determiner
verb		

In many cases punctuation is also encoded as a major class. This is desirable for two reasons. First of all, it leads to a higher degree of consistency in the annotation of the corpus if each token receives a label. Also, more importantly, punctuation in some languages (e.g. German) has strong implications for the interpretation of an utterance and as such for its grammatical analysis. It is also important for parsing systems which operate on sequences of tags rather than on text.

Very often tagsets which are used for the annotation of corpora also include some kind of *open* or *unique* wordclass, which can be used to account for the encoding of unique grammatical elements that are of significant importance for the further annotation or for the interpretation of an utterance (e.g. infinitive markers, negative particles, existential particles, etc.). Finally, there may be a *residual* or *miscellaneous* wordclass, used for the encoding of elements that do not fit into any of the other wordclasses (e.g. foreign expressions, symbols, abbreviations or unclassifiable tokens). Alternatively, the classes for unique and miscellaneous items may be combined into a single *special* class.

[1] For extensive examples of all these types of annotation, see Garside *et al.* (eds.) (1997).

The overview above seems to indicate that there is a consensus as to what one should tag as far as major wordclasses are concerned. However, a class as *pronoun/determiner* reveals that the information in a morphosyntactic tagset depends heavily on the research context the tagset was designed for. Another important factor is the existence of various descriptive linguistic traditions or 'schools', even for individual languages. For example, there are languages which in their descriptive tradition do not identify the class of determiner. For English it was shown by Aarts and Aarts (1982) that the determiner can also be regarded as a syntactic function that can be realized by a series of categories (e.g. pronouns, numerals, articles, etc.).

Such differences may not be significant at the level of the major wordclasses, but they certainly are at the level of features. Here, we first of all find the *subclassifications* of the major wordclasses, such as distinctions between common and proper nouns, or between the different types of pronoun. For verbs we find a global distinction between main verbs and auxiliaries. Auxiliaries can be further subdivided into various types, such as general, passive or semi-auxiliary. Adverbs are often subclassified on semantic grounds, whereas adpositions vary in their position (preposition or postposition). Other types of subdivision are possible too, depending on one's research aim. For example, in the CLAWS2a tagset, we find a special, detailed subdivision of nouns, such as *noun of style*, *noun of organization* and *noun of title*. This tagset emerged from the cooperation between Lancaster University (UCREL, Prof. G. Leech) and IBM and was designed to encode the information needed for the parsing of computer manuals.

After subclassification, a possible further step is the addition of *morphological* information, in particular for nouns, verbs and adjectives. Features containing this type of information are for example: number (singular, plural), person (1, 2, 3), gender (feminine, masculine, neuter), degree (positive, comparative, superlative). For the more synthetic languages such as German and Russian case information (nominative, genitive, dative, accusative, ablative, vocative) also forms part of the features that can be included in a morphosyntactic tagset.

Finally we may find features which go beyond the purely morphological level of annotation. A morphosyntactic tagset usually establishes a link between what are traditionally called the morphological and syntactic levels of description. This link is of special importance with respect to a future extension of the linguistic enrichment of the corpus texts, for example in the form of automatic syntactic analysis. Typical features which establish this link are the *verbal subcategorizations* (indicating the possible functional structures of an utterance) and features like prepositional case in German, where the case which a certain preposition requires for its complement is encoded. Other information of this kind refers to the treatment of compounded expressions or discontinuous elements. This is discussed more extensively in 4.3.

The information contained in a tag can usually be ordered (partially) hierarchically, beginning with the major wordclass and continuing with subtype, morphological features, etc. For example, in the Mecolb annotation of a verb we could get the following structure (cf. Cloeren 1994):

```
VRB                                             (major wordclass)
'subcat', pas, npas                             (vtype)
infin, ppart, ger/indic, subj, imp, cond        (vform/mood)
pres, imperf, pret, fut                         (tense)
```

Here the traditional distinction between lexical (main) verbs and auxiliaries is encoded by the combination of the values of the *vtype* and *vform/mood* features. This kind of hierarchical information structure and the corresponding increase in degree of detail has implications for the notation of the labels, as we will see in 4.4.

4.2.2 Syntactic tags

So far, we have only dealt with the morphological and morphosyntactic level of annotation. One can also think of an extension in the form of the labelling of *syntactic* information. We already discussed the morphosyntactic nature of some tagsets and gave some examples of typical features. If morphosyntactic features establish the link between the morphological and syntactic levels of annotation, the syntactic features give a description in terms of how a certain token relates to others. In other words, an indication is given of the functional role of this token. It is clear that at this point we are about to leave the scope of this book, since this area is normally considered as syntactic parsing. There is, however, one approach which integrates morphosyntactical and syntactical tagging.

The approach, which emerged some years ago, is based on the *Constraint Grammar* formalism (cf. Chapter 14; also Karlsson *et al.* (eds.) 1995). In the optimal case, the Helsinki parser, which was based on this approach, generates a fully disambiguated sentence analysis, with one morphological reading and one syntactic function label for each wordform. In addition to the morphological analysis, the result of the sentence analysis is a 'flat' functional syntactic labelling of each wordform. The kind of syntactic information that is encoded consists of verb chain labels (covering all verbal uses of verbs), head labels (subject, object, indirect object, subject complement, adverbial, etc.) and modifier labels (premodifier, postmodifier, determiner as premodifier, etc.)[2]. In other words: no constituent structure is assigned. A typical analysis would look like the following:[3]

```
her            @GN>
hand           @SUBJ
```

[2] A full list of ENGCG tags can be found in Appendix 17.6.

[3] The example given here is from Karlsson *et al.* (eds.) (1995). For illustration purposes we look at the syntactic labels only.

```
had        @+FAUXV
come       @-FMAINV
to         @INFMARK>
rest       @-FMAINV
on         @ADVL
that       @DN>
very       @AN>
book       @<P
```

Although this kind of syntactic analysis lacks a description in terms of constituents, it certainly constitutes an additional and valuable enrichment of a corpus, while it can be carried out fully automatically (cf. Karlsson *et al.* (eds.) 1995).

4.2.3 Semantic and discourse tags

Semantic tagging and discourse labelling have also gained in popularity over the past few years. *Semantic tagging* generally concerns the process of determining the *sense* of a certain wordform in a specific context. One possibility of achieving this is to provide the token with some semantic feature information, as for example verb aspect (static, progressive, resultative) or noun features (colour, animateness, human, etc.).

This type of information can be useful for a further analysis, where semantic features can guide the analysis process and give clues by imposing selection criteria for the interpretation of a certain utterance. Of course it is impossible to sketch a full overview of the different features and values one might want to include in this kind of annotation. However, the problem of semantic ambiguity is crucial for the whole range of natural language processing applications, where 'understanding' of what was said plays a central role (e.g. in machine translation, topic-identification, etc.). Several methods have been proposed in order to identify the various senses of a word automatically (e.g. the various senses of "bridge" in English). It certainly goes beyond the scope of this chapter to give a detailed overview. See for example the work of Hearst (1991), Zernik and Jacobs (1990), Veronis and Ide (1990), Yarowsky (1992) or the efforts of the NMSU-team around Louise Guthrie and Jim Cowie (Cowie *et al.* 1992). Instead we discuss some problems and present general characteristics of techniques that are used to tackle the issue of automatic sense tagging. To begin with the most serious problem: the question of which senses should be selected. There is no (or only very little) agreement as to the degree of detail a sense tagging scheme should have. Very often researchers use broader classifications than conventional dictionaries do. This is because they feel this to be adequate from their experience with printed textual materials in which they do not encounter the oldish or rare senses which are described in the dictionaries. So, sense classification by one researcher generally cannot be compared to that of another. In order to overcome this problem, researchers very often use definitions from the well-known standard dictionaries of language use, such as LDOCE (1978) or the Collins COBUILD dictionary (1987). Besides the factor of comparability for those using the same dictionary, the reason for this lies in the fact that one avoids the sisyphean task of having to define a 'complete' set of sense classifications from scratch (which would actually imply the creation of an entirely new dictionary).

In line with the sense classifications of conventional dictionaries, various methods have been proposed for the discrimination of word senses in real text. Basically, these methods take co-occurrence information, dictionary subject-codes and wordclass tagging, thesauri and also foreign language equivalents into account. Subsequently, rules or statistical models have been developed for the word-sense disambiguation process. The future will show which implementation of word sense disambiguation technique will be the most successful. However, it is more than likely that with the existence of larger quantities of sense-tagged corpus data, the progress in the field of sense tagging could be significantly accelerated as happened with morphosyntacic tagging. An example of how one can come to coherent semantic classification of lexical entries is the recent WordNet online-lexical database (Miller *et al.* 1993). The setup for this database was inspired by psycholinguistic theories about the mental lexicon. As such, the lexical information is not organized according to alphabetical criteria (lists), but in a net-like fashion in which English nouns, verbs and adjectives are organized into sets of synonyms. Each of these sets represents an underlying concept, while different types of relations link the synonym sets to each other. Resnik (1995) demonstrates the usefulness of WordNet for the identification and tagging (discrimination) of noun senses. He describes a method for automatic sense disambiguation of nouns appearing in the context of related nouns. The tagging was carried out with WordNet senses. The basic idea behind the algorithm is that the most informative subsumer of two related polysemous words indicates the relevant sense or at least provides some information on which sense is the relevant one. For example, the words "doctor" and "nurse" are both polysemous ("doctor": health professional, holder of a Ph.D.; "nurse": health professional, nanny). Here the most informative subsumer is *health professional* and determines the relevant sense.

With respect to *discourse tagging*, the situation is even more nebulous because of the various possible research aims, approaches and understandings within the field. The span reaches from the various attempts on text-grammar to the semantico-syntactic borderline within one single utterance, where discourse elements cause a distortion of the propositional structure of the utterance (e.g. summarizing repetitions, hesitations, juxtapositions, etc.). A description of phenomena of this kind can be found in Aarts and Oostdijk (1997). It would go far beyond the scope of this section to try to give a coherent and meaningful description of what should be tagged in relation to discourse.

4.2.4 Distributional similarity tags

Yet another way to tag tokens in a text is the technique of *word type clustering*, which originates from information retrieval. It has been applied for the annotation of text corpora by Eric Atwell and John Hughes (Hughes 1992). It is mainly a statistical procedure, in which the determination of the tagset is left to a program. The basic idea behind this technique is that words belonging to certain categories have specific positions in the sentence. Determiners, for example, can never occur as the last word of a sentence. Another source of information for a basic classification is the relative position of one word in relation to other words (individually or in groups). For example, prepositions rarely appear immediately before a verb but rather frequently follow them, whereas the opposite is true for the category of pronouns. So, according to its position it is more likely that one word belongs to one class than to another. The algorithm proposed for the automatic word classification starts with the grouping of the most common words according to their absolute sentence positions, such that only the closely matching words are allowed to merge. This basic process is repeated until all the frequent words in the corpus have been classified. To classify the remainder of the words (i.e. to determine whether a word belongs to a certain group or class), their relative positions are used. The actual word is then compared with one of the other classes. If it is sufficiently similar, it becomes a member of this class; if not, it may become the first member of a new group. This process converges until all the words in the corpus have been classified and subsequently all different groups have been interpreted in terms of a certain wordclass tag. Note that the resulting groups do not necessarily correspond to the traditionally accepted wordclasses. For the main applications of the technique, such as information retrieval or word prediction, this is unimportant. However, it has been shown that this technique works quite satisfactorily even for identifying (some of) the traditionally accepted classes (Hughes and Atwell 1993), especially pronouns, auxiliaries, copulas and prepositions. Of course the method also has its drawbacks. One problem, not so different from other stochastic approaches to wordclass tagging, is the resolution of ambiguity in cases where words are dominantly associated with one specific tag and the choice for an alternative tag cannot be made because the positional histograms are not finegrained enough. To take an extreme example, in "¿Es la torre alta? No, es la *baja*." ("Is the tower high? No, it is low.") the word "baja" clearly should be tagged as an adjective, but its position, sentence final and following an article, leads the software to conclude that it must be a noun. Another problem with the method is that it may encounter some problems with the analysis of spoken (transcribed) language, which consists of syntactically more or less complete utterances, but only rarely of complete sentences. As a consequence, the calculation of coherent word-position histograms for the utterances is likely to be quite an undertaking.

4.3 SPECIAL PROBLEMS IN THE APPLICATION OF TAGSETS

4.3.1 Multi-unit tokens and multi-token units

One of the most obvious problems in the morphosyntactic annotation of texts is that the linguistic units one would like to tag do not necessarily coincide with the *orthographically* marked tokens. Usually, a token is defined as a unit consisting of one or more characters, separated by a clear delimiter, usually a blank or a punctuation symbol. However, there are cases where one single token would receive more than one tag (e.g. contractions), where a group of more than one token receives only one tag (e.g. compounds) or where two tokens, which form one unit, are separated from each other (e.g. discontinuities).

Contractions are typically referred to as tokens which, due to their nature, are two tokens merged into one, while they preserve their grammatical identities. As such they require two tags instead of one. For example, the Spanish form of "del" = "de" + "el" would receive two labels: that of preposition ("de") and that of definite article ("el"). The same procedure applies to enclitic forms in English (for example "he's" or "you'd"). Here as well, the units may keep their forms, but receive two labels, that of pronoun ("he" and "you") and that of verb ("'s" and "'d"). Alternatively, such a token may be split into two units, each tagged separately. Both methods can lead to problems. When keeping the tokens together further processing becomes more complicated. Also, inappropriate taggings may result, e.g. in "the father of my friend's car" the token "friend's" would be tagged as a genitive noun while the genitive actually applies to "father". Extreme use of token separation produces some quirks as well: splitting off the enclitic "not" from forms such as "didn't" seems harmless at first, but then one encounters the sequence "wo"+"n't".

A *compounded expression* or *multi-token unit* is an element that is composed of two or more tokens, which together form a strong lexical unit. The best known examples are *nominal compounds*, such as the French "pomme de terre" ("apples of the earth", i.e. "potatoes") and the English "House of Lords". The automatic identification and appropriate tagging of those units is not a trivial task since we face a lot of possible ambiguities. For example "know how" can be a common noun or a legal sequence of a verb plus an adverb. Of course one could opt for the solution not to make any effort in analysing compounded expressions. This would certainly reduce the complexity of the task of tagging the corpus. On the other hand, it would cause a considerable loss of information and would certainly complicate subsequent stages of corpus analysis, such as syntactic parsing, tremendously. Sometimes an attempt is made to identify these expressions and to treat them as single items only. This is certainly more defensible than the former approach. However, in this case as well there is loss of information. This loss is probably not important for subsequent stages of corpus analysis, but it blocks the exploration of the internal structures of multi-token units. In order to avoid such loss of information it has been proposed at various occasions to encode compounded expressions in two ways. In the ideal situation one would get two labels for each element of the compounded expression, one for each individual token and one for the whole expression. To indicate the cohesion between the tokens of the compounds, a ditto tag could be added to the labels, which indicates the number of the compounded elements and the position of the token in the compound. For example, "pommes de terre" would receive some labelling like[4]

```
pommes_NOU(plu,...)_NOU(plu,...,1/3)
de_ADP(pre)_NOU(plu,...2/3)
terre_NOU(sg,...)_NOU(plu,...,3/3)
```

Another common problem with multi-token expressions is that of *discontinuity*. Normally a discontinuous structure consists of two or more tokens belonging to one wordclass (no matter whether compounded or not), which are separated by one or more other tokens. For example discontinuous conjunctions, such as in English "*provided* after all *that*", or discontinuous verbal expressions as in German "ich *hole* ihn vom Bahnhof *ab*" (where the infinitive is "abholen"). Again, it is obvious that automatic identification of these elements is not an easy task, but the added value for following stages in the analysis of the corpus is also evident. If one tries to encode this sort of constructions, one probably marks those elements as discontinuous (for example to trigger a syntactic parser) and encodes the cohesion of certain discontinuous elements at the same time. An example would be:

```
provided_CON(sub,disc,1/2) after all that_CON(sub,disc,2/2)
```

If there happens to be more than one discontinuous element with the same tag in the same text unit, one would have to include an additional feature, indicating which part belongs to which. Hardly any tagset today provides for this eventuality, mostly because the situation is an extremely unlikely one. A related, more frequent phenomenon is the combination of compounding and coordination, e.g. "oil and gas profits". The tagging problem here is usually circumvented by viewing this as one long compound rather than two interleaved ones.

Identification of cases like the above is normally not possible with one single dictionary only. Instead it is common practice to cover these phenomena in separate lists. So, normally one has lists for person names, place names, names and acronyms of organizations, foreign expressions, formulae and also compounded expressions. In addition, it can be very useful to have a list of verbal expressions (periphrases), for the identification of special verbal constructions, such as in Spanish the verb "quedar" (English: "stay") followed by the preposition "en" (English: "in") followed by an infinitive, meaning "to arrange sth. together". These lists need not only be inventories of

[4] Note, however, that following this strategy to its logical extreme leads to quite complicated effects for nested compounds, such as "New York Times".

fixed expressions. Sometimes it can be a good idea to apply some additional heuristics (cf. Leech *et al.* 1994). For example if we come across a form of address (e.g. “Prof.”, “Mr.”, “Herr”, “Frau”, “Sr.”, etc.), it is very likely that one of the following tokens will be a proper noun. Hence, the construction of a template would certainly be useful for the identification of rare proper names.

4.3.2 Underspecification and ambiguity

Underspecification and ambiguity are two descriptively incomplete phenomena, where some information which the tagset could in principle provide is not provided. But the reasons for this lack of information are quite different in the two cases, which should therefore be clearly distinguished.

Underspecification (sometimes called *neutralization*) is a phenomenon that normally occurs at the feature level of the tags rather than at the wordclass level. We can speak of underspecification whenever a certain value of a feature that is foreseen in the annotation scheme is irrelevant for or even not applicable to a certain token. In other words, in this situation the whole feature of the tag becomes obsolete for the token to be described. The reason for this phenomenon lies mainly in the ‘underspecification’ of the respective tagset or in the morphosyntactic ‘underspecification’ of the token itself. The first type of underspecification originates from the definition of the tagset, e.g. a tagset that does not account for the tagging of case in German, although it is a feature that would apply perfectly to various wordclasses. In the second type a feature is generally applicable to a word belonging to a certain class, but for one specific word it is not marked or is neutralized. There are many examples, such as French adjectives like “superbe”, “facile”, “difficile”, which are not marked for gender or the German article “die”, where, when plural, gender is not marked and, when feminine, number is not marked.

In relation to the efforts of various research projects funded by the European Commission (e.g. EAGLES, MECOLB, MULTEXT) which, among other objectives, try to develop guidelines for interchangeable, cross-linguistic morphosyntactic tagsets, a third type of underspecification can be observed: underspecification of the language. This is the case when a certain feature is suggested for the cross-linguistic tagset, but does not apply for a certain language. An example is the case feature for adjectives in Italian, French and Spanish.

Underspecification is normally signalled, in a tagset, simply by the absence of any indicator of the attribute. Alternatively, a special value may be used to make the absence of an attribute explicit.

There is room for different viewpoints on whether morphological syncretism should lead to underspecification of values, or whether values, even where they are not morphologically signalled, should be specified on the basis of context. There is also room for difference of opinion about whether the unmarked value of a binary attribute should be applied to the absence of the marked value, e.g. should we mark all verbs which are not passive in Danish as active? Or should we leave voice unspecified, except with those verbs for which the passive is an option? The final decisions are usually based on the projected use of the material, on whether or not further processing is either aided or hampered by underspecification.

Another important factor in these decisions is the feasibility of removing the underspecification. In case a feature is absent in a language (underspecification of the language), it is clear that there is no point in including it in the tagset. In the case of underspecification of the tagset one could consider the re-design of the annotation scheme when one feels that the actual tagset is not fine-grained enough. In the case of underspecification of a certain token, the underspecification can usually be resolved on the basis of contextual information. Parsing systems based on unification (e.g. Koster 1991) or constraints (cf. Voutilainen *et al.* 1992) usually offer suitable mechanisms for resolving feature underspecification. The following example, based on AGFL compliant coding (Koster 1991), illustrates this:

```
die_ART( def , sg|pl , m|f|n , nom|acc )
gute_ADJ( sg , m|f|n , nom|acc )
Frau_NOU( com , sg , f , nom|acc )
```

In this example of a noun phrase there is a great deal of underspecification in all the tokens. The individual tokens underspecify for number (singular or plural), gender (masculine, feminine or neutral) and/or case (nominative or accusative). By means of unification at least the features for number and gender can be resolved. The case feature would remain underspecified and could only be resolved in a following step. If the main verb is copulative, the only possible value for case would be nominative, since both the subject and the subject complement carry this case. For any transitive verb, there is only the choice between subject, direct object and object complement.

The example above indicates at least three things. First, it demonstrates that quite a lot of underspecification can already be resolved when one looks at the immediate context. It can therefore be implemented in a tagging system. Secondly, there are cases of underspecification that can only be treated at a higher level of linguistic analysis. But most importantly, for further linguistic analysis, it is desirable to have more, even underspecified information, rather than no information on certain features at all.

Ambiguity, as contrasted with underspecification, is the phenomenon of lack of information, where there is uncertainty between two or more alternative descriptions. Several different senses of 'ambiguity' can be distinguished in wordclass tagging.

In *grammatical homonymy*, one wordform, when isolated from its context, belongs to more than one morphosyntactic class. For example the Spanish word "bajo" has at least five readings:

- adjective ("low")
- adverb ("quiet")
- verb (1st person singular of "bajar", "going down")
- noun ("apartment on the ground floor")
- preposition ("under")

Normally, this type of ambiguity, if it is considered such, does not occur in an annotated corpus, since the ambiguity is resolved, either by automatic methods, where constraints, likelihoods and context play an important role, or by human disambiguation.

A further type of ambiguity may arise where even a human annotator cannot decide on a single appropriate tag. The problem is that the categories used in the tagset may have unclear boundaries – not because of any human failing, but because that is what linguistic categories are like. This problem is compounded by the fact that different linguists may have different opinions on the same data. This is usually due to the different theoretical backgrounds of the linguists and it reflects another serious problem, namely the incompleteness and incompatibility of present linguistic descriptions. The only solution is not to count on the linguistic knowledge of the annotators, but to support each tagset with extensive documentation on how it is supposed to be applied to actual text. However, such documentation can never be complete and even the most extensive manual is bound to fail to provide criteria for disambiguation in some cases.

Finally, there are also *genuine* textual ambiguities, by which we mean cases where the text simply does not provide enough information for disambiguation between two or more clearly defined categories. For example, it may be unclear whether in a given case the exclamatory word "Fire!" is a verb or a noun.

With respect to the encoding of ambiguity only little has been proposed. However, ambiguity is certain to occur in any annotated corpus. Especially for the annotation of large corpora, as they do emerge in present day corpus linguistics (e.g. the Bank of English (Collins COBUILD), the British National Corpus BNC (Oxford Computing Services), the Spanish Reference Corpus CREA, the Mannheim corpora (cf. al Wadi 1994), without being exhaustive), fully manual correction of an automatic morphosyntactic tagging can be discarded for the sake of consistency and for the practical reason of it being a too labour-intensive task. If no special measures are taken, i.e. the ambiguity problem is ignored, each token receives only a single tag. The removal of the alternative tags leads to a lower recall rate for some subsequent processes, i.e. some potentially interesting tokens can no longer be retrieved.

Sometimes, as for example was done in the tagging of the BNC, it is more sensible to provide more than one tag for those tokens where the automatic system (or the human disambiguator) proves to have insufficient evidence for successful ambiguity resolution. This allows later users of the material access to all possible interpretations, i.e. it improves recall at the cost of precision. This strategy is sometimes supported by the definition of a set of so-called *portmanteau* tags, tags that represent a set of other, 'normal' tags. An example from the BNC is the tag VVD-VVN, which means "either the past tense or the past participle of a lexical verb". In effect, this introduces underspecification into the tagset to reduce the problem of ambiguity. The BNC example uses a long but user-friendly notation. More extreme would be shorter mnemonic tags, e.g. VVND for the example above, or even tags completely unrelated in appearance. Using this kind of pre-defined tags is supposed to be reserved for situations where it can be assumed that a human reader/analyst would have no problem in coping with the ambiguity, but in practice this reservation may be conveniently forgotten.

4.4 NOTATION

In the development of an annotation scheme, one usually has to consider three things. First of all, one has to determine the linguistic content of the tagset, as we described in the sections above, secondly one will have to decide on the representation of the labels and thirdly one has to determine the most appropriate data-format of the annotations. In this section we discuss the more formal aspects of a tagset: the possible ways of representation. In addition, we address some terminological issues. Finally, we present a current proposal by the *Text Encoding Initiative* (TEI) for the encoding of wordclass tags using *Standard Generalized Markup Language* (SGML).

4.4.1 Class and feature value names

From the comparison of quite a number of well-established tagsets it appears that there is a consensus among linguists about a common linguistic terminology for wordclasses, even across languages (cf. Cloeren 1994). So, by adopting this terminology, no special linguistic training would be required to recognize the content of a certain tag. However, it is very often the case that even a fully-trained linguist is not able to recognize the meanings of certain codes and abbreviations without referring to some form of documentation, which is usually entirely separate from the annotated texts. The problem is one of tag representation. Generally speaking, there are several different ways of encoding – ranging from fully written out names through mnemonic letter-digit sequences to completely numerical labels. For reasons of readability there is a preference for *mnemonic* tags, as shown in all the examples above. Full-length names may be clearer individually, but make the annotated text virtually unreadable. Numerical labels can be used as well, since it is not difficult to transform one well-defined form into another. The advantage of numerical labels is that they are relatively compact compared to the mnemonic tags and therefore can be stored and manipulated more efficiently. The EAGLES cross-linguistic tagging guidelines (cf. Chapter 5; also Leech and Wilson 1994) provide a good example, where a nearly completely numerical tagset is used as an *intermediate tagset* in order to assist the automatic mapping of one annotation scheme (e.g. one for a certain language) onto another or for the mapping of dictionary contents onto a tagset.

To focus on mnemonic coding, the question is how one should encode the information in such a way that the result is recognizable, without the necessity of frequent consultation of manuals or tag lists. As we said earlier, there seems to be a common linguistic terminology for wordclasses. For this reason this terminology should find its way into the tags. For English, however, there is a longstanding tradition in the encoding of corpora. It originates from the annotated Brown and LOB corpora (Francis and Kučera 1982; Johansson 1986), where this terminology has not been encoded in a very accessible way. An example is the JJ tag for a general adjective, or II for a simple preposition. For the tagging of the British National Corpus a tagset was developed that for its linguistic content shows a close resemblance to the LOB/Brown and the CLAWS-schemes (although the tagset is limited to approximately 60 tags). However, it uses much more mnemonic abbreviations for otherwise similar tags, for example the JJ from above has become AJO, so that the readability has improved significantly.

4.4.2 Structure of tags

The tagset of the BNC, although more mnemonic in its labels can (just as its predecessors) be characterized as a 'flat' annotation scheme, in that there is no clear distinction between major wordclass information and additional feature information. The various tagsets developed by the Nijmegen TOSCA-group (Oostdijk 1991; Greenbaum 1992), the MECOLB Minimal Tagset (Cloeren 1994) or the tagset developed at the Lexicographical Institute of the Royal Spanish Academy (Santalla and Cloeren 1995) make use of hierarchical data-field-oriented encoding schemes. The hierarchy in the tags is represented as follows: the level of detail increases from left to right and the different entries are separated by unique delimiters. More specifically, the tag starts on the left with a mnemonical letter code for the major wordclass, followed by a series of feature values, which are separated by commas, where the feature sequence is enclosed between brackets. Examples from the MECOLB Tagset are PRN(pos,pl,1) for a possessive pronoun, plural, 1st person and VRB(copula,indic,pres) for copula verb, indicative, present tense.

Since this way of encoding corresponds directly with the input format required by a two-level, unification-based grammar formalism, it is sometimes referred to as *two-level encoding*. Two-level grammars consist of a set of context-free rewrite rules and a set of so-called *affix* or *feature* rules. The feature level plays a double role in grammars of this kind. First of all, it allows for linguistic generalizations over the first level; secondly it invokes the unification mechanism in the grammar. This is based on the condition that, within one rule of the grammar, all features with the same name must have the same value. As such, the two-level tags operate as the terminal symbols of a syntactic grammar, where the features form a direct interface between the morphosyntactic tagging and the subsequent syntactic parsing. Another, very important advantage of a hierarchically data-field-oriented encoding scheme is its convertibility into other formats. In other words, if the representation of the tag is well-defined (i.e. is unambiguous), its actual format does no longer matter as a conversion from one format into another is then an easy task. One could consider the conversion into numerical codes for database-internal treatment of the tags. Or, if one adheres to other types of formalisms for linguistic analysis, one could convert the above examples into a more general feature structure like:

```
{ {cat:  VRB}
  {head: {form:     finite}
         {subject: {agreement: {number: plural}}}
         {subcat:   copula}
  }
}
```

4.4.3 Positioning of tags

Once we have defined the linguistic contents of the tagset and the representation of its labels, we are practically ready to incorporate it and start tagging. However, we should decide how we want to integrate our tags with the text. Generally, one can think of three ways of applying the tags: horizontally (in the text between the tokens or on a second line beneath the token), vertically (in columns) or in a separate file. It is obvious that the placement of the tags depends heavily on one's personal opinion and taste. Apart from that, it is possible to transform one variant into another. With respect to further processing of the data, it is probably better to have some vertical representation (in columns); with respect to readability and manual correction, a horizontal representation may be preferred, since in most parts of the world people are used to read horizontal lines of texts and because one can focus more intensely on word–tag combinations in their context.

Keeping the analyses in a separate file reflects the difference between the base text and the interpretation dependent annotation. It can also be useful for the storage of the tagged corpus in a database. Normally, a corpus database system separates the text from the tags and establishes a link in the indices by means of the numerical position of each token in the text in order to enhance efficient exploration. When the results of a query are presented on the screen or are stored in a file, the tokens and tags appear together for the sake of readability.

4.4.4 SGML/TEI guidelines for tags

The reader will have noticed that in all sections of this chapter, the various aspects of the development of a tagset, ranging from its contents structure to the more formal issues, were governed by a series of principles: the use of common linguistic terminology and of hierarchically oriented data-field encoding schemes in order to smoothen the interaction between linguistic contents and formal representation, resulting in a large amount of flexibility with respect to the concrete data format of the tags. We have also paid attention to the maxims of care and accuracy in tagset development. One of the reasons for all this was that corpus-tagging is still a very time-, labour- and money-consuming task and it should be avoided that a tagged corpus can only be exploited once and for one very specific purpose. It is our strong conviction that the provider of annotated corpora should design his resources to be as reusable as possible, for the benefit of several research communities, consisting, among others, of linguists, translators and language engineers.

In order to further improve the interchangeability of (tagged) corpora, the Text Encoding Initiative (TEI) has come up with a set of recommendations and proposals for SGML-encoded texts and corpora (cf. TEI-P3; Sperberg-McQueen and Burnard 1994; van Herweijnen 1993). Generally, the TEI publishes guidelines which enable electronic texts to be shared among researchers even from different disciplines. The markup of electronic text according to the TEI-guidelines applies to the documentation of text and text sources (author, editor, person responsible for electronization, person responsible for TEI-markup, etc.). It allows the encoding of structural features of the text (paragraph structure, chapters, typographical markup, etc.). Also – and this is important for the linguistic annotation of corpora – the TEI-guidelines allow for the encoding of the results of analysis and interpretation of texts, as e.g. wordclass tagging.

For the encoding of wordclass tags a *Feature Structure Library Framework* has been proposed, which describes how tags and feature-structures should be encoded (for details, see TEI-P3, chapter 16).

First of all, one has to encode the tags as they appear in the text. To stick to the more analytical way of encoding in the style of the Nijmegen TOSCA group or that of the MECOLB-tagsets (Bodmer 1994; Cloeren 1994), we would encode the possessive pronoun in the phrase "Our house" as

```
<w ana=PRN(pos,pl,1)>Our</w> house.
```

Here the SGML-element <w> is used as the code for the wordform. Within this element the attribute *ana* (ana-lysis) serves to express the wordclass tag as a literal string, thereby not making any difference at this level between major wordclass and additional features. In case of portmanteau-tags the *ana*-attribute may also contain a list of different tags, for example:

```
<w ana='VRB(main,imp)_1 NOU(com,sg)_2'>Fire</w> !
```

The values which *ana* can adopt have to be declared in the so-called *Feature Structure Library*, which normally belongs to the document description, called *header*, for example:

```
<fsLib ... type='MECOLB Minimal Tagset'>
  ...
  <fs type=morfsynt.tag id=PRN(pos,pl,1) feats='PRN POS PL 1'>
  <fs type=morfsynt.tag id=PRN(pos,pl,2) feats='PRN POS PL 2'>
  ...
</fsLib>
```

The names for the types of the feature structures can be chosen freely. Each feature structure (or in our case: each tag) which appears in the document has to be declared in the feature structure library by means of the SGML label *id*. Finally, the components of each tag should be specified, as can be seen in the declaration of *feats*. The feature names and values have to be declared too. The place where this should be done is the so-called *Feature Library*:

```
<fLib ... type='morphosyntactic information of
                         the MECOLB Minimal Tagset'>
  ...
  <f name=major_wordclass id=PRN><sym value=Pronoun>
  <f name=PRNType id=POS><sym value=possessive>
  ...
  <f name=Number id=PL><sym value=plural>
  <f name=Number id=SG><sym value=singular>
  ...
  <f name=Person id=1><sym value=first>
  <f name=Person id=2><sym value=second>
  ...
</fLib>
```

Within this declaration of features and feature values we can observe that a distinction is made between major wordclasses and corresponding features. However, in a strict sense, this difference remains an issue for the interpretation by the human analyst, since within this proposal for the encoding of feature structures, the major wordclass level and the feature level are treated in the same way.

We close this section with an example of how lexically more complex tokens can be encoded within this framework. Let us have another look at the compound expression "pommes de terre", which would ideally receive the following labels:

```
pommes_NOU(plu,...)_NOU(plu,...,1/3)
de_ADP(pre)_NOU(plu,...2/3)
terre_NOU(sg,...)_NOU(plu,...,3/3)
```

In line with the framework proposed by the Text Encoding Initiative, this expression could be encoded as follows:

```
<w ana='NOU(plu,...)_1 NOU(plu,...)_2'
            id=25 n=1-3 form=comp>pommes</w>
<w ana='ADP(pre)_1 NOU(plu,...)_2'
            id=25 n=2-3 form=comp>de</w>
```

```
<w ana='NOU(sg,...)_1 NOU(plu,...)_2'
          id=25 n=3-3 form=comp>terre</w>
```

Each single token of the compound expression is identified as belonging to the lexical item of the corpus with *id* 25. This indicates the lexical value of the expression. The *n* attribute refers to the number of tokens that make up the expression and to the position of the token in the sequence. Furthermore, each token of the expression has two tags, where the first indicates its morphosyntactic classification, abstracted from its context, and the last tag indicates the classification of the compound as a whole. And, finally, we introduced an attribute *form*, which indicates the type of expression we are dealing with.[5] Fortunately, the framework allows such freedom. Other types that could be expressed in this way are, for example, discontinuous expressions and contractions.

[5] Note that the notation is the same as that for portmanteau tags and the attribute *form* also serves to distinguish the type of multiple tag that is used. However, more intricate means would be needed if both portmanteau and compound tags would be needed for any token in the text. An extreme case of this would be when a token can be part of more than one compound expression and we would be forced to use multiple *id* attributes as well, somehow linked to the multiple tags.

5 STANDARDS FOR TAGSETS*

Geoffrey Leech and Andrew Wilson

5.1 INTRODUCTION

We normally add annotations to a text so that they can be re-used as a general research resource, by varied end-users other than the annotators themselves. This implies that the choice of linguistic categories for annotation should take account of the need for annotations which are as far as possible theoretically neutral, so that their re-use is not limited to those who have adopted a particular theoretical framework. Although this ideal of 'theoretical neutrality' is itself controversial and probably unattainable, it may be realistically seen as one of the goals of annotation. This is because we need to minimize the amount of automatic or manual adaptation that would have to be undertaken for the annotated corpus to be successfully used by research groups taking different theoretical positions. In the interests of interchangeability and re-usability of annotated corpora, it is important to avoid a 'free-for-all', or a 'reinvention of the wheel' every time a new project begins. A possible strategy to accomplish this is to strive for some kind of standardization.

Another aim of standardization is that annotations should as far as possible be capable of integration with NLP components such as lexicons and grammars, and with tasks such as parsing. For example, if wordclass tags are to be used for lexical knowledge acquisition, the information in the corpus must be assimilated to the information already in the lexicon. This implies that, as far as possible, whatever standards are to be employed for the categorization of items in a lexicon should also be employed

*The following text is a reduced and edited version of a document that was produced for the Text Corpus Working Group of EAGLES. The work is in the Public Domain.

H. van Halteren (ed.), Syntactic Wordclass Tagging, 55-80.

in corpus annotation. For this reason, the EAGLES morphosyntactic guidelines were developed in close collaboration with the EAGLES Lexicon Working Group.

At the cross-linguistic level, annotations used for one language should as far as possible be compatible with annotations used for another. Compatibility here means that where there are descriptive categories in common between different languages, these should be recognized in the annotation scheme and recoverable from the annotations applied to texts in different languages. This is likely to prove an increasingly important issue in multilingual engineering. Whereas, in the past, there has been a tendency to produce software that concentrates on single languages or bilingual pairs, a growing need for multilingual NLP applications may required a different approach, and there is already some evidence for this. One might characterize this as a 'plug and play' approach: a basic 'engine' or algorithm is developed, which will work with all the relevant languages, and the resources (e.g. lexical) for each language are simply slotted into that framework. Such an approach requires basic agreement on the categories to be encoded in the language resources, so that the language-independent engine can be developed. One example might be anaphoric resolution: here, all the languages would need to encode the number, gender and case of pronouns, as far as is relevant to each individual language. The basic engine could not be applied to a given language if number were relevant to its pronoun system but were not encoded in the tagset.

Since 1993, moves have been made to develop standards to be followed by anyone undertaking wordclass tagging of a corpus. These attempts are, to our knowledge, so far limited to the European Union, which has supported an initiative called EAGLES (*Expert Advisory Groups on Language Engineering Standards*), aiming towards the standardization of practices in NLP (or Language Engineering) in general. However, there is expectation that this standardization tendency will spread more widely through the continents of the world as it is already spreading to parts of Europe (currently) outside the EU.

It may be asked why the initiative to standardise practices in language engineering has been so far largely limited to Europe. Two likely reasons for this are that (a) the EU is one of the technologically advanced regions of the world and (b) the EU is very much a multilingual community, in which the different major languages of member nations are given equality of status. Consequently, the EU has invested heavily in language engineering programmes (such as the EUROTRA machine translation programme) and has laid great emphasis on the need to develop comparable, interchangeable technologies for the main languages of the union: Dutch, English, French, German, Italian and so on. Obviously, in this context, it makes good economic sense to ensure that the resources developed for different languages (corpora, grammars, lexicons, tools) can 'talk to one another'. A central aspect of this impetus is the need to develop multilingual resources, such as bilingual and multilingual lexicons and parallel corpora. The need to annotate corpora for use across national and linguistic boundaries also leads to a requirement for common *de facto* standards in areas such as wordclass tagging. At a recent EAGLES workshop at Pisa (January 1999), it was noted that some 20 users and institutions in various European countries had already adopted the morphosyntactic guidelines (as outlined in this chapter) as a model for developing tagsets for their own languages.

A second part of the answer to the above question is: why have other regions of the world not invested such effort in language engineering standards? Looking beyond Europe, to the international scene, we see that the two technologically most powerful influences are the USA and Japan, both of which are predominantly monolingual societies, where the need for such standards is correspondingly less urgent. As nation states, the USA and Japan have other institutions promoting collaboration within their boundaries (such as centralised state sponsorship of research). Nevertheless, it is noteworthy that both Japan and the USA have shown strong interest in the work of EAGLES and recently funding has been provided by the American NSF (National Science Foundation) to enable American computational linguists to collaborate with EAGLES. Indeed, this was the main theme of the workshop at Pisa mentioned above, which was attended by a considerable number of American speech and language scientists.

Standardization, understood as convergence towards uniformity of practice, arouses strong feelings of both a positive and a negative kind. The benefits of standardization, as an aid to technological progress, are obvious: it promotes interchange and reusability of resources far and wide and provides a basis for development of products which can command an unrestricted international market. The disadvantages of standardization lie in its imposition of a 'straight-jacket' on scientific and intellectual endeavour. In linguistics, where the immense diversity of human languages continues to challenge and baffle the research community, any attempt to regiment the use of terms, categories or theories is likely to be anathemized. But somehow the requirements of science and of technology need to be reconciled. One compromise, for example, lies in setting up provisional *guidelines* (rather than standards in the strict sense), in the expectation that, over time, convergence will take place. It is in this spirit that EAGLES provides the following guidelines for wordclass annotation (or, to use the EAGLES terminology, *morphosyntactic annotation*).

The text below is a reduced and slightly edited version of a document that was produced for the Text Corpus Working Group of EAGLES. The work for this document (*Morphosyntactic Annotation*: EAGLES Document EAG-CSG/IR-T1.3) was undertaken by a Corpus Annotation Subgroup and was disseminated within EAGLES as *work in progress* in October 1994. Since then, the guidelines have been applied to a number of EAGLES projects. However, the guidelines remain provisional and open to further discussion and refinement.[1] At the time, the membership of the EU was more restricted than it is today and the languages taken into consideration, in working towards these guidelines, were the following: Danish, Dutch, English, French, German, Greek, Italian, Portuguese and Spanish. Inevitably, the recommendations were influenced by the fact that all these languages belong to the Indo-European family and all but one belong to the Germanic or Romance language groups. It remains to be seen how far these guidelines can be extended, without substantial revision, to other languages.

Section 5.2 first investigates the reasonable goals for standardization and then goes on to present the tagset guidelines themselves. Section 5.3 describes the EAGLES intermediate tagset for interchange between different local tagsets and illustrates its use (and that of the guidelines in general) with example tagsets for English and Italian.

5.2 RECOMMENDATIONS FOR MORPHOSYNTACTIC (WORDCLASS) CATEGORIES

5.2.1 Reasonable goals for standardization

Some kind of standardization is becoming urgent, particularly in the area of wordclass annotation. This is an area in which most annotation has been done and wordclass tagging is likely to be undertaken for many different languages in the next few years. In the interests of interchangeability and reusability of annotated corpora and particularly for the development of multi-lingual corpora, it is important to avoid a free-for-all in tagging practices.

But the varied needs and constraints which govern any annotation project, or which might govern such projects in the future, urge caution in setting out to achieve a rigid standardization. Where possible, it is important to offer a default specification which can be adopted where there are no overriding reasons for departing from it. In this way, invariance will establish itself across different projects and languages and a *de facto* standard will progressively come into being.

However, the need to go beyond a preferred standard — a principle of *extensibility* — should also be recognized. There will be a need to extend the specification to new phenomena and sometimes a need to represent different perspectives on the same data. *Extensibility* means, on the one hand, the ability to extend the specification to language-

[1] We acknowledge the help and advice of the following in drawing up the preliminary EAGLES guidelines presented here: Susan Armstrong, Gerardo Arrarte, Fernanda Bacelar de Nascimento, Nuria Bel, Anna Braasch, Nicoletta Calzolari, Pieter de Haan, K.A.C. Depuydt, Tilly Dutilh-Ruitenberg, Paula Guerreiro, Ramia Hatzidaki, Ulrich Heid, Dirk Heylen, Liliane Khouri, Philip King, Jean-Marc Langé, Lothar Lemnitzer, Christine Meunier, Monica Monachini, Wolfgang Paprotté, Anne Schiller, Helena Soares, Petra Steiner, Simone Teufel, Jean Veronis, Marta Villegas and Annie Zaenen.

specific phenomena and, on the other, the ability to vary the degree of granularity for this or that annotation task.

The use of the term *guidelines*, in reference to the documentary specification of annotation standards, is salutary in suggesting that there is no absolute normative prescription of annotation practices, but at most a set of recommendations, from which the annotator may justify departures or extensions for particular purposes. Even the term *recommendations* is too strong a word in some cases: often we can only point out the range of practices which exist, without offering advice to prefer one to another.

We consider, in the following three sections, the feasibility of achieving a measure of standardization in three important areas.

5.2.1.1 Common standards for representation and encoding of annotations of texts. At face value, the most trivial aspect of annotation guidelines is in the recommendation of *visible* character-coded means to represent this or that linguistic phenomenon. Any device for encoding a given linguistic phenomenon is (in the last resort) arbitrary and, so long as it is distinctive, can automatically be converted into a different device. We propose that the criteria of *unambiguity*, *compactness*, *readability* and *processibility* be given priority, although different degrees of priority may be assigned to such criteria for different projects. On the other hand, we suggest below (5.3) that local specifications should be translatable into a common *EAGLES standard* automatically, by a regular mapping.

5.2.1.2 Common standards in describing and representing linguistic categories or structures. The specification of common standards for linguistic categories/structures is more serious and challenging. If a common standard implies the recognition of invariants across different languages or different descriptions of the same language, then the extent to which this is feasible depends on the extent to which such invariants are recognized by those already working in the field. This may be unproblematic in the case of the grossest categories such as Noun, Prepositional Phrase, etc., but as one moves toward (a) greater granularity of description and (b) more abstract levels of linguistic annotation, the degree of consensus is likely to decline. The level of wordclass tagging is the one most favourable to a reasonable degree of standardization in this sense and is also the level for which the urgency of establishing common standards is greatest. In 5.2.2–5.3, below, this will be dealt with in some detail and in close relation to the standards for wordclass categorization in the lexicon (cf. Chapter 11).

5.2.1.3 Common standards for specifying annotation schemes and their application to texts. The final area for standardization that we consider, here, appears to be the most difficult to achieve, if it is to be equated with laying down rules for *consistency* in the application of tags to texts. To restrict our attention to the apparently favourable area of wordclass annotation: the ideal need is to specify an annotation scheme so precisely that a different annotator, applying the same annotation scheme to the same text, would arrive at exactly the same result. This would mean that each word-token, in a given text, would end up with the same tag, even if done independently by two analysts. But in practice, there are always 'fuzzy boundaries' between categories, such as the uncertainty (in English) of whether to regard "gold" in "a gold watch" as an adjective or a noun.[2] Decisions on such matters have to be specified in the annotation scheme, which should also deal with such general issues as whether functional or formal definitions of the use of tags are to be adopted, or whether both function and form have to be represented in the annotation. Individual words may need to be discussed, where their recognition as members of this or that category is problematic. But new phenomena, not covered by existing guidelines, are always liable to occur, however detailed the annotation scheme.

Such issues as these cannot be decided in the abstract, in a way which generalizes across languages and across annotation tasks. This kind of standardization is best met, not by laying down detailed specifications of how this or that category is applied in the tagging of this or that word, but by recommending that a sufficiently detailed annotation scheme be made available to users of the annotated corpus. There is little possibility of seeking detailed agreement between different annotators on matters of how to apply tags to texts, particularly if different languages are involved. But, at least, one can ensure that the user be provided with information, as detailed as possible, about how annotations have been applied to texts.

5.2.1.4 Conclusion: manageable levels of achievement in specifying standards. The following degrees of standardization may be thus be recommended at the current stage:

1. **Representation/encoding:** observance of general principles of unambiguity, compactness, readability and processibility; translatability of annotation devices into a set of language-generic conventions.

2. **Identifying categories/subcategories/structures:** agreement on common categories, etc., across different languages, where these can be justified by linguistic analysis and descriptive tradition; allowance for variation, subject to three degrees of constraint: *obligatory*, *recommended* and *optional* specifications.

3. **Annotation schemes and their application to texts:** agreement merely on the requirement that annotation schemes should be made available to end-users and to other annotators, and should be as detailed as possible.

---

[2] This uncertainty can also be seen as just one of the forms of *ambiguity* (cf. 4.3.2). The encoding of ambiguity in wordclass annotation has so far received little attention and we make no recommendations except to propose that, in principle, all the kinds of ambiguity listed in 4.3.2 should be distinguishable by different mark-up.

5.2.2 Word categories: tagset guidelines

Four degrees of constraint are recognized in the description of word categories by means of wordclass tags:

1. **Obligatory attributes or values** have to be included in any wordclass tagset. The major parts-of-speech (Noun, Verb, Conjunction, etc.) belong here, as obligatorily specified.

2. **Recommended attributes or values**. These are widely-recognized grammatical categories which occur in conventional grammatical descriptions (e.g. *Gender, Number, Person*).

3. **Special extensions:**

   (a) *Generic attributes/values*. These are not usually encoded, but may be included by anyone tagging a corpus for any particular purpose. For example, it may be desirable for some purposes to mark semantic classes such as temporal nouns, manner adverbs, place names, etc. But no specification of these features is made in the guidelines, except for exemplification purposes. They are purely optional.

   (b) *Language-specific attributes/values*. These may be important for a particular language, or maybe for a few languages at the most, but do not apply to the majority of languages under consideration.

In practice, 3a and 3b cannot be clearly distinguished.

Type 3 (*special extensions*) is an acknowledgement that the guidelines are not closed, but allow modification according to need. The four types, 1, 2, 3a and 3b, correspond to the four types of constraint applied to word categorization in the lexicon.

5.2.2.1 Obligatory attributes/values. Only one attribute is considered obligatory: that of the major word categories, or *parts-of-speech*:

Major categories

1.	N	[noun]	8.	C	[conjunction]
2.	V	[verb]	9.	NU	[numeral]
3.	AJ	[adjective]	10.	I	[interjection]
4.	PD	[pronoun/determiner]	11.	U	[unique/unassigned]
5.	AT	[article]	12.	R	[residual]
6.	AV	[adverb]	13.	PU	[punctuation]
7.	AP	[adposition]			

Of these, the last three values are in need of explanation.

The *unique* value (U) is applied to categories with a unique or very small membership, such as *negative particle*, which are unassigned to any of the standard part-of-speech categories. The term *unique* cannot always be strictly applied, since (for example) Greek has three negative particles, $\delta\epsilon\nu$, $\mu\eta(\nu)$, and $o\chi\iota$.

The *residual* value (R) is assigned to classes of textword which lie outside the traditionally accepted range of grammatical classes, although they occur quite commonly in many texts and very commonly in some, e.g. foreign words or mathematical formulae. It can be argued that these are on the fringes of the grammar or lexicon of the language in which the text is written. Nevertheless, they need to be tagged.

Punctuation marks (PU) are (perhaps surprisingly) treated here as a part of wordclass annotation, as it is very common for punctuation marks to be tagged and to be treated as equivalent to words for the purposes of automatic tag assignment.[3]

The symbols used to represent the major categories (above) and the attributes and values of other categories (below) will be used again in 5.3 for a method of language-neutral representation called the *Intermediate Tagset*.

5.2.2.2 Recommended attributes/values. These are specified below under part-of-speech headings. The set of values for each attribute is definitely not a closed set and will need to be augmented to handle the peculiar features of individual languages (see 5.2.2.3). Not all languages will instantiate all attributes or all values of an individual attribute. For such cases of *underspecification* (cf. 4.3.2) 0 designates a zero value, meaning "this attribute is not applicable" for the particular language or for a particular textword in that language.[4] The standard requirement for these *recommended* attributes/values is that, if they occur in a particular language, then it is advisable that the tagset of that language should encode them.

1. Nouns (N)

(i)	Type:	1. Common	2. Proper	
(ii)	Gender:	1. Masculine	2. Feminine	3. Neuter
(iii)	Number:	1. Singular	2. Plural	
(iv)	Case:	1. Nominative	2. Genitive	3. Dative
		4. Accusative	5. Vocative	

Inflection type is omitted as an attribute, as it is purely morphological.

2. Verbs (V)

(i)	Person:	1. First	2. Second	3. Third
(ii)	Gender:	1. Masculine	2. Feminine	3. Neuter
(iii)	Number:	1. Singular	2. Plural	
(iv)	Finiteness:	1. Finite	2. Non-finite	

[3] Punctuation marks account for c. 15% of the tokens in a text (calculated on the basis of the tagged Lancaster-Oslo/Bergen Corpus of written British English; see also Nunberg 1990).

[4] However, there are some cases where underspecified values have been given a name of their own, e.g. the *oblique* case for pronouns/determiners.

(v)	Verb form/Mood:	1. Indicative	2. Subjunctive	3. Imperative
		4. Conditional	5. Infinitive	6. Participle
		7. Gerund	8. Supine	
(vi)	Tense:	1. Present	2. Imperfect	3. Future
		4. Past		
(vii)	Voice:	1. Active	2. Passive	
(viii)	Status:	1. Main	2. Auxiliary	

Attribute (v) has two names because of different traditions, for different European languages, regarding the use of the term *Mood.* In fact, the first four values (v) 1–4 are applicable to Finite Verbs and the last four (v) 5–8 to Non-finite Verbs.

Attribute (vii) *Voice* refers to the morphologically-encoded passive, e.g. in Danish and in Greek. Where the passive is realized by more than one verb, this does not need to be represented in the tagset.

The same applies to compound tenses (Attribute (vi)). In general, compound tenses are not dealt with at the wordclass level, since they involve the combination of more than one verb in a larger construction.

3. Adjectives (AJ)

(i)	Degree:	1. Positive	2. Comparative	3. Superlative
(ii)	Gender:	1. Masculine	2. Feminine	3. Neuter
(iii)	Number:	1. Singular	2. Plural	
(iv)	Case:	1. Nominative	2. Genitive	3. Dative
		4. Accusative		

Attribute (i) *Degree* applies only to inflectional comparatives and superlatives. In some languages, e.g. Spanish, the number of such adjectives is very small.

4. Pronouns and Determiners (PD)

(i)	Person:	1. First	2. Second	3. Third
(ii)	Gender:	1. Masculine	2. Feminine	3. Neuter
(iii)	Number:	1. Singular	2. Plural	
(iv)	Possessive:	1. Singular	2. Plural	
(v)	Case:	1. Nominative	2. Genitive	3. Dative
		4. Accusative	5. Non-genitive	6. Oblique
(vi)	Category:	1. Pronoun	2. Determiner	3. Both
(vii)	Pron.-Type:	1. Demonstrative	2. Indefinite	3. Possessive
		4. Int./Rel.	5. Pers./Refl.	
(viii)	Det.-Type:	1. Demonstrative	2. Indefinite	3. Possessive
		4. Int./Rel.	5. Partitive	

The parts-of-speech Pronoun, Determiner and Article heavily overlap in their formal and functional characteristics and different analyses for different languages entail separating them out in different ways. For the present purpose, we have proposed placing Pronouns and Determiners in one 'super-category', recognizing that for some descriptions it may be thought best to treat them as totally different parts-of-speech.

There is also an argument for subsuming Articles under Determiners. The present guidelines do not prevent such a realignment of categories, but do propose that articles (assuming they exist in a language) should always be recognized as a separate class, whether or not included within determiners. The requirement is that the descriptive scheme adopted should be automatically mappable into the present one (see 5.3).

Attribute (iv) accounts for the fact that a possessive pronoun or possessive determiner may have two different numbers. This attribute handles the number which is inherent to the possessive form (e.g. Italian "(la) mia", "(la) nostra" as first-person singular and first-person plural) as contrasted with the number it has by virtue of agreeing with a particular noun (e.g. Italian "(la) mia", "(le) mie").

Under Attribute (v) *Case*, the value *Oblique* applies to pronouns such as "them" and "me" in English and equivalent pronouns such as "dem" and "mig" in Danish. These occur in object function and also after prepositions.

Under Attributes (vi) and (vii), the subcategories *Interrogative* and *Relative* are merged into a single value *Int./Rel.* it is often difficult to distinguish these in automatic tagging, but they may be optionally distinguished (see 5.2.2.3) at a more delicate level of granularity.

Similarly, under Attribute (vi), *Personal* and *Reflexive* pronouns are brought together as a single value *Pers./Refl.*. Again, they may be optionally separated as a more delicate level (5.2.2.3).

5. Articles (AT)

(i)	Article-Type:	1. Definite	2. Indefinite	
(ii)	Gender:	1. Masculine	2. Feminine	3. Neuter
(iii)	Number:	1. Singular	2. Plural	
(iv)	Case:	1. Nominative	2. Genitive	3. Dative
		4. Accusative		

6. Adverbs (AV)

(i)	Degree:	1. Positive	2. Comparative	3. Superlative

There are many possible subdivisions of adverbs on syntactic and semantic grounds, but these are regarded as optional rather than recommended (see 5.2.2.3 below).

7. Adpositions (AP)

(i)	Type:	1. Preposition

In practice, the overwhelming majority of cases of adpositions we have to consider in the European languages under consideration are prepositions. Hence only this one value needs to be recognized at the *recommended* level. Other possibilities, such as Postpositions and Circumpositions, are dealt with at the *optional* level.

8. Conjunctions (C)

(i)	Type:	1. Coordinating	2. Subordinating

9. Numerals (NU)

(i)	Type:	1. Cardinal	2. Ordinal	
(ii)	Gender:	1. Masculine	2. Feminine	3. Neuter
(iii)	Number:	1. Singular	2. Plural	
(iv)	Case:	1. Nominative	2. Genitive	3. Dative
		4. Accusative		
(v)	Function:	1. Pronoun	2. Determiner	3. Adjective

In some languages (e.g. Portuguese) this category is not normally considered to be a separate part-of-speech, because it can be subsumed under others (e.g. cardinal numerals behave like pronouns/determiners; ordinal numerals behave more like adjectives). We recognize that in some tagsets *Numeral* may therefore occur as subcategory within other parts-of-speech. (Compare the treatment of articles under 5. above). At the same time, it is possible to indicate the part-of-speech function of a word within the numeral category by making use of Attribute (v).

10. Interjections (I)

No subcategories are recommended.

11. Unique/Unassigned (U)

No subcategories are recommended, although it is expected that tagsets for individual languages will need to identify such one-member wordclasses as *Negative particle, Existential particle, Infinitive marker*, etc. (see 5.2.2.3).

12. Residual (R)

(i)	Type:	1. Foreign word	2. Formula	3. Symbol
		4. Acronym	5. Abbreviation	6. Unclassified
(ii)	Number:	1. Singular	2. Plural	
(iii)	Gender:	1. Masculine	2. Feminine	3. Neuter

The *Unclassified* category applies to word-like text segments which do not easily fit into any of the foregoing values, for example: incomplete words and pause fillers such as "er" and "erm" in transcriptions of speech or written representations of singing such as "dum-de-dum".

Although words in the Residual category are on the periphery of the lexicon, they may take some of the grammatical characteristics, e.g., of nouns. Acronyms such as "IBM" are similar to proper nouns; in most languages, symbols such as alphabetic characters can vary for singular and plural (e.g. "How many Ps are there in PSYCHOPATH?") and are in this respect like common nouns. In some languages (e.g. Portuguese) such symbols also have gender. It is quite reasonable that in some tagging schemes some of these classes of word will be classified under other parts-of-speech.

13. Punctuation marks

Word-external punctuation marks, if treated as words for wordclass tagging, are sometimes assigned a separate tag (in effect, an attribute value) for each main punctuation mark:

(i)	1. Period	2. Comma	3. Question mark	... etc. ...

An alternative is to group the punctuation marks into positional classes:

(i)	1. Sentence-final	2. Sentence-medial	3. Left-Parenthetical
	4. Right-Parenthetical		

Under 1 are grouped ".", "?" and "!". Under 2 are grouped ",", ";", ":" and "—". Under 3 are placed punctuation marks which signal the initiation of a constituent, such as "(", "[" and "¿" in Spanish. Under 4 are grouped punctuation marks which conclude a constituent the opening of which is marked by one of the devices in 3, e.g. ")", "]" and Spanish "?". We make no recommendation about choosing between these two sets of punctuation values.[5]

5.2.2.3 Optional attributes/values. Here we deal with aspects of wordclass annotation which are optional and may be included in the annotation scheme according to need. Many of them go beyond morphosyntax and are of a syntactic or semantic nature. There is decidedly no claim to completeness. We do not recommend any of these features, but simply present them as having illustrative value. We start with optional features which are application- or task-specific.

1. Nouns

One might wish to introduce semantically and syntactically oriented attributes such as countability:

(v)	Countability:	1. Countable	2. Mass

2. Verbs

Additional optional attributes:

(ix)	Aspect:	1. Perfective	2. Imperfective
(x)	Separability:	1. Non-separable	2. Separable
(xi)	Reflexivity:	1. Reflexive	2. Non-reflexive
(xii)	Auxiliary:	1. Have	2. Be

Attribute (ix) is needed for Greek and Slavonic languages. It corresponds also to the *Past Simple/Imperfect* distinction of Romance languages.

Attribute (x) is relevant for German compound verbs ("fängt ... an", "anfangen") and also for phrasal verbs in Danish and English.

Attribute (xii) is applied to main verbs in French, German, Dutch, etc., and determines the selection of "avoir" or "être", etc., as auxiliary for the *Perfect*.

[5]The punctuation category is clearly the most peripheral of the above categories as regards relevance to morphosyntax. There is also a scale of peripherality within the punctuation category. For example, on the boundary between punctuation and the mark-up of a text are such features as highlighting, whether realized by italics, bold-face or capitals, which according to one view, should be included within tagging schemes.

Additional optional values for attributes recommended in 5.2.2.2 above:

(viii)	Status:	3. Semi-auxiliary

In addition to main and auxiliary verbs, it may be useful (e.g. in English) to recognize an intermediate category of semi-auxiliary for such verbs as "be going to", "have got to", "ought to".

3. Adjectives

Additional optional attributes:

(v)	Inflection-type:	1. Weak-Flection	2. Strong-Flection	3. Mixed
(vi)	Use:	1. Attributive	2. Predicative	
(vii)	NP Function:	1. Premodifying	2. Postmodifying	3. Head-function

Weak and *Strong* (attribute (v)) are values for adjectival inflection in the Germanic languages German, Dutch and Danish. The syntactic attribute (vi) makes a distinction between, for example, "main" (*Attributive*) and "asleep" (*Predicative*) in English.

4. Pronouns and Determiners:

Additional optional attributes:

(ix)	Special Pronoun Type:	1. Personal	2. Reflexive	3. Reciprocal
(x)	Wh-Type:	1. Interrogative	2. Relative	3. Exclamatory
(xi)	Politeness:	1. Polite	2. Familiar	

Attribute (xi) is limited to second-person pronouns. In some languages (e.g. French) it is possible to treat *polite* and *familiar* simply as pragmatic values encoded through other attributes — especially person and number. In languages where there are special polite pronoun forms (e.g. Dutch "u" and Spanish "usted"), the additional *Politeness* attribute is required.

6. Adverbs

(ii)	Adverb-Type:	1. General	2. Degree	
(iii)	Polarity:	1. Wh-type	2. Non-wh-type	
(iv)	Wh-type:	1. Interrogative	2. Relative	3. Exclamatory

Attribute (ii) allows the tagset to distinguish degree adverbs, which have a distinctive syntactic function (such as "very", "so", "too"), from others. Attribute (iv) enables the tagset to mark separately the "Wh-" or "Qu-" adverbs which are interrogative, relative or exclamatory in function. The relevant adverbs, in English, are "when", "where", "how" and "why".

7. Adposition

(i)	Type:	2. Fused prep-art

The additional value is for the benefit of those who do not find it practical to split fused words such as French "au" (= "à" + "le") into two textwords (cf. 4.3.1). This very common phenomenon of a fused preposition + article in West European languages can now be handled by using *Fused prep-art*, in effect assigning two tags to the same orthographic word (one for the preposition and one for the article).

8. Conjunctions

(ii)	Coord-Type:	1. Simple	2. Correlative	3. Initial
		4. Non-initial		

This attribute subclassifies coordinating conjunctions. It is easier to assign one tag to one orthographic word and it is therefore suggested that the four values are assigned as follows: *Simple* applies to the regular type of coordinator occurring between conjuncts: German "und", for example. When the same word is also placed before the first conjunct, as in French "ou... ou...", the former occurrence is given the *Correlative* value and the latter the *Simple* value. When two distinct words occur, as in German "weder... noch...", then the first is given the *Initial* value and the second the *Non-initial* value.

There are also examples of special extensions of the tagset which may be needed for specific languages. Again, the examples are purely illustrative and there is certainly no claim to completeness. In some cases, they derive from work already done on tagsets and their applications to texts. In other cases, they derive from specialist research, or from comments on an early draft of these guidelines, supplied by specialists in particular languages.

1. Nouns

An additional language-specific attribute is:

(vi)	Definiteness:	1. Definite	2. Indefinite	3. Unmarked	[Danish]

This is to handle the suffixed definite article in Danish: e.g. "haven" ("the garden"); "havet" ("the sea")

Additional values for recommended attributes:

(ii)	Gender:	4. Common		[Danish, Dutch]
(iv)	Case:	6. Vocative	7. Indeclinable	[Greek]

The *Common* gender contrasts with *Neuter* in a two-gender system.

2. Verbs

An additional attribute:

(xiii)	Aux.-function:	1. Primary	2. Modal	[English]

The primary (non-modal) auxiliaries are "be", "have" and "do".

An additional value to the non-finite category of verbs is arguably needed for English, because of the merger in that language of the gerund and participle functions. The -ing form does service for both and the two traditional categories are not easily distinguishable.

(v)	Verb-form/Mood:	9. -Ing form	[English]

3. Adjectives
Additional values for *Case*:

(iv)	Case:	5. Vocative	6. Indeclinable	[Greek]

4. Pronouns and Determiners
An additional value for *Gender* and for *Case*:

(ii)	Gender:	4. Common	[Danish]
(v)	Case:	7. Prepositional	[Spanish]

An additional attribute:

(xii)	Strength	1. Weak	2. Strong	[French, Dutch, Greek]

Weak and *Strong* distinguish, for example, "me" from "moi" in French and "me" from "mij" in Dutch.

5. Articles
Again, additional values for Article-Type, Gender and Case are:

(i)	Article-Type:	3. Partitive		[French]
(ii)	Gender:	4. Common		[Danish]
(iv)	Case:	5. Vocative	6. Indeclinable	[Greek]

6. Adverbs
Additional values for Adverb Type:

(ii)	Adverb-Type:	3. Particle	4. Pronominal	[English, German, Dutch]

In some tagging schemes, especially for English, a particle such as "out", "off" or "up" counts as a subclass of adverb. In other tagging schemes, the particle may be treated under *Residual* as a special wordclass. German and Dutch have pronominal adverbs such as German "daran", "davon", "dazu".

7. Adpositions
Values for adposition Type, in addition to 1. *Preposition* and 2. *Fused prep-art*:

(i)	Type:	3. Postposition	4. Circumposition	[German, English]

German "entlang" is a *Postposition* and arguably the "'s" which forms the genitive in English is no longer a case marking but an *enclitic Postposition*, as in "the Secretary of State's decision", "in a month or so's time". German "auf . . . hin" is an example which can be analysed as a *Circumposition*.

8. Conjunctions
An additional attribute, applying to subordinating conjunctions only:

(iii)	Subord.-type:	1. With-finite	2. With-infin.	3. Comparative	[German]

For example, in German, "weil" introduces a clause with a finite verb, whereas "ohne (zu...)" is followed by an infinitive and "als" is followed by various kinds of comparative clause (including clauses without finite verbs).

11. Unique/Unassigned
The following miscellaneous values may occur:

(i)	Unique-type:	1. Infinitive marker	[German "zu", Danish "at", Dutch, English]
		2. Negative particle	[English "not", "n't"]
		3. Existential marker	[English "there", Danish "der"]
		4. Second negative particle	[French "pas"]
		5. Anticipatory "er"	[Dutch]
		6. Mediopassive voice marker "se"	[Portuguese]
		7. Preverbal particle	[Greek]

5.3 INTERMEDIATE TAGSET

5.3.1 Basic Structure

For any tagset designed for the annotation of texts in a given language, the guidelines do not impose any particular set of choices to be used in distinguishing and representing grammatical categories. But it is important that the tagset should be mappable (automatically, if possible) onto a set of attribute-value pairs in conformity with the guidelines presented in 5.2.2. This includes the possibility (indeed the probability) that the annotator will need to define optional values other than those presented in 5.2.2.3 above.

This mapping will have the additional value that it will enable the annotator to transfer the information in a wordclass tagged corpus to the wordclass component of a lexicon (e.g. in order to record frequencies of word–tag pairs). It will also enable a lexicon of the given language (if one exists in conformity with the EAGLES guidelines for morphosyntactic information in the lexicon; cf. Chapter 11) to be used as a major input to automatic tagging.

To aid this mapping, and to test out its efficacy, we suggest that an *Intermediate Tagset* can be used as a language-neutral representation of a set of attribute-value pairs, based on the word categorization presented in 5.2.2 above. This can act as an intermediate stage of mapping between the tags assigned to textwords in corpus annotation and the labels assigned to words in a lexicon. Another important function of this Intermediate Tagset is to act as a basis for interchange between different local tagsets for particular corpora and particular languages. Note that the intermediate tagset is not intended to be used by human users, but only to serve as an interchange mechanism and a mechanism for validating tagsets.

A convenient linear method of representation is arrived at as follows:

(i) Represent the obligatory part-of-speech attribute value by using one or more letters, as indicated in 5.2.2.1:

N = noun	C = conjunction
V = verb	NU = numeral
AJ = adjective	I = interjection
PD = pronoun/determiner	U = unique/unassigned
AT = article	R = residual
AV = adverb	PU = punctuation
AP = adposition	

(ii) Represent the whole tag as a linear string of characters, each attribute (roman number (i), (ii), (iii), (iv), ...) representing the first, second, third, fourth, ... place in a string of digits.

(iii) Represent each value of each attribute by employing the arabic digits used in 5.2.2.2. Thus the interpretation of the string of digits will vary according to the part-of-speech category. (The optional attributes and values in 5.2.2.3 may also be used, but have to be specially defined for each tagset.)

Here are some examples: A common noun, feminine, plural, countable, is represented: N122010. A 3rd-person, singular, finite, indicative, past-tense, active, main-verb, non-phrasal, non-reflexive, verb: V3011141101200. A comparative, general adjective: AJ2000000. A coordinating conjunction, simple: C110. An interjection: I. A plural symbol (as in "two Bs"): R320.

5.3.2 Underspecification

An attempt is made in the Intermediate tagset to always make underspecification (cf. 4.3.2) of attribute values explicit. Wherever an attribute is inapplicable to a given word in a given tagset, the value 0 fills that attribute's place in the string of digits. When the 0's occur in final position, without any non-zero digits following, they could be omitted without loss of information. Thus a comparative general adjective could simply be represented: AJ2. However, for clarity, the 0's should be added.

There may be cases where a category needed for tagging in a specific language (given current limitations of automatic tagging) cuts across two or more values in the optional categories of the guidelines and may even cut across different attributes as well. If no special attribute value has been defined for such a case (as has been for the *oblique* value of the case attribute), it is possible to describe the category by means of the *OR* operator (|) and brackets to identify the arguments of this operator. Another operator we can use is the negative operator, signalled by the minus (-), so that -4 means "all values of this attribute except the 4th".

A good example (because of its complexity) is the base form of the English verb. The finite base form in English can be specified by using a disjunction "[finite indicative present tense [plural OR [first person OR second person] singular] OR imperative OR subjunctive]". This is spelled out, using the intermediate tagset, as follows:

V[[-301|002]111|000121|000130]0200001

Even this leaves out the non-finite use of the base form, as an infinitive. This example, awkward as it is, has an explanatory value: the relation between tagsets and a language-neutral representation can be very indirect. Although such cases as this are unusual, they show that the mapping between a lexicon and a tagged corpus is not always an easy one to automate.

5.3.3 Example tagsets

To illustrate the method of converting a tagset into this type of language-neutral labelling, we present a rendering into an *Intermediate Tagset* of a tagset for English and a set of dictionary codes for Italian based, for the former, on the English implementation of the above guidelines and, for the latter, on the DMI codes, provided by Nicoletta Calzolari, ILC, University of Pisa. For English, with its simple morphology, we find a complex interrelation between the wordclass guidelines and the requirements of a particular language. With other languages under consideration, the mapping from the language-specific tagset to the Intermediate Tagset is likely to be more straightforward.

5.3.3.1 English tagset, with intermediate tags.

Tag	*Description of word category*	*Example(s)*	*Intermediate Tag*
AJ	(Positive) adjective, general	big	AJ10000000
AJR	Comparative adjective	bigger	AJ20000000
AJT	Superlative adjective	biggest	AJ30000000
APR	Preposition	at, of	AP1
APO	Postposition	's	AP3
ATD	Definite article	the	AT1000
ATIs	Indefinite article, singular	a, an	AT2010
AV	(Positive) adverb, general	soon	AV1120
AVD	(Positive) adverb of degree	very, so	AV1220
AVDR	Comparative adverb of degree	more, less	AV2220
AVDT	Superlative adverb of degree	most, least	AV3220
AVDWQ	Adverb of degree, other wh-type	how	AV021[1|3]
AVR	Comparative adverb, general	sooner	AV2120
AVT	Superlative adverb, general	soonest	AV3120
AVWQ	General adverb, other wh-type	when, why	AV011-2
AVWR	General adverb, relative	where, why	AV0112
CC	Coord. conjunction, simple	and	C110
CCI	Coord. conjunction, initial	both (...and)	C130
CCM	Coord. conjunction, medial	neither (...nor)	C140
CSC	Subord. conjunction, comparative	than	C203
CSF	Subord. conjunction, with finite	if, while	C201
CSN	Subord. conjunction, with non-finite	in order (to)	C202

Tag	*Description of word category*	*Example(s)*	*Intermediate Tag*
DDs	Singular demonstrative determiner	this, that	PD001002010000
DDp	Plural demonstrative determiner	these	PD002002010000
DI	Indefinite det., neutral for number	no, some	PD000002020000
DIs	Indefinite determiner, singular	every, much	PD001002020000
DIp	Indefinite determiner, plural	both, many	PD002002020000
DVs1	Possessive det., 1st pers. sing	my	PD100102030000
DV2	Possessive det, 2nd person	your	PD200002030000
DV3sF	Possessive det, 3rd pers. sg. fem.	her	PD320102030000
DV3sM	Possessive det, 3rd pers. sg. masc.	his	PD310102030000
DV3sU	Possessive det, 3rd pers. sg. neut.	its	PD330102030000
DVp1	Possessive det, 1st pers. plur	our	PD100202030000
DVp3	Possessive det, 3rd pers. plur	their	PD300202030000
DWR	Relative determiner	which	PD000002040200
DWQ	Other wh-determiner	what	PD000002040-200
IJ	Interjection	Oh, Yes	I
NCs	Singular common noun	book, man	N101000
NCp	Plural common noun	books, men	N102000
NPs	Singular proper noun	Mary	N201000
NPp	Plural proper noun	Rockies	N202000
NUC	Cardinal numeral, neutral for number	two	NU10000
NUCs	Singular cardinal numeral	one	NU10100
NUCp	Plural cardinal numeral	fifties	NU10200
NUOs	Singular ordinal numeral	seventh	NU20100
NUOp	Plural ordinal numeral	sevenths	NU20200
PDs	Singular demonstrative pronoun	this	PD001001100000
PDp	Plural demonstrative pronoun	those	PD002001100000
PI	Indefinite pronoun, neutral for number	all, none	PD000001200000
PIs	Singular indefinite pronoun	anyone	PD001001200000
PIp	Plural indefinite pronoun	few, many	PD002001200000
PPs1N	Personal pronoun, 1st pers. sg. nom.	I	PD101011501000
PPs1O	Pers. pron., 1st pers. sg. obl.	me	PD101061501000
PP2	Pers. pron., 2nd pers.	you	PD2000[1|6]1501000
PPs3NF	Pers. pron., 3rd pers. sg. nom. fem.	she	PD321011501000
PPs3NM	Pers. pron., 3rd pers. sg. nom. masc.	he	PD311011501000
PPs3U	Pers. pron., 3rd pers. sing. neuter	it	PD3310[1|6]1501000
PPs3OF	Pers. pron., 3rd pers. sg. obl. fem.	her	PD321061501000
PPs3OM	Pers. pron., 3rd pers. sg. obl. masc.	him	PD311061501000
PPp1N	Pers. pron., 1st pers. pl. nom.	we	PD102011501000
PPp1O	Pers. pron., 1st pers. pl. oblique	us	PD102061501000
PPp3N	Pers. pron., 3rd pers. pl. nom.	they	PD302011501000
PPp3O	Pers. pron., 3rd pers. pl. oblique	them	PD302061501000
PRs1	Reflexive pronoun, 1st pers. sg.	myself	PD101001502000
PRs2	Reflexive pronoun, 2nd pers. sg.	yourself	PD201001502000
PRs3F	Reflexive pronoun, 3rd pers. sg. fem.	herself	PD321001502000
PRs3M	Reflexive pronoun, 3rd pers. sg. masc.	himself	PD311001502000
PRs3U	Reflexive pronoun, 3rd pers. sg. neut.	itself	PD331001502000
PRp1	Reflexive pronoun, 1st pers. plural	ourselves	PD102001502000
PRp2	Reflexive pronoun, 2nd pers. plural	yourselves	PD202001502000
PRp3	Reflexive pronoun, 3rd pers. plural	themselves	PD302001502000

Tag	*Description of word category*	*Example(s)*	*Intermediate Tag*
PVs1	Possessive pronoun, 1st pers. singular	mine	PD100101300000
PV2	Possessive pronoun, 2nd pers.	yours	PD200001300000
PVs3F	Possessive pronoun, 3rd pers. fem.	hers	PD320101300000
PVs3M	Possessive pronoun, 3rd pers. masc.	his	PD310101300000
PVs3U	Possessive pronoun, 3rd pers. neut.	its	PD330101300000
PVp1	Possessive pronoun, 1st pers. plural	ours	PD100201300000
PVp3	Possessive pronoun, 3rd pers. plural	theirs	PD300201300000
PWQ	Other wh-pronoun, neutral for case	what, which	PD000001400-200
PWQN	Other wh-pronoun, nominative	who	PD000011400-200
PWQO	Other wh-pronoun, oblique	whom	PD000061400-200
PWR	Relative pronoun, neutral for case	which	PD000001400200
PWRN	Relative pronoun, nominative	who	PD000011400200
PWRO	Relative pronoun, oblique	whom	PD000061400200
RFO	Formula	X/21	R200
RFW	Foreign word	mawashi	R100
RSY	Symbol, neutral for number	$, '	R300
RSYs	Symbol, singular	A, b	R310
RSYp	Symbol, plural	As, b's	R320
RUN	Unclassified	er, um	R600
UI	Infinitive marker	to (eat)	U1
UN	Negative particle	not, -n't	U2
UX	Existential "there"	there	U3
VM	Modal auxiliary verb	can, will	V0001100200002
VPB	Finite base form, primary auxiliary	be, do, have	V[[-301|002]111|000121|000130]0200001
VPD	Past tense, prim. aux.	did, had	V0001140200001
VPDR	Past tense -re form, prim. aux.	were	V[[201|002]11|00012]40200001
VPDZ	Past tense -s form, prim. aux.	was	V-2011140200001
VPG	-Ing form, prim. aux.	being, having	V0002900200001
VPI	Infinitive, prim. aux.	(to) be/have	V0002500200001
VPM	Pres. tense 1st pers. sg, prim. aux.	am	V1011110200001
VPN	Past participle, prim. aux.	been	V0002640200001
VPR	Pres. tense -re form, prim. aux.	are	V[201|002]1110200001
VPZ	Pres. tense -s form, primary auxiliary	is, does, has	V3011110200001
VVB	Finite base form, main verb	eat, have	V[[-301|002]111|000121|000130]0100000
VVD	Past tense, main verb	ate, had	V0001140100000
VVDR	Past tense -re form, main verb	were	V[[201|002]11|00012]40100000
VVDZ	Past tense -s form, main verb	was	V-2011140100000
VVG	-Ing form, main verb	leaving, being	V0002900100000
VVI	Infinitive, main verb	(to) leave/do	V0002500100000
VVM	Present tense 1st pers. sg., main verb	am	V1011110100000
VVN	Past participle, main verb	eaten, left	V0002640100000
VVR	Present tense -re form, main verb	are	V[201|002]1110100000
VVZ	Present tense -s form, main verb	is	V3011110100000

5.3.3.2 Italian DMI codes, with Intermediate Tags.

Tag	*Description of word category*	*Example(s)*	*Intermediate Tag*
AFN	Adj.pos.femm.inv.	carta/e valore	AJ12[1|2]0
ANS	Adj.pos.comm.sing.	dolce	AJ1410
ANP	Adj.pos.comm.plur.	dolci	AJ1420
AMN	Adj.pos.masc.inv.	complemento/i oggetto	AJ11[1|2]0
AFSS	Adj.sup.femm.sing.	grandissima, massima	AJ3210
AFPS	Adj.sup.femm.plur.	grandissime, massime	AJ3220
AMPS	Adj.sup.masc.plur.	grandissimi, massimi	AJ3120
AMSS	Adj.sup.masc.sing.	grandissimo, massimo	AJ3110
ANSC	Adj.com.comm.sing.	maggiore	AJ2410
ANPC	Adj.pos.comm.plur.	maggiori	AJ2420
ANNC	Adj.pos.comm.inv.	meglio, peggio	AJ24[1|2]0
ANN	Adj.pos.comm.inv.	pari, dappoco	AJ14[1|2]0
AFS	Adj.pos.femm.sing.	vera	AJ1210
AFP	Adj.pos.femm.plur.	vere	AJ1220
AMP	Adj.pos.masc.plur.	veri	AJ1120
AMS	Adj.pos.masc.sing.	vero	AJ1110
B	Adv.pos.	forte	AV1000
BC	Adv.com.	maggiormente	AV2000
BS	Adv.pos.mann.	fortemente	AV1600
BSS	Adv.sup.mann.	fortissimamente	AV3600
C	Conj.subord.	perche'	C200
CC	Conj.coord.	e	C100
DDMS	PrAdj.dem.masc.sing.	quello, quel	PD01100201
DDMP	PrAdj.dem.masc.plur.	quelli	PD01200201
DDFS	PrAdj.dem.femm.sing.	quella	PD02100201
DDFP	PrAdj.dem.femm.plur.	quelle	PD02200201
DDNS	PrAdj.dem.comm.sing.	cio'	PD04100201
DDNP	PrAdj.dem.comm.plur.	costoro	PD04200201
DIMS	PrAdj.ind.masc.sing.	alcuno, alcun	PD01100202
DIMP	PrAdj.ind.masc.plur.	alcuni	PD01200202
DIFS	PrAdj.ind.femm.sing.	qualcuna	PD02100202
DIFP	PrAdj.ind.femm.plur.	poche	PD02200202
DINS	PrAdj.ind.comm.sing.	ogni	PD04100202
DINP	PrAdj.ind.comm.plur.	tali, altrui	PD04200202
DEMS	PrAdj.escl.masc.sing.	quanto!	PD0110020003
DEMP	PrAdj.escl.masc.plur.	quanti!	PD0120020003
DEFS	PrAdj.escl.femm.sing.	quanta!	PD0210020003
DEFP	PrAdj.escl.femm.plur.	quante!	PD0220020003
DENS	PrAdj.escl.comm.sing.	quale!	PD0410020003
DENP	PrAdj.escl.comm.plur.	quali!	PD0420020003
DENN	PrAdj.escl.comm.inv.	che!	PD04[1|2]0020003

Tag	*Description of word category*	*Example(s)*	*Intermediate Tag*
DPMS1	PrAdj.poss.1p.masc.sing.	mio	PD11100201
DPMP1	PrAdj.poss.1p.masc.plur.	miei	PD11200201
DPFS1	PrAdj.poss.1p.femm.sing.	mia	PD12100201
DPFP1	PrAdj.poss.1p.femm.plur.	mie	PD12200201
DPMS2	PrAdj.poss.2p.masc.sing.	tuo	PD21100201
DPMP2	PrAdj.poss.2p.masc.plur.	tuoi	PD21200201
DPFS2	PrAdj.poss.2p.femm.sing.	tua	PD22100201
DPFP2	PrAdj.poss.2p.femm.plur.	tue	PD22200201
DPMS3	PrAdj.poss.3p.masc.sing.	suo	PD31100201
DPMP3	PrAdj.poss.3p.masc.plur.	suoi	PD31200201
DPFS3	PrAdj.poss.3p.femm.sing.	sua	PD32100201
DPFP3	PrAdj.poss.3p.femm.plur.	sue	PD32200201
DPMS1	PrAdj.poss.1p.masc.sing.	nostro	PD11100201
DPMP1	PrAdj.poss.1p.masc.plur.	nostri	PD11200201
DPFS1	PrAdj.poss.1p.femm.sing.	nostra	PD12100201
DPFP1	PrAdj.poss.1p.femm.plur.	nostre	PD12200201
DPMS2	PrAdj.poss.2p.masc.sing.	vostro	PD21100201
DPMP2	PrAdj.poss.2p.masc.plur.	vostri	PD21200201
DPFS2	PrAdj.poss.2p.femm.sing.	vostra	PD22100201
DPFP2	PrAdj.poss.2p.femm.plur.	vostre	PD22200201
DPNP3	PrAdj.poss.3p.comm.plur.	loro	PD34200201
DPNN	PrAdj.poss.comm.inv.	altrui	PD04[1|2]00201
DTMS	PrAdj.int.masc.sing.	quanto?	PD0110020001
DTMP	PrAdj.int.masc.plur.	quanti?	PD0120020001
DTFS	PrAdj.int.femm.sing.	quanta?	PD0210020001
DTFP	PrAdj.int.femm.plur.	quante?	PD0220020001
DTNN	PrAdj.int.comm.inv.	che?	PD04[1|2]0020001
DTNS	PrAdj.int.comm.sing.	quale?	PD0410020001
DTNP	PrAdj.int.comm.plur.	quali?	PD0420020001
DRNN	PrAdj.rel.comm.inv.	che	PD04[1|2]0020002
DRNS	PrAdj.rel.comm.sing.	quale	PD0410020002
DRNP	PrAdj.rel.comm.plur.	quali	PD0420020002
I		oh!	I
SFN	Noun comm.femm.inv.	attivita' (la/le)	N12[1|2]
SFP	Noun comm.femm.plur.	case	N122
SFS	Noun comm.femm.sing.	casa	N121
SMN	Noun comm.masc.inv.	re, caffe' (il/i)	N11[1|2]
SMP	Noun comm.masc.plur.	libri	N112
SMS	Noun comm.masc.sing.	libro	N111
SNN	Noun comm.comm.inv.	sosia (il/la, i/le)	N14[1|2]
SNP	Noun comm.comm.plur.	insegnanti (gli/le)	N142
SNS	Noun comm.comm.sing.	insegnante (un/una)	N141
SPFP	Noun prop.femm.plur.	Marie	N222
SPFS	Noun prop.femm.sing.	Maria	N221
SPMP	Noun prop.masc.plur.	Borboni	N212
SPMS	Noun prop.masc.sing.	Mario	N211

Tag	*Description of word category*	*Example(s)*	*Intermediate Tag*
PDMS3	Pron.dem.masc.sing.3	costui	PD31100110
PDMS	Pron.dem.masc.sing.	quello	PD01100110
PDMP	Pron.dem.masc.sing.	quelli	PD01200110
PDFS	Pron.dem.femm.sing.	quella	PD02100110
PDFP	Pron.dem.femm.plur.	quelle	PD02200110
PDNS	Pron.dem.comm.sing.	cio'	PD04100110
PDNP	Pron.dem.comm.plur.	tali	PD04200110
PEMS	Pron.escl.masc.sing.	quanto!	PD0110010003
PEMP	Pron.escl.masc.plur.	quanti!	PD0120010003
PEFS	Pron.escl.femm.sing.	quanta!	PD0210010003
PEFP	Pron.escl.femm.plur.	quante!	PD0220010003
PENS	Pron.escl.comm.sing.	che (vedo!)	PD0410010003
PENN	Pron.escl.comm.inv.	chi!	PD04[1|2]0010003
PIMS	Pron.ind.masc.sing.	uno	PD01100120
PIMP	Pron.ind.masc.plur.	alcuni	PD01200120
PIFS	Pron.ind.femm.sing.	una	PD02100120
PIFP	Pron.ind.femm.plur.	alcune	PD02200120
PINS	Pron.ind.comm.sing.	chiunque	PD04100120
PINP	Pron.ind.comm.plur.	tali, quali	PD04200120
PPMS1	Pron.poss.1p.masc.sing.	mio	PD11100130
PPMP1	Pron.poss.1p.masc.plur.	miei	PD11200130
PPFS1	Pron.poss.1p.femm.sing.	mia	PD12100130
PPFP2	Pron.poss.1p.femm.plur.	mie	PD12200130
PPMS2	Pron.poss.2p.masc.sing.	tuo	PD21100130
PPMP2	Pron.poss.2p.masc.plur.	tuoi	PD21200130
PPFS2	Pron.poss.2p.femm.sing.	tua	PD22100130
PPFP2	Pron.poss.2p.femm.plur.	tue	PD22200130
PPMS3	Pron.poss.3p.masc.sing.	suo	PD31100130
PPMP3	Pron.poss.3p.masc.plur.	suoi	PD31200130
PPFS3	Pron.poss.3p.femm.sing.	sua	PD32100130
PPFP3	Pron.poss.3p.femm.plur.	sue	PD32200130
PPMS1	Pron.poss.1p.masc.sing.	nostro	PD11100130
PPMP1	Pron.poss.1p.masc.plur.	nostri	PD11200130
PPFS1	Pron.poss.1p.femm.sing.	nostra	PD12100130
PPFP1	Pron.poss.1p.femm.plur.	nostre	PD12200130
PPMS2	Pron.poss.2p.masc.sing.	vostro	PD21100130
PPMP2	Pron.poss.2p.masc.plur.	vostri	PD21200130
PPFS2	Pron.poss.2p.femm.sing.	vostra	PD22100130
PPFP2	Pron.poss.2p.femm.plur.	vostre	PD22200130
PPNP3	Pron.poss.3p.comm.plur.	loro	PD34200130
PTNS	Pron.int.comm.sing.	chi?	PD0410010001
PTNN	Pron.int.comm.inv.	che?	PD04[1|2]0010001
PTMS	Pron.int.masc.sing.	quanto?	PD0110010001
PTMP	Pron.int.masc.plur.	quanti?	PD0120010001
PTFS	Pron.int.femm.sing.	quanta?	PD0210010001
PTFP	Pron.int.femm.plur.	quante?	PD0220010001

Tag	*Description of word category*	*Example(s)*	*Intermediate Tag*
PRNN	Pron.rel.comm.inv.	che, chi, cui	PD04[1|2]0010002
PRNS	Pron.rel.comm.sing.	quanto	PD0410010002
PRMS	Pron.rel.masc.sing.	quanto	PD0110010002
PRMP	Pron.rel.masc.plur.	quanti	PD0120010002
PRFP	Pron.rel.femm.plur.	quante	PD0220010002
PQNS1	Pron.pers.comm.sing.1	io	PD141001001
PQNS2	Pron.pers.comm.plur.2	tu	PD241001001
PQMS3	Pron.pers.masc.sing.3	egli, lui, esso	PD311001001
PQFS3	Pron.pers.femm.sing.3	ella, lei, essa	PD321001001
PQNP1	Pron.pers.comm.plur.1	noi	PD142001001
PQNP2	Pron.pers.comm.plur.2	voi	PD242001001
PQNP3	Pron.pers.comm.plur.3	loro	PD342001001
PQMP3	Pron.pers.masc.plur.3	essi	PD312001001
PQFP3	Pron.pers.femm.plur.3	esse	PD322001001
PQNS1	Pron.pers.comm.sing.1	me	PD141001001
PQNS2	Pron.pers.comm.sing.2	te	PD241001001
PQMS3	Pron.pers.masc.sing.3	lui, esso	PD311001001
PQFS3	Pron.pers.femm.sing.3	lei, essa	PD321001001
PQNP1	Pron.pers.comm.plur.1	noi	PD142001001
PQNP2	Pron.pers.comm.plur.2	voi	PD242001001
PQNP3	Pron.pers.comm.plur.3	loro	PD342001001
PQMP3	Pron.pers.masc.plur.3	essi	PD312001001
PQFP3	Pron.pers.femm.plur.3	esse	PD322001001
PQNS1	Pron.pers.comm.sing.1	mi	PD141001001
PQNS2	Pron.pers.comm.sing.2	ti	PD241001001
PQMS3	Pron.pers.masc.sing.3	gli	PD311001001
PQNP1	Pron.pers.comm.plur.1	ci	PD142001001
PQNP2	Pron.pers.comm.plur.2	vi	PD242001001
PQNP3	Pron.pers.comm.plur.3	loro	PD342001001
PQMP3	Pron.pers.masc.plur.3	li	PD312001001
PQFP3	Pron.pers.femm.plur.3	le	PD322001001
PFNS1	Pron.refl.comm.sing.1	mi (me stesso)	PD141001002
PFNS2	Pron.refl.comm.sing.1	ti (te stesso)	PD241001002
PFNN3	Pron.refl.comm.inv. 3	se', si	PD311001002
PFNP1	Pron.refl.comm.plur.1	ci	PD142001002
PFNP2	Pron.refl.comm.plur.2	vi	PD242001002
PFNP3	Pron.refl.comm.plur.3	loro	PD342001002
VFY	Verb aux. inf.pres.	avere	V00025101
VGY	Verb aux. ger.pres.	avendo	V00027102
VF	Verb main inf.pres.	amare	V00025101
VG	Verb main ger.pres.	amando	V00027102
VP1IFY	Verb aux. 1pl.ind.fut.	avremo	V10211302
VP2IFY	Verb aux. 2pl.ind.fut.	avrete	V20211302
VP3IFY	Verb aux. 3pl.ind.fut.	avranno	V30211302
VS1IFY	Verb aux. 1sg.ind.fut.	avro'	V10111302
VS2IFY	Verb aux. 2sg.ind.fut.	avrai	V20111302
VS3IFY	Verb aux. 3sg.ind.fut.	avra'	V30111302

Tag	*Description of word category*	*Example(s)*	*Intermediate Tag*
VP1IF	Verb main 1pl.ind.fut.	ameremo	V10211301
VP2IF	Verb main 2pl.ind.fut.	amerete	V20211301
VP3IF	Verb main 3pl.ind.fut.	ameranno	V30211301
VS1IF	Verb main 1sg.ind.fut.	amero'	V10111301
VS2IF	Verb main 2sg.ind.fut.	amerai	V20111301
VS3IF	Verb main 3sg.ind.fut.	amera'	V30111301
VP1CIY	Verb aux. 1pl.subj.impf.	avessimo	V10212202
VP2CIY	Verb aux. 2pl.subj.impf.	aveste	V20212202
VP3CIY	Verb aux. 3pl.subj.impf.	avessero	V30212202
VS1CIY	Verb aux. 1sg.subj.impf.	avessi	V10112202
VS2CIY	Verb aux. 2sg.subj.impf.	avessi	V20112202
VS3CIY	Verb aux. 3sg.subj.impf.	avesse	V30112202
VP1CI	Verb main 1pl.subj.impf.	amassimo	V10212201
VP2CI	Verb main 2pl.subj.impf.	amaste	V20212201
VP3CI	Verb main 3pl.subj.impf.	amassero	V30212201
VS1CI	Verb main 1sg.subj.impf.	amassi	V10112201
VS2CI	Verb main 2sg.subj.impf.	amassi	V20112201
VS3CI	Verb main 3sg.subj.impf.	amasse	V30112201
VP1IIY	Verb aux. 1pl.ind.impf.	avevamo	V10211202
VP2IIY	Verb aux. 2pl.ind.impf.	avevate	V20211202
VP3IIY	Verb aux. 3pl.ind.impf.	avevano	V30211202
VS1IIY	Verb aux. 1sg.ind.impf.	avevo	V10111202
VS2IIY	Verb aux. 2sg.ind.impf.	avevi	V20111202
VS3IIY	Verb aux. 3sg.ind.impf.	aveva	V30111202
VP1II	Verb main 1pl.ind.impf.	amavamo	V10211201
VP2II	Verb main 2pl.ind.impf.	amavate	V20211201
VP3II	Verb main 3pl.ind.impf.	amavano	V30211201
VS1II	Verb main 1sg.ind.impf.	amavo	V10111201
VS2II	Verb main 2sg.ind.impf.	amavi	V20111201
VS3II	Verb main 3sg.ind.impf.	amava	V30111201
VP1CPY	Verb aux. 1pl.subj.pres.	abbiamo	V10212102
VP2CPY	Verb aux. 2pl.subj.pres.	abbiate	V20212102
VP3CPY	Verb aux. 3pl.subj.pres.	abbiano	V30212102
VS1CPY	Verb aux. 1sg.subj.pres.	abbia	V10112102
VS2CPY	Verb aux. 2sg.subj.pres.	abbia	V20112102
VS3CPY	Verb aux. 3sg.subj.pres.	abbia	V30112102
VP1CP	Verb main 1pl.subj.pres.	amiamo	V10212101
VP2CP	Verb main 2pl.subj.pres.	amiate	V20212101
VP3CP	Verb main 3pl.subj.pres.	amino	V30212101
VS1CP	Verb main 1sg.subj.pres.	ami	V10112101
VS2CP	Verb main 2sg.subj.pres.	ami	V20112101
VS3CP	Verb main 3sg.subj.pres.	ami	V30112101
VP1DPY	Verb aux. 1pl.cond.pres.	avremmo	V10214102
VP2DPY	Verb aux. 2pl.cond.pres.	avreste	V20214102
VP3DPY	Verb aux. 3pl.cond.pres.	avrebbero	V30214102
VS1DPY	Verb aux. 1sg.cond.pres.	avrei	V10114102
VS2DPY	Verb aux. 2sg.cond.pres.	avresti	V20114102
VS3DPY	Verb aux. 3sg.cond.pres.	avrebbe	V30114102

Tag	*Description of word category*	*Example(s)*	*Intermediate Tag*
VP1DP	Verb main 1pl.cond.pres.	ameremmo	V10214101
VP2DP	Verb main 2pl.cond.pres.	amereste	V20214101
VP3DP	Verb main 3pl.cond.pres.	amerebbero	V30214101
VS1DP	Verb main 1sg.cond.pres.	amerei	V10114101
VS2DP	Verb main 2sg.cond.pres.	ameresti	V20114101
VS3DP	Verb main 3sg.cond.pres.	amerebbe	V30114101
VP1IPY	Verb aux. 1pl.ind.pres.	abbiamo	V10211102
VP2IPY	Verb aux. 2pl.ind.pres.	avete	V20211102
VP3IPY	Verb aux. 3pl.ind.pres.	hanno	V30211102
VS1IPY	Verb aux. 1sg.ind.pres.	ho	V10111102
VS2IPY	Verb aux. 2sg.ind.pres.	hai	V20111102
VS3IPY	Verb aux. 3sg.ind.pres.	ha	V30111102
VP1IP	Verb main 1pl.ind.pres.	amiamo	V10211101
VP2IP	Verb main 2pl.ind.pres.	amate	V20211101
VP3IP	Verb main 3pl.ind.pres.	amano	V30211101
VS1IP	Verb main 1sg.ind.pres.	amo	V10111101
VS2IP	Verb main 2sg.ind.pres.	ami	V20111101
VS3IP	Verb main 3sg.ind.pres.	ama	V30111101
VP2MPY	Verb aux. 2pl.imp.pres.	abbiate	V20213102
VS2MPY	Verb aux. 2sg.imp.pres.	abbi	V20113102
VP2MP	Verb main 2pl.imp.pres.	amate	V20213101
VS2MP	Verb main 2sg.imp.pres.	ama	V20113101
VNPPPY	Verb aux. comm.pl.part.pres.	aventi	V04226102
VNSPPY	Verb aux. comm.sg.part.pres.	avente	V04126102
VNPPP	Verb main comm.pl.part.pres.	amanti	V04226101
VNSPP	Verb main comm.sg.part.pres.	amante	V04126101
VP1IRY	Verb aux. 1pl.ind.past	avemmo	V10211402
VP2IRY	Verb aux. 2pl.ind.past	aveste	V20211402
VP3IRY	Verb aux. 3pl.ind.past	ebbe	V30211402
VS1IRY	Verb aux. 1sg.ind.past	ebbi	V10111402
VS2IRY	Verb aux. 2sg.ind.past	avesti	V20111402
VS3IRY	Verb aux. 3sg.ind.past	ebbe	V30111402
VP1IR	Verb main 1pl.ind.past	amammo	V10211401
VP2IR	Verb main 2pl.ind.past	amaste	V20211401
VP3IR	Verb main 3pl.ind.past	amarono	V30211401
VS1IR	Verb main 1sg.ind.past	amai	V10111401
VS2IR	Verb main 2sg.ind.past	amasti	V20111401
VS3IR	Verb main 3sg.ind.past	amo'	V30111401
VFPPRY	Verb aux. femm.pl.part.past	avute	V02226402
VFSPRY	Verb aux. femm.sg.part.past	avuta	V02126402
VMPPRY	Verb aux. masc.pl.part.past	avuti	V01226402
VMSPRY	Verb aux. masc.sg.part.past	avuto	V01126402
VFPPR	Verb main femm.pl.part.past	amate	V02226401
VFSPR	Verb main femm.sg.part.past	amata	V02126401
VMPPR	Verb main masc.pl.part.past	amati	V01226401
VMSPR	Verb main masc.sg.part.past	amato	V01126401

6 PERFORMANCE OF TAGGERS

Hans van Halteren

6.1 INTRODUCTION

One of the most important properties of a tagger is how well it does its work, i.e. how many of the tags it assigns are correct and how many superfluous tags are eliminated. As a result, reports on wordclass taggers invariably list numbers, usually percentages, that are supposed to indicate this property. Unfortunately, the manner in which these numbers are determined differs widely between reports, so that a straightforward comparison of taggers by a simple arithmetical comparison of these numbers is not possible.

This chapter first examines what is being measured and how useful the measures are. It then examines factors which influence the measurements themselves and which therefore ought to be kept invariant or at the very least described in reports.

6.2 PERFORMANCE MEASURES

In the literature one finds several measures which try to indicate the accuracy of the output of taggers. The most popular ones are two pairs of measures: *correctness and ambiguity*, which are the easiest to understand, and *precision and recall*, which are the best suited in an information retrieval environment.

H. van Halteren (ed.), Syntactic Wordclass Tagging, 81-94.

6.2.1 Definitions of measures

Correctness is a straightforward measurement of how many tokens receive a correct, i.e. *contextually appropriate*, tag. It is defined as the number of correctly tagged tokens, divided by the total number of tokens. It is usually expressed as a percentage. Correctness has many aliases in the literature, such as *success rate* (or its inverse, *error rate*), *percentage of correct tags* or the simple but rather inexact *score*. It is most often found in the description of taggers which yield a single tag for each token. When used for taggers which are allowed to yield multiple tags per token it is accompanied by a second measure, the *average number of tags per token* or *ambiguity*. This is calculated as the total number of tags, divided by the total number of tokens.

An alternative pair of measures, more geared towards the description of ambiguous taggings, is the combination *precision and recall*, which finds its origin in information retrieval. There, *precision* describes how many of the retrieved objects are correct matches for the user's query. For tagging, this means that precision measures how many of the tokens tagged X are tagged X correctly. *Recall* describes how many of the objects matching the user's query are retrieved. For tagging this means that recall measures how many of the tokens that ought to be tagged X are indeed tagged X. With these descriptions we get precision and recall values for each individual tag. In order to get a single value for the tagger output as a whole the descriptions are generalized to the following definitions: *precision* is the number of correct token–tag pairs that is produced, divided by the total number of token–tag pairs that is produced; *recall* is the number of correct token–tag pairs that is produced, divided by the number of correct token–tag pairs that is possible. Both precision and recall are usually expressed as percentages.

The two pairs of measures are related, but not equal. Recall is equal to correctness, except that, in the case of a genuinely ambiguous token, recall demands all correct tags to be present while correctness is satisfied with the presence of only one correct tag. This difference can probably be explained by the fact that correctness was designed for the description of single tag taggers and recall for multiple tag taggers. Note, by the way, that in the annotation of actual text, the effect of genuine ambiguity is negligible. As a result, the difference between recall and correctness is often ignored (or perhaps not even perceived) in the literature, leading to calculations of 'recall' in which unambiguously tagged corpora such as Brown are used as a benchmark, something which is not possible according our definition of recall since Brown undoubtedly contains genuine ambiguity. Ambiguity and precision both attempt to indicate the amount of superfluous tags, but precision includes a reference to correctness. If a text without

Table 6.1 Precision and recall for selected individual tags.

Tag	Description	Percentage of LOB Corpus	Recall	Precision
Overall		100	97.6	97.5
BEDZ	3rd sing. past of "be"	0.9	100	100
CS	subordinating conjunction	1.5	91.2	89.4
IN	preposition	10.6	98.3	97.1
NN	'normal' common noun	13.0	98.8	96.0
RP	adverbial particle	0.8	81.9	96.8
TO	infinitival "to"	1.4	98.5	97.7

Table 6.2 Confusion table for selected tags.

Proposed Tag	Correct Tag				
	CS	IN	RP	TO	Other
CS	16231	901	–	–	1014
IN	1111	120346	1272	197	860
RP	–	194	7123	20	21
TO	–	357	–	15460	4
Other	293	292	285	–	988380

genuine ambiguity is tagged with a single tag per token, ambiguity becomes 1, while precision becomes equal to correctness and recall.[1]

6.2.2 Usefulness of measures

Despite the popularity of these four measures, we have to ask ourselves if they are really all that useful. The information they provide only describes exactly how much work it will be to postprocess the tagged material: ambiguity and precision let us calculate how many superfluous tags have to be removed and correctness and recall let us calculate how many tags will have to be corrected. And even then we do not get all the information we need, because none of the measures tells us how the superfluous and erroneous tags are distributed, something which may well influence the amount of work involved.

[1] If a benchmark tagging is used to check the tagger output, this assumes that the tokenization by the tagger equals that of the benchmark.

One would expect that precision and recall at least would be suited for determining the quality of a tagger for information retrieval purposes. However, the values as used for taggers are no more than indications. As already stated, precision and recall are primarily measures which stand in relation to a specific query and a specific query refers to specific tags. Now, for most taggers the precision and recall fluctuates rather a lot from tag to tag. An example can be found in table 6.1, which shows a number of precision and recall measurements for the TOSCA/LOB tagger (producing a tagset almost identical to that described in Johansson 1986) when tested on the full LOB corpus.[2] Some tags have better scores than the overall one,[3] e.g. BEDZ with a perfect score but also, more surprisingly, the tag TO, which is usually deemed a difficult case. On the other hand, there are also tags, e.g. CS, with much lower scores.

However, even tag-specific scores lack information which might be vital to users. First of all, they do not show what type of errors are made, information which might indicate which other tag(s) should be involved in queries as well. A more useful presentation for such a purpose is the *confusion matrix*. An example is table 6.2, which, among other things, shows us that the errors for CS most often derive from a confusion with IN. Even more useful would be information about the context in which the errors are made, but it is hardly possible to represent this in a limited amount of space.

Furthermore, even tag-specific scores are only generalizations, namely over all words with that tag. If the end-users are linguists, their queries tend to concentrate on the 'interesting' words, which all too often correspond to problematic tags and/or problematic words within tags. As an extreme example we give the word "after", which leads to the precision and recall scores in table 6.3 and the confusion table in table 6.4.

There is another, maybe even more fundamental, problem with the current measures in an information retrieval environment. Hardly any user involved in information retrieval blindly follows the indexing information (here: tags) that is offered. Users know that generally the information that is offered at most approximates the information they want. With regard to tagging, it is likely that there are differences between the tagset that is provided and the one that is desired. And even when the tagset that is present in the data coincides with the one desired, there is a chance that the interpretation of the tags differs between tagger and end-user. Given the resulting likelihood of reinterpretation of tags, there is a marked decrease in the value of the performance measures, which only refer to the original interpretation of the tags. For information

[2]The scores given here cannot be extrapolated to tagging accuracy on new text as parts of this tagger (such as the Hidden Markov Model used) were also trained on the LOB corpus.

[3]Even though the TOSCA/LOB tagger produces a single tag for each token, overall precision is not equal to recall. This is because the tokenizer in the tagger arrived at a different tokenization than that found in the tagged LOB corpus.

Table 6.3 Precision and recall for the word "after".

Tag	Description	Percentage in Benchmark Tagging	Recall	Percentage in Tagger Output	Precision
Overall		100	82.4	100	82.4
CS	subordinating conjunction	15.1	86.4	27.6	47.2
IN	preposition	83.2	82.2	70.9	96.4
JJB	attributive-only adjective	0.7	50.0	0.4	100
RI	adverb (homograph of preposition)	1.0	63.6	1.2	53.8

Table 6.4 Confusion table for the word "after".

Proposed	Correct Tag			
Tag	CS	IN	JJB	RI
CS	146	160	–	3
IN	23	766	4	1
JJB	–	–	4	–
RI	–	6	–	7

retrieval it is more important that the tagging is consistent, i.e. that the same word in the same context receives the same tag. Obviously, correctness and consistency are linked, since for any meaningful tagset correct tagging generally implies consistent tagging. However, they are certainly not the same and a correctness measure cannot easily be translated into a consistency measure. In fact, consistency is much harder to measure than correctness. Possibly for this reason, it usually remains unmeasured. As a result, nothing much can be said about the consistency of existing taggers, apart from the obvious fact that rule based systems tend to produce a more consistent tagging than probabilistic systems.

Yet another problem with the measures is that they refer to individual tokens. For some uses of the tagged text other units are more appropriate. A good example is parsing: the presence of even one incorrect tag in a sentence can prevent a parser from producing a correct parse. Here, the sentence (or rather utterance or text unit) would be a better unit for measurement of correctness. In other uses, such as the identification of technical terms in a text, the optimal unit for measurement is harder to determine, especially since tokens may be ambiguous as to the exact unit or units they belong to.

Table 6.5 Material used for tagging experiments.

Ident	Text Type	#Samples	#Utterances	#Tokens
A	Newspaper	810	3355	68654
B	Newspaper	810	3283	68521
C	Scientific	236	2799	69080
D	Scientific	256	2894	68722

We can conclude that there are circumstances under which the existing measures, correctness/ambiguity and precision/recall, are not optimally useful. However, for general purposes they suffice. In addition, they are well-defined and easily understood. As a result, we support the use of these measures and will use them ourselves in this book.

6.3 PERFORMANCE MEASUREMENTS

Since we now have reasonably good measures of the quality of tagger output, does this mean that we can always judge the relative quality of two taggers by comparing measurements, e.g. correctness values? The answer to this question must, unfortunately, be negative. First of all, the translation from the correctness percentage to a quality judgement has to consider the amount and the type of the information conveyed by the tagset used by the tagger. Secondly, the correctness percentages themselves may not be comparable due to differences in the methods of measuring.

In this section we will attempt to demonstrate the potential pitfalls of performance measurements, not only theoretically but also by way of a number of experiments. We hope that this demonstration will not only serve as a warning to the user but will also inspire authors of future reports to clearly describe what exactly they have been measuring.

6.3.1 Experimental setup

In the experiments we use a number of taggers to tag Dutch text with the WOTAN tagset (Berghmans 1994). This tagset, of some 240 tags, encodes not only the part-of-speech and morphological features, but also some more syntactic information. An example is "V(trans,ott,1_of_2_of_3,mv)" which indicates a verb ("V"), in present tense ("ott"), 1st, 2nd or 3rd person plural ("1_of_2_of_3,mv") and used as a transitive ("trans"). In the experiments we use four text sets taken from the Eindhoven corpus (Uit den Boogaart (ed.) 1975), annotated with manually corrected WOTAN tags. The properties of the text sets are listed in table 6.5. The union of all four is used as well and referred to as U.

The taggers used in the experiments are all built with three components. The first is a lexical lookup component which is used to determine all possible tags for known tokens. The second is a lexical heuristics component based on word endings which proposes possible tags if the token is unknown, i.e. if it is not in the lexicon. The third is a *Hidden Markov Model* (HMM; cf. Chapter 16) which determines which of the possible tags is the contextually most likely. Of each of the three components there exist five versions, one derived from or trained on each of the tagged text sets A, B, C, D and U. Note that the size of the text sets is insufficient for the training tasks, i.e. one would not expect a training set this size to yield a tagger with such a detailed tagset which can be generally applied. In these experiments, however, this insufficiency will lead to exaggerated differences which help demonstrate the points we want to make.

Each complete tagger consists of specific combinations of components and is indicated by the identifications of its components, e.g. tagger BUA uses lexicon B, heuristics U and HMM A. To keep the complexity of the experiments within bounds, the texts are not retokenized if they are used as test material. In particular, multi-token units (e.g. proper names) are marked as such and treated as if they were single tokens.

6.3.2 Effects of the tagset

The first reason for the difficulty of correctness comparison, then, is that the meaning of the correctness percentage number is dependent on the tagset used by the tagger. A tagger with a correctness of 95 percent with tagset X might well be better than a tagger with a correctness of 97 percent with tagset Y. This is because the tagger performance must be measured against the difficulty of its task. For this we should first look at the ambiguity of the tokens in the text. The more different choices there are for each position, the more problematic the selection of the contextually correct one will be. In addition, there may be differences in the type of ambiguity. Ambiguity in some features (e.g. tense) can often be resolved with a small context window. For other features (e.g. transitivity) much more information is needed. As a consequence, presence of such features in a tagset usually leads to lower correctness scores. Note that the size of the tagset only has an indirect influence and cannot be used to express the difficulty of the tagger's task. As an example, consider the practice of adding *portmanteau tags* (cf. 4.3.2), which increases the size of the tagset but reduces ambiguity and hence makes the tagging task easier.

To demonstrate the relation between difficulty and correctness, we use tagger AAA to tag all text sets. The resulting correctness values, measured by comparing with the original tagging of the material, are listed in the first row (P240) of table 6.6. The corresponding ambiguity values are listed in table 6.7. The correctness and ambiguity vary considerably from text set to text set but this is not important in this experiment (see below for a discussion of this aspect). Instead, the difference between tagsets is of interest: if we do not measure complete equality of the tags but just equality of

Table 6.6 Correctness of tagger AAA at different resolutions.

Tagset	Text Set A	B	C	D	U
P240	97.9	88.8	85.8	87.0	89.9
P13	99.1	95.5	92.7	93.8	95.3
P112	97.9	89.3	86.3	87.5	90.3
P127	98.6	92.3	89.5	90.5	92.7

Table 6.7 Ambiguity in tagger AAA at different resolutions.

Tagset	Text Set A	B	C	D	U
P240	2.08	2.41	2.45	2.42	2.34
P13	1.40	1.52	1.55	1.55	1.51
P112	2.02	2.30	2.34	2.32	2.24
P127	1.69	1.95	1.96	1.97	1.89

the parts-of-speech (13 different ones), this leads to the values listed in the second row (P13). In effect, this simulates tagging the material with a tagset of 13 tags. The correctness is much higher and the ambiguity much lower. That it is ambiguity and not size which is the most important factor is demonstrated with two intermediate tagsets P112 and P127. The translation from P240 to P127 removes a number of 'difficult' features, such as transitivity. For P112, on the other hand, the translation removes a number of 'easy' features, such as number. P112 is the smaller of the two (112 vs. 127 different tags) but has a higher ambiguity and, as expected, correctness for P112 is lower than for P127.

At this point we want to repeat that the tokenization and especially the recognition of multi-token units in the experiment is already catered for in the input to the system. This means that the current series of experiments cannot be used to demonstrate the influence of attempted automatic recognition of multi-token units. However, earlier experiments (van Halteren 1996) have shown that in tagging English with the TOSCA tagset fully one third of the errors consisted of misassignment of multi-token unit membership. Considering that these direct errors can be assumed to have caused additional indirect errors, we can conclude that the difficulty of the tagging task is increased considerably if the tagger attempts to determine by itself whether or not tokens should be grouped into larger units. A good way of measuring this is the use of *ditto tags* (cf. 4.3.1) to indicate membership of a larger unit and to include ditto tags in the ambiguity count.

6.3.3 Effects of the method of comparison

Even when comparing two taggers which use the same tagset one has to be careful in the interpretation of published results. The reason is that the method used in the evaluation can greatly influence the results.

On designing an evaluation we first have to decide how we are going to determine whether or not a tag is correct. There are basically two ways of doing this. The easiest approach is to take the output of the tagger and go through it, marking the tags one thinks are not correct. There are two problems with this approach. First, a human checker tends to miss things, especially if there is a large amount of text to go through and/or the tagset is a detailed one. Secondly, even when checking very thoroughly, the tagger will often get the benefit of the doubt, e.g. when the checker is not completely certain what the correct tag should be. Another, more objective, approach is to prepare a benchmark tagging with which the output of the tagger can be compared. Here, too, there are problems. Obviously, since it is probably prepared by humans, the benchmark may itself contain mistakes. Optimally, several people, preferably experienced in the use of the tagset, should independently tag the text, after which they should examine the differences and try to reach a consensus on the choice of the correct tag. However, it may well be that there is not just a single correct tag for each token. With some tagsets (and extensive and specific tag application manuals) a single correct choice seems to be possible: Voutilainen and Jarvinen (1995) show that full agreement can be reached for the EngCG tagset.[4] Marcus *et al.* (1993), on the other hand, show that even for the small Penn tagset (48 tags, including punctuation) annotators on average disagree about the tagging of 3-4 percent of the text, a percentage which is comparable to the percentage of errors made by some automatic taggers.[5] It would appear that a report on the level of agreement between human annotators, both before and after negotiation, is a welcome addition to reports on tagger correctness for the same tagset.

To examine the effects of the choice of comparison method the author manually checked a number of samples from text set B (972 words in all), tagged with tagger AAA. The check was done as it would have been for a larger amount of text, i.e. contemplation of each token took seconds rather than minutes. 83 tags were marked as erroneous, yielding a correctness score of 91.5 percent. Afterwards an automatic comparison method was used to check the tagger output against the original tagging of the material. This produced an error count of 128 and a correctness score of 86.8

---

[4] The full agreement was reached after negotiation; before negotiation agreement was not full, but still over 99%. A joint examination by the annotators showed the remaining cases to be mainly clerical errors, with only a few initial differences that were due to an incomplete manual. None of the differences turned out to be due to a 'genuine difference of opinion' (cf. Church 1992).

[5] Ratnaparkhi (1996) even argues that in a set of features that could be used by a tagger to determine the correct tag in a context, one of the most useful features is the name of the human annotator responsible for annotating the particular sentence which the tagger output is to be compared to.

percent. Closer examination shows several interesting facts about the differences. First of all, 9 of the 128 tags marked by the automatic comparison are actually correct and are errors in the original annotation. Next, the manual check missed 29 certain errors. However, the automatic check was not flawless either for the manual check uncovered 3 errors not found in the automatic comparison. It may well be that there are further errors unrecognized by both methods. Finally, there are 12 tags where either the automatic (11) or the manual check (1) marks an error but where closer examination reveals that both proposed tags can be defended as being correct. Strict rules as to how tags should be assigned, not available during the experiment, are needed to decide the issue in these cases. Assuming that all errors have been spotted by at least one of the methods, the correctness of the tagging is 88.6 percent, with 1.3 percent allowing more than one correct tag. Neither method yields perfect results but the benchmarking method is closer to the combined score and, especially in the light of the large amount of errors missed in the manual check, appears to be a better choice.[6]

6.3.4 *Effects of choice of tokens measured*

We not only have to ask ourselves how to compare tags, but also for which tokens we want to do so. A text contains words and punctuation and may in addition contain markup. If we do not take special precautions we are likely to get a measurement based on all three types of tokens. Now it is hard to imagine a sensible syntactic wordclass tag for markup, so this should certainly be excluded from measurement. If it is not – and which author ever specifies this? – a heavily marked up text will have a remarkably high tagging correctness.[7] For punctuation, matters are slightly more complicated, for they do have a syntactic meaning and may even be ambiguous. As a result they are usually given a tag and are also included in the measurement. Generally, however, punctuation tends to be tagged as being itself and hence is unambiguous. If this is the case, punctuation tags are always correct and boost the correctness. Table 6.8 shows what correctness percentages calculated on words plus punctuation translate to when considering only the words. It shows the recalculated score of popular correctness percentages if the punctuation forms 10, 15 or 20 percent of the total number of tokens. The differences may not be overwhelming, but they do help to reach that magical number one is aiming for.

[6] For a more optimistic view of human annotators, see Baker (1997).

[7] Alternatively, markup may have been removed on purpose before the measurement, because some taggers have problems with SGML tags and entities, e.g. splitting "é" in three tokens and tagging each of them.

Table 6.8 Recalculated correctness if punctuation is excluded.

Correctness	Reported Percentage of Punctuation 10	15	20
97	96.7	96.5	96.3
95	94.4	94.1	93.8
90	88.9	88.2	87.5
85	83.3	82.3	81.3

6.3.5 Effects of separation of test and training material

Another major influence on measurements is the relation between training material and test material. The best results are achieved if the test material is equal to the training material. In such a test, training is optimally tuned for the task at hand: all information that is needed is present and no superfluous information hinders the work. This also means that the smaller the size of the material, the better. The strictest test, and the one most likely to give an indication of what will happen if the tagger is used on new texts, is to have no overlap between test and training material whatsoever. Intermediate strategies involve having a partial overlap, e.g. training set is included in the test set or the test set is taken from the same sources as the training set so that at least the lexicon is more attuned.

The influence of the choice of test material is best shown by using taggers derived completely (i.e. lexicon, heuristics as well as HMM) from a single text set. The results with these taggers are listed in table 6.9. There is an enormous difference between the correctness if the test set is the training set (consistently around 97.9 percent for the component text sets) and if it is not (between 85.6 and 88.8 percent). As predicted earlier, the small size of the training sets serves to produce such exaggerated differences. Furthermore, the differences are a compounded effect. On the one hand they are caused by the appropriateness of the lexicon to the test set, on the other hand by the appropriateness of the HMM. We will try to separate these two influences in some more experiments.

The appropriateness of the lexicon can be measured directly. Its completeness can be expressed in terms of the percentage of tokens that are known. However, this percentage does not yield enough information by itself, since a token may be known but the correct tag may not be among those provided and, on the other hand, a token may not be known but the heuristics can provide the correct token. Instead, or in addition, we can use the percentage of tokens for which the correct tag is present. The amount of superfluous information, the lexicon's overcompleteness, can be expressed in a measure we already have: ambiguity. In order to examine the influence of the lexicon in practice we use

Table 6.9 Correctness for single training set taggers.

Tagger	Text Set A	B	C	D	U
AAA	97.9	88.8	85.8	87.0	89.9
BBB	88.7	97.9	85.6	87.0	89.8
CCC	85.6	85.7	97.9	87.7	89.3
DDD	86.0	86.4	87.8	98.0	89.6
UUU	96.4	96.8	96.7	96.9	96.7

Table 6.10 Token known by lexicon in xUU.

Tagger	Text Set A	B	C	D	U
AUU	99.4	85.6	82.9	83.3	87.8
BUU	85.7	99.4	82.7	83.4	87.8
CUU	81.3	81.4	99.8	84.8	86.8
DUU	81.4	82.0	85.8	99.6	87.2
UUU	99.5	99.5	99.8	99.7	99.6

Table 6.11 Correct tag present for xUU.

Tagger	Text Set A	B	C	D	U
AUU	100	97.3	95.2	96.2	97.2
BUU	97.4	100	95.1	96.2	97.2
CUU	96.8	96.7	100	97.0	97.6
DUU	96.7	96.8	96.8	100	97.6
UUU	100	100	100	100	100

five taggers in which the lexicon is derived from different text sets but in which the lexical heuristics and the HMM are derived from all our material.

The measurements are listed in tables 6.10 (token in lexicon), 6.11 (correct tag present), 6.12 (ambiguity) and 6.13 (correctness of output). Note that even for the text sets from which the lexicon was derived the percentage of known tokens is not 100, because there are tokens which are potentially unknown for reasons of capitalization. It is obvious from the tables how influential a factor the completeness of the lexicon is. Furthermore, the effect of overcompleteness becomes clear on comparison of the table entries where the percentage of available correct tags is 100: on the individual

Table 6.12 Ambiguity in xUU.

Tagger	Text Set A	B	C	D	U
AUU	2.10	2.48	2.43	2.44	2.36
BUU	2.56	2.16	2.47	2.49	2.42
CUU	2.90	2.90	2.37	2.72	2.72
DUU	2.79	2.77	2.55	2.26	2.59
UUU	2.96	2.96	2.92	2.95	2.95

Table 6.13 Correctness of output of xUU.

Tagger	Text Set A	B	C	D	U
AUU	97.0	92.1	89.8	91.0	92.5
BUU	91.8	97.2	89.6	91.0	92.4
CUU	90.3	90.6	97.2	91.8	92.5
DUU	90.3	90.9	91.8	97.4	92.6
UUU	96.4	96.8	96.7	96.9	96.7

text sets tagger UUU consistently scores 0.5 percent lower than the tagger using the correct tuned lexicon. Obviously, these effects need attention especially if the lexicon is derived from training material. If the lexicon is acquired in another manner, e.g. derived from an existing machine readable dictionary, they become less important.

The other side of the relation between training and test material is the appropriateness of the Hidden Markov Model. To examine this aspect we use an experiment similar to the previous one. This time we use the full lexicon U and lexical heuristics U, but the five different Hidden Markov Models derived from each of the text sets. This means that the correct tag is present for all tokens. Ambiguity varies minimally between text sets, as can be seen in the UUU row of table 6.12, which corresponds to the current choice of lexicon and heuristics. The resulting correctness measurements are listed in table 6.14. They vary only slightly less than those in table 6.13 and show the same pattern: the more overlap and the smaller the sets, the better the results are. This means that even for taggers which only use training material for context statistics, we should be extremely wary of extrapolating measurements made with overlapping test and training sets, especially for small and fully overlapping sets.

Table 6.14 Correctness of output of UUx.

Tagger	Text Set				
	A	B	C	D	U
UUA	97.6	93.1	92.1	92.4	93.8
UUB	92.9	97.6	92.0	92.3	93.7
UUC	92.6	92.7	97.6	92.8	93.9
UUD	92.7	92.9	92.5	97.7	93.9
UUU	96.4	96.8	96.7	96.9	96.7

6.3.6 Effects of representativity of test material

The final aspect of evaluation we wish to discuss is the representativity of the test material. As we have seen above, the completeness of the lexicon and the statistics greatly influence the results of a tagger. Assuming that there are marked differences between language varieties, we can expect that test results for one type of text cannot be easily extrapolated to another type (cf. Weischedel *et al.* 1993; Biber 1993). If we reexamine table 6.9, we see that on top of the enormous distance between training set and test set there is another clearly visible gap of 2 to 3 percent between same variety text and different variety text. Since all four text sets are of the same general type, i.e. written non-fiction, we can expect even greater differences when using these taggers on fiction or spoken material. Interestingly, there are also differences between the results for text sets of the same type, especially the scientific sets C and D. This means we should be careful in the interpretation of test results even if they have been measured on the desired text type.

7 SELECTION AND OPERATION OF TAGGERS

Hans van Halteren

7.1 INTRODUCTION

Discussions about taggers tend to focus on the software involved. However, this software should always be considered in relation to the task it is used for and the people it is used by. This chapter examines the role of the user in the tagging process, both in the preparatory phase, during selection of a tagger to use (7.2), and in the operational phase, during the actual use of the tagger (7.3). The user obviously has a further important role to play in the exploitation of the tagged material, but this role is so dependent on the exact use of the material (cf. Chapter 3) that we will not attempt to describe it here.

7.2 SELECTION OF A TAGGER

For each kind of text research one must first decide whether or not tagged text will be more useful than raw text. If this decision turns out in favour of tagging, one is faced with a new and possibly much harder choice: which tagger to use? First and foremost, obviously, the choice of tagger is determined by the language which is to be tagged. And this is really the only *sine qua non*; all other factors must be weighed and, hence, can be outweighed.

H. van Halteren (ed.), Syntactic Wordclass Tagging, 95-104.

We see all too often that a selection is made simply on the basis of availability. Possibly a tagger is already available at one's institute. Or a new tagger is chosen because it happens to run on the computer platform one uses. Or the tagger just happens to be the cheapest, possibly even free and in the public domain. Now availability, platform and cost are certainly factors which should be considered but they ought to be only secondary ones. The prime consideration should always be that the tagger is suited for the job it is supposed to do. This means that it is not possible to award any tagger with the predicate *best tagger for language X*. The perfect tagger for one job may be completely useless for another.

7.2.1 Tagset

The most important factor in the suitability of a tagger is that the form of its output (its *tagset*; cf. Chapters 4 and 5) is suitable. We advise every user to write down the specifications for the tagset he needs for his work before examining the available taggers, as this tends to lead to a more objective comparison. The major decision here relates to the fundamental nature of the tagset: should it be a linguistic tagset and, if so, should it adhere to a specific linguistic tradition? Further decisions focus on the finegrainedness of the tagset and on exactly which groups of words are to be distinguished. Here, one should not assume that more is better, but one should rather aim for just enough. In that way the other requirements for the tagger, such as acceptable performance, are not impeded by unnecessary demands.

After the specifications have been set up and taggers are being examined, one should keep in mind that suitability of a candidate tagset is not the same as equality with the specified tagset. It is often possible to translate between tagsets, e.g. by way of the EAGLES intermediate tagset (5.3), so that compatibility ought to be sufficient. Also, when we suggest a specification beforehand, we do this under the assumption that the tagger is to be acquired for a single job. If the possible future uses of the tagged text are unknown and the user is therefore unable to provide clear specifications, it is advisable to opt for conformance with standards such as EAGLES.

7.2.2 Documentation

The next factor to consider is the documentation accompanying the tagger. For a tagger to be useful, it must be amply documented. Documentation on installation and operation of the software are usually well provided for. Equally, or possibly even more, important, however, is documentation on the in- and output of the tagger.

As for the input, most taggers make assumptions about the form of the text they are going to tag. In extreme cases, tokens must be separated and decapitalized beforehand. Most taggers are more robust than this, but even these may yet run into problems, e.g. when faced with binary files from a word-processor rather than with pure ASCII text. Documentation should specify what the tagger can and cannot handle in terms of working at all (e.g. "this tagger needs SGML entity representations for diacritics and will crash when encountering 8-bit MSDOS codes") as well as in terms of working well (e.g. "the user should put mentioned words in <mention> markers if he wants them tagged as nouns instead of as themselves").

As for the output of a tagger, we dare say that tagged text becomes all but useless if there is no clear description of what the individual tags mean and how they are applied to actual text.[1] In a so called *tagset manual*, an explanation should be present of each tag in terms of major wordclass and subcategorizations and/or other features. For possibly unclear (if not all) distinctions, exact criteria should be specified. For *open* classes these criteria should be supported by a number of examples and for *closed* classes complete lists should be provided. Apart from examples per tag, the documentation should also contain contrastive examples for ambiguous words, showing what choices are to be made in which context and why. Finally, the presence of a large piece of actual tagged text can be useful for illustration and additional exemplification.

When examining the suitability of a tagger, the user can best judge the documentation by attempting to manually tag some text (preferably the example text in the documentation as well as some of the text the tagger is supposed to be tagging later on), following the criteria laid down in the tagset manual. If the information provided is insufficient for this task, it is very likely that the use of the tagged text produced by the tagger will be problematic as well.[2]

7.2.3 The tagging process

After judging whether the functionality of a tagger is sufficient, it is also necessary to examine if the tagging can be done in the amount of time that is available. This is determined by two factors: the *processing speed* of the automatic tagger and the amount of *user interaction* that is required.[3]

Of these, the speed of the tagger is the easiest to measure. Preferably one should perform a test run on the machine on which the tagger will have to do its work. If this is not possible, measurements on other machines can be made. However, when comparing speeds, one should take into account the platform on which the measurements were made, noting not just the processor but also size of memory, speed of harddisk, etc.

[1] Also, good documentation may allow reinterpretation of a text tagged with a tagset originally meant for a different purpose. Obviously, it is a good sign if the documentation contains a translation table to the EAGLES intermediate tagset.

[2] If the tagset manual is inadequate, the user may have already had trouble judging the suitability of the tagset itself.

[3] A further factor which may be of importance is the type of user interaction that is required. It may well be that in the work environment in which the tagger is to be used, some of the types of interaction (described in 7.3) are not feasible. Note, however, that opting for a fully automatic tagger to avoid all human involvement can turn out to be a delusion: performance requirements may well force a human correction phase after all.

A quantification of the time-investment by the user is much more elusive. There are many modes of human–machine interaction (see 7.3 below) and in most of these one should count on a substantial training period before optimal speed is reached (cf. Marcus *et al.* 1993). Since the training time tends to be longer than one wants to invest just for selection, it is impossible to get exact measurements. All one can do is attempt to get a feel for the user-friendliness of the interaction process (difficulty of the task, support by documentation and by special tools such as disambiguation environments) and an impression of the order of magnitude of the time involved.[4]

The importance of speed is much more variable than that of the functionality of the tagger. For some purposes, e.g. the one time tagging of a single book, all but the lowest speeds are acceptable. For others, e.g. the daily tagging of all of the day's newspapers, it is of the utmost importance. Every user will have to decide for himself how important speed is. Again, one should attempt to decide on the desired and/or acceptable speed before examining individual taggers.

7.2.4 Performance

The final factor we want to address is the quality of the output. Chapter 6 describes in detail how this is measured and how difficult it can be to translate the resulting measurements into a judgement on the relative quality of a tagger. Again, we can only suggest that the user should decide beforehand what is expected of the tagger, both in terms of the minimally acceptable and in terms of the preferred quality. Measurements should be based on the use that is to be made of the tagged material and hence should concentrate on the tags/features that are important for the task at hand rather than on the full tagset. It should also be noted that the way in which the quality is measured should be related to any correction phase envisaged, e.g. consistent errors are preferable to inconsistent ones.

7.2.5 Combining the factors

The selection of a tagger, then, ideally consists of three steps. First, one determines what the tagger is supposed to do. One specifies criteria for all the relevant factors and attaches a relative importance to each factor. Then one examines each individual candidate tagger, preferably with one's own test data on one's own machine, and determines how they score for each factor. Any tagger which scores below the minimum criteria for even one factor can be eliminated from the selection process. Finally, one compares the remaining taggers, using the compound score determined by the factor scores and their weight, and selects the tagger with the highest score.

---

[4] The information gained by this exercise is not just useful for the selection of a tagger. It can also serve as the basis of planning for the actual annnotation work.

In practice, of course, the procedure is never as mathematically precise as it is presented here. However, the methodology ought to be clear and should be usable even if it is not possible to determine criteria which are expressed as numbers precise to four digits after the decimal point. The main point is that one should not let the tagger determine what one can do with one's data but that one lets one's plans with the data determine which tagger one uses.

There is one special case, which completely thwarts our selection methodology. This is when all taggers are rejected because they all fall short of either individual or compound criteria. In this case, there are only two options. The first is to try to build one's own tagger (cf. Part II of this book). This is methodologically preferable, but may well be practically impossible, for any number of reasons. The alternative is to lower one's standards. Care should be taken, however, that the lowered standards are still sufficient for the planned use of the tagged text. If they are not, one should not proceed on the assumption that any tagging is better than no tagging at all, but should rather re-evaluate the necessity of tagging.

7.3 USER INTERACTION

Although much of the literature gives the impression that tagging software does its work completely automatically, there is usually a certain amount of work that the user is required and/or allowed to contribute. Since tagging software generally lacks access to semantics, pragmatics and world knowledge, it cannot be expected to select the contextually appropriate tag for each and every token all by itself. For some purposes a small percentage of mistagged tokens is acceptable, but if the error rate is too high, an additional source of information must be consulted. For the time being, that source can only be a human, i.e. the user.[5]

From the user's point of view, user interaction is a two-edged sword. On the one hand, it can be expected to improve the quality of the output. On the other, it only works if one puts in a certain amount of effort, an amount which may become impossible to provide if one wants to tag, say, millions of words of text. Most likely, the user is best served with a system in which he is allowed but not required to contribute information. How close each individual tagger comes to this ideal depends on the way the user interaction is integrated in the tagging process. The freedom of the user to do as much or as little as he wants is complete if the interaction takes the form of *pre- or post-processing* (as long as the tagger does not just assume all the pre-processing has been done and may crash if it has not). Freedom is completely absent if the interaction is

[5] This section might as well have been placed in Part II of the book, since it describes how user interaction can help in the tagging process. We decided to place it here, however, because it directly concerns the user, something which the other subjects of Part II do not.

implemented as *co-processing*, i.e. if the system treats the user as if he were simply one of the software modules, available to its every beck and call.

From the system's point of view, however, co-processing is the optimal solution. The information the user can contribute relates to every phase of the tagger's work: tokenization, assignment of potential tags and selection of the contextually appropriate tag. Furthermore, the earlier the information is present the better, for errors in the earlier phases are unlikely to be corrected automatically in subsequent phases (cf. the discussion on lexicon completeness in 6.3.5). Post-processing is simply too late to be of use except for the single tag being corrected. Pre-processing is certainly early enough but can only be used for very specific aspects, such as text unit separation, since in general the user cannot be expected to predict which additional information the tagger will require.

There is clearly a conflict between what is best for the user (pre- or post-processing) and what is best for the system (co-processing). Fortunately, there is room for compromise. It is possible to split the tagging process into phases and to alternate automatic phases and (optional) user interaction phases. This approach we would like to call *inter-processing*. Several examples of inter-processing are given below, where we discuss the contributions the user can make to each of the tagger's subtasks.

7.3.1 Tokenization

The first subtask in which the tagger may need help is tokenization (cf. Chapter 9). Examples of token separation problems are (for English at least) the interpretation of the period (sentence end, abbreviation marker or both) and of the single quote (opening quote, closing quote, genitive marker, enclitic marker, marker for substandard pronunciation or simply part of a word). Problems with text unit separation (which is often necessary for later modules such as Hidden Markov Models) occur with the abovementioned period or when punctuation is missing altogether, e.g. in newspaper headlines or in spoken text.

The information needed at this point is often expected to be inserted in a pre-processing phase, e.g. in the form of SGML markers. Markers can be used to indicate sentence delimitation. Problematic characters can be replaced by more explicit SGML entities, e.g. quotes can be separated out into left quotes and right quotes. Most of this markup is easy to explain and to use and, it is assumed, its insertion does not place too much of a burden on the user.

Unfortunately, pre-processing is more of a burden than assumed, because the user is forced to make the decision whether or not markers have to be inserted. In addition, much of the pre-processing is superfluous. Even with a limited amount of information, e.g. a list of known abbreviations, a tokenizer should be able to make a correct decision in the majority of the cases (as shown in Chapter 9). In the remaining cases the user can be prompted for a decision, either in a co-processing or in an inter-processing manner.

For inter-processing, the prompts take the form of special markers indicating uncertain tokenization, which can lead the user to the places where his information is needed. Co-processing is probably superior to inter-processing here, since the ability of the tokenizer to handle all effects and side-effects allows the decision to take the form of a simple answer to a question rather than explicit and possibly extensive marker editing in the tokenized text. Note, however, that this stance is only acceptable because of the expected sparseness of decisions.

7.3.2 Classification of unknown tokens

Another possible job for the user is the classification of unknown tokens. As we have seen in 6.3.5, correct information here can prevent a lot of errors. The task allows either a co- or an inter-processing implementation. Inter-processing is superior, because a preceding automatic phase can check the whole text first. Ideally, this allows it to find the answer itself, e.g. a word starting with a capital letter in the first position of a sentence can be assumed to be capitalized due to its position if that same word is observed elsewhere, in non-initial positions and uncapitalized. And even when the user has to be consulted after all, a full sweep through the text serves to collect more contextual information, allowing the user to make a more informed decision.

The optimal solution, then, is to have a special phase before the actual assignment of potential tags which scans the text for unknown tokens (and possibly also potential compounds), orders them and presents them to the user so that he can indicate their potential tags. Ordering should take into account that the user may not wish to solve all problems and should start off with the most frequent or most problematic cases. Care must also be taken that the user provides all potential tags, which means that he must be able to consult the context of all instances and that morphological and/or heuristic support should be very useful.

The possibilities and impossibilities of user interaction here will determine whether or not users will invest the effort to extend the lexicon for each new text. If the user does not help out, the tagger has to fall back on heuristics to assign potential tags to unknown tokens (cf. Chapter 13), the effects of which on the quality of the output are described in Chapter 6. Of course, a tagger using a co-processing approach, which always forces the user to provide tags for unknown tokens, runs the risk of being abandoned altogether if the user's task is too awkward or too time-consuming. All in all, the quality of the user interface at this point should be reckoned as a major factor in the determination of the quality of a tagger as a whole.

7.3.3 Selection of the contextually appropriate tag

Finally, most taggers can certainly use some help in the selection of the contextually appropriate tag for each token. Again, the main possibilities for interaction are co-processing and inter-processing. In a co-processing procedure, the user's help can be called upon the moment the automatic selection phase knows it is in trouble. This seems most likely in a rule-based system, where this knowledge can be included in the rules. In a probabilistic system one can imagine the co-processing module to be activated when there is no single best tag, e.g. when the difference in likelihood of the best tags falls below some threshold. For the user, co-processing at this point is most probably unacceptable. He is reduced to playing a module in a computer system and has no control over what and how much he wants to do. Furthermore, the information provided by the user is of value only locally, for tag selection of the specific token in question. This means that a separate selection phase following the automatic one is only slightly less efficient.

The better choice, then, is inter-processing. However, since tag selection is the last subtask, inter-processing is in fact the same as post-processing for the tagging process as a whole, which we will discuss below. Viewing the tagging process as a whole rather than the tag selection phase separately also leads to the conception of a pre-processing strategy where the user is expected to pre-tag specific types of tokens on which the automatic tagger is known to perform badly. This can certainly improve the performance of the tagger but brings back the basic problem of pre-processing: the user having to decide what is useful to pre-tag and what is not.

7.3.4 Post-processing of tagged text

Whether or not the user has been helping the tagger along, there will always be errors in the final output. This means that it must always be possible to change that output. It is certainly a point in favour of a tagger if output adjustments are supported by a tag selection tool accompanying the tagger (see e.g. van Halteren and Oostdijk 1993). Obviously, as for the taggers themselves, there are differences in quality between tag selection tools. Also, again as for taggers themselves, which design features are best may depend on the tagset being used and the user using it. This means that we will just have to repeat our adage that one should always try a tool for oneself and we will have to restrict ourselves here to some more general points (see also McEnery and Rayson 1997).

First and foremost, a tag selection tool should use a clear presentation of the text and the tags. Whether tokens should be presented vertically or horizontally, with tags connected or on a separate row/column and with only the selected tag or with all possible tags depends on the situation, including factors such as the size of the individual tags, the ambiguity of the tagset and what the user is most comfortable with. The only general statement that can be made here is that the ideal system allows the user a choice in all these aspects of the layout.

For the individual tags, we can be more specific. Here, clarity of presentation means a general preference for readable names over mnemonics (cf. Chapter 4). Note, however, that for expert users of the tagset in question mnemonics can be preferable since they are more space-efficient. Even so, the use of mnemonics should be available to such experts only and probably only if they explicitly ask for it. Numerical notation for tags is anathema. Expecting humans to notice that that a 381 should be a 382 or a 163 is an invitation to produce an annotation riddled with errors.

Next, the user should never have to actually type the name of a tag. Manually typed tags are too error-prone. Preferably, the user should select from the list of possible tags for the token at hand. The use of such a shortlist means less work for the user and in addition increases the consistency of the output. If the correct tag is not on the list there should be a selection procedure to pick any tag from the tagset, possibly by way of a hierarchically structured menu. Note that this not only prevents errors in tag names but it also allows the tool to use long form tag names without increasing the work load on the user.

Then, available information should be used in support of the selection process wherever possible. We already saw an example of this in the previous paragraph, where the list of possible tags for a particular token was used to help the user decide on a tag. Another example is the use of known relations between tokens and their tags: the selection of a ditto tag for part of a multi-token unit should automatically result in the selection of the corresponding ditto tags for the other parts of the same unit. A third and final example is the use of knowledge about consistent errors by the tagger: if there are tags with a below average correctness, it might be useful to provide a command in the selection tool which lets the user search for specific tag/alternative-tag pairs.

Finally, it is useful if the selection tool provides some trace abilities. First of all, the information gained by observing the error correction process may be useful for implementing improvements in the automatic tagger.[6] Also, such information might be used to monitor consistency (cf. 6.2). However, neither of these uses appears to be straightforward. A more likely example of a trace function is a timing device to keep track of the time spent by the user, which can help improve the estimates for the correction time of the remaining material, possibly leading to adjustments in the time schedule and/or committed manpower.

Appendix: NOT an inventory of taggers

Originally, it was envisaged that this book would also contain an inventory of existing automatic taggers and possibly even test results for each. This plan has been abandoned for several reasons. The most important of these is that the number of taggers is very large already and still growing. We could never hope to be complete, even with regard to taggers in existence now, as most of them are still being further developed. Furthermore, as we hope to have shown above, judging the usefulness of a tagger is very much an

[6] Although some of the more obvious improvements, such as additional items to be put in the lexicon, can also be derived from the corrected material.

exercise that must be done in a context. Rather than filling a very large amount of space with potentially out-of-date and hence useless information, we just want to advise the reader to scourge the internet for the most recent news about taggers and examine them for usefulness in the right context.

II The Implementer's View

8 AUTOMATIC TAGGERS: AN INTRODUCTION

Hans van Halteren and Atro Voutilainen

8.1 GENERAL ARCHITECTURE

In the first part of the book we examined tagging from the point of view of the user. Among other things, we discussed what can be expected from automatic tagging software. We did not discuss in very much detail how such software works or which methods are used to arrive at an acceptable quality output. In this second part of the book, we will do just that: describe some popular techniques in detail.

Conceptually, the tagging task can be divided into a number of subtasks:

- segmentation of the text into tokens;
- assignment of potential tags to tokens, usually resulting in ambiguity;
- determination of the contextual appropriateness of each potential tag, usually in order to remove the less appropriate tags and thus resolve the ambiguity.

Although not all tagging systems need actually use this subdivision, we will assume it as the basis of a general architecture for taggers and use it to structure this part of the book.

H. van Halteren (ed.), Syntactic Wordclass Tagging, 109-115.

8.1.1 Tokenization

The first task, then, is identifying the tokens which are to be tagged, a process generally referred to as *tokenization*. Since most taggers (and other NLP aplications) operate on individual utterances rather than full texts, tokenization usually also tries to identify utterance boundaries. It is described in detail in Chapter 9.

8.1.2 Assignment of potential tags

Once the tokens are identified, we have to determine for each token which tags we have to choose from.

In the best case, the token is a *known* wordform (or symbol) and we can either find it in a *wordform lexicon* or derive it from a base form, by using *morphological rules* and a *base-form lexicon*. Matters pertaining to lexical lookup can be found in Chapter 10 and to morphological analysis in Chapter 12.

If the token is not known, we have several options. The first is to assign all existing tags to the token and let subsequent modules figure out which are the most likely. However, most of the time the form of the token will give sufficient clues to make an educated guess about the set of tags which are relevant. For languages with a strong morphology, it may be possible to apply the morphological rules even though the base form of the word is unknown (cf. Chapter 12). But even for other languages highly useful clues can be found, as shown in Chapter 13.

8.1.3 Determination of the most likely tag

The final and most difficult task is to select the correct tag from among the potential ones. Since *correct* should be read as *contextually appropriate*, this involves an examination of the context of the token. As yet, we cannot expect software to understand the context and to base its selection on such understanding. As a result, any software module instead bases its selection on a limited *model* of language use, be it a linguistic or a statistical one. It is in these models, and hence within this subtask, that the highest variation exists. Two important dimensions of variation are the way information is expressed in the model (linguistic rules, neural networks, Markov models, etc.) and the main source of the information (linguists' knowledge or corpus data). The most popular approaches are described below, in Chapters 14 to 17.

8.2 CORPUS RESOURCES

As you will see in the following chapters, practically all activities related to automatic wordclass tagging benefit from the availability of annotated corpus material (see also Smith 1997:139–140). Some activities, e.g. supervised training of statistical models, are impossible without access to an annotated corpus. For other activities, such access may not be strictly necessary but will still be highly useful. A writer of a disambiguation grammar, e.g., requires a prior understanding of what is the correct analysis of any acceptable input text in the object language (the variants of the language the grammar is supposed to describe). In addition to documentation of the individual tags and their application principles this can be accomplished by means of a so-called *grammar definition corpus* (cf. Voutilainen 1994).

In general, annotated corpora can serve as:

- an illustration and partial definition of the tagging scheme;
- training material for a corpus-based model or inspiration for a hand-crafted one;
- testing material during development or for final quality determination.

8.2.1 Form of corpus resources

In order to be useful in the ways listed above, especially the first one, an annotated corpus will have to have certain properties. First of all, the annotation should show the 'correct' tags. In addition, but this is not common practice, the annotation can include the other potential tags as well. Consider the following sentence, taken from an English corpus and annotated with EngCG tags (cf. Chapter 14):[1]

```
"<Holland>"
        "holland" <Correct!> <*> <Proper> N NOM SG
"<and>"
        "and" <Correct!> CC
"<France>"
        "france" <Correct!> <*> <Proper> N NOM SG
"<have>"
        "have" <SVO> <SVOC/A> V SUBJUNCTIVE VFIN
        "have" <Correct!> <SVO> <SVOC/A> V PRES -SG3 VFIN
        "have" <SVO> <SVOC/A> V INF
        "have" <SVO> <SVOC/A> V IMP VFIN
"<both>"
        "both" CC
        "both" <Quant> DET PRE PL
        "both" <Correct!> <NonMod> <Quant> PRON NOM PL
"<made>"
        "make" <Correct!> <SVOC/N> <SVOC/A> <SVO> PCP2
        "make" <SVOC/N> <SVOC/A> <SVO> V PAST VFIN
"<tough>"
        "tough" <Correct!> A ABS
"<deals>"
        "deal" <SVO> <SVOO> V PRES SG3 VFIN
        "deal" <Correct!> <Count> N NOM PL
```

[1] A full list of EngCG tags can be found in Appendix 17.6.

```
"<with>"
        "with" <Correct!> PREP
"<industry>"
        "industry" <Correct!> N NOM SG
"<to>"
        "to" PREP
        "to" <Correct!> INFMARK>
"<encourage>"
        "encourage" <SVO> V SUBJUNCTIVE VFIN
        "encourage" <SVO> V IMP VFIN
        "encourage" <Correct!> <SVO> V INF
        "encourage" <SVO> V PRES -SG3 VFIN
"<more>"
        "much" ADV CMP
        "much" <Quant> PRON CMP SG
        "much" <Correct!> <Quant> DET POST CMP SG
        "many" <Quant> PRON CMP PL
        "many" <Quant> DET POST CMP PL
"<recycling>"
        "recycle" <Correct!> <SVO> PCP1
"<of>"
        "of" <Correct!> PREP
"<packaging>"
        "package" <Correct!> <SVO> PCP1
"<$.>"
```

Basically, this sample is the output of a morphological analyser used in the EngCG tagger (Karlsson *et al.* (eds.) 1995). On the left-hand margin we find the running text wordform and punctuation marks (e.g. "<France>", "<$;>"). Each indented line is a morphological analysis consisting of a base form (e.g. "retailer") and a number of morphosyntactic tags indicating part-of-speech (e.g. N), inflection (e.g. NOM, SG), derivation (e.g. <DER:er>) and typical syntactic properties of the word (e.g. <SVO> for monotransitive verbs). The feature <Correct!> is added manually to those readings that are regarded as appropriate. For example, "both" is analysed as a pronoun in the above sentence (rather than as a coordinator or a determiner).

Furthermore, comment lines can be added to an annotated corpus, e.g. to document analyses that may not seem obvious at first glance. Here is an example:[2]

```
"<Jane>"
        "jane" <Correct!> <*> <Proper> N NOM SG
"<is>"
        "be" <Correct!> <SV> <SVC/N> <SVC/A> V PRES SG3 VFIN
"<older>"
        "old" <Correct!> A CMP
;; a NP is introduced by PREP rather than CS. See Doc:3.4.12
```

[2] A special notation will have to be used for such comment lines, in order to avoid problems with further processing, e.g. by a semantic annotator.

```
"<than>"
        "than" <**CLB> CS
        "than" <Correct!> PREP
"<John>"
        "john" <Correct!> <*> <Proper> N NOM SG
"<$.>"
```

Without a predefined, at times perhaps somewhat arbitrary, descriptive policy, it might not be at all clear how a prenominal "than" should be analysed. For instance, one could argue for the CS (subordinating conjunction) analysis by pointing out that CSs introduce clauses and "John" is a clause, though an elliptical one. On the other hand, the PREP (preposition) analysis could be defended by arguing that a preposition is followed by a noun phrase, which "John" is, at least in terms of a surface-oriented view of grammar. As is done in this example, comments may include pointers to a written documentation where the problematic phenomenon may be discussed in more detail.

8.2.2 Size of corpus resources

How much annotated corpus material is needed for testing and improving taggers? At least the following factors are relevant:

- desired quality of the intended tagger;
- degree of domain specificity of the intended tagger;
- nature of the tagset: number and quality of distinctions;
- (for hand-crafted models) expertise of the grammarian;
- (for corpus-based models) learning strategy of the method;
- (for testing) desired certainty of quality measurement.

To serve as a linguistically motivated test bench, this kind of corpus should be representative at least in the modest sense that (for each of the target domains) it represents all the grammatical distinctions introduced by the tagset. For instance, if the tagset includes ambiguity-forming imperative verb readings, the corpus should contain a reasonable number of cases where the imperative is the correct reading. If this is not the case, it would be difficult to determine the context conditions under which an imperative tag is the desired choice.

As to the quantity of the test corpus, it seems obvious that there can never be too much of it; the limits are probably set by the available human resources and annotation techniques.

8.2.3 Creation of corpus resources

How, then, does one go about creating corpus resources? For the annotation shown above, the following simple routine for annotating corpora manually has actually been used:

1. Analyse the text with the tokenizer and lexical analyser.

2. Take the text into a text editor (e.g. Gnu Emacs).

3. Moving the cursor with a mouse, select the correct alternative analysis line and insert a unique string somewhere in it, e.g. "*********" in

```
"<to>"
        "to" PREP
        "to" INFMAR*********K>
"<return>"
        "return" <SVO> V SUBJUNCTIVE VFIN
        "return" <SVO> V IMP VFIN
        "return" <SVO> V INF*********
        "return" <SVO> V PRES -SG3 VFIN
        "return" N NOM SG
```

4. Using e.g. a keyboard macro, convert the string into a unique symbol:

```
"<to>"
        "to" PREP
        "to" <Correct!> INFMARK>
"<return>"
        "return" <SVO> V SUBJUNCTIVE VFIN
        "return" <SVO> V IMP VFIN
        "return" <Correct!> <SVO> V INF
        "return" <SVO> V PRES -SG3 VFIN
        "return" N NOM SG
```

Using this routine, 2,000-3,000 words can be processed in an hour, according to experiences with English. Spending a work month (150 hours) would thus result in at least 300,000 words of annotated corpus. A technique which can further improve the efficiency of the corpus resource creation process is *bootstrapping*: first annotate a small corpus, e.g. using the above method, then create and test a tagger on the basis of the annotated material and use this tagger to prepare more texts, so that less work has to be done by hand on the later texts.

An obvious problem with manually created corpus resources is that annotators make misanalyses even with well-motivated and documented tagsets, e.g. because of lapses in attention (cf. 6.3.3). One method to correct misanalyses is to have the same text individually annotated by different people and then reconsider the cases where the annotators chose a different analysis.[3] A less expensive but also somewhat less reliable method is to correct the misanalyses made by an annotator when testing and improving the tagger.

Whatever method is used, one should always be aware that there will be differences between annotated corpus material and the 'Linguistic Truth' (if such a thing exists at all). However, most of the contributors to this book, even though they acknowledge this fact, do not let themselves be hampered by it. They continue to make use of corpus resources, probably assuming that as long as the Linguistic Truth remains elusive, they may as well settle on a well-annotated corpus being close enough.

[3] There may well be cases where there is no single 'correct' tag, but more than one tag is equally defendable.

9 TOKENIZATION*

Gregory Grefenstette

9.1 INTRODUCTION

The linguistic annotation of naturally occurring text can be seen as a progression of transformations of the original text, with each step abstracting away surface differences. *Tokenization* is one of the earliest steps in this transformation during natural language processing.

Tokenization means dividing up the input text, which to a computer is just one long string of characters, into subunits, called *tokens*.[1] These tokens are then fed into subsequent natural language processing steps such as morphological analysis, wordclass tagging and parsing. Since these subsequent treatments are usually designed to work on individual sentences, a subsidiary task of tokenization is often to identify sentence boundaries as well as token boundaries.[2] Though rarely discussed, and quickly dismissed, tokenization in an automated text processing system poses a number of thorny questions, few of which have completely perfect answers.

Tokenization is not the first step in the abstraction process. If you consider an original printed document, most typesetting distinctions (e.g. font size, font style, page layout, pictures and graphics) are filtered out of the text that will be tokenized

[1] Here a *token* means the individual appearance of a word in certain position in a text. For example, one can consider the wordform *dogs* as an instance of the word *dog*. And the wordform *dogs* that appears in, say, line 13 of page 143 as a specific *token*.
[2] Sentence identification is not always considered as a part of tokenization.

*The following work was accomplished at Xerox Limited.

H. van Halteren (ed.), Syntactic Wordclass Tagging, 117-133.

and further analysed. It is not that these elements do not carry meaning (cf. Holstege *et al.* 1991) that could be exploited by a machine, but rather that the great variety of different typesetting conventions make it difficult for any general processing system to take them into account. Preferably, however, the information is not deleted altogether but rather translated into *markup*, so that the possibility remains to exploit it during later processing.

The input to a tokenizer, then, is a stream of characters which consists of *graphic tokens* separated by layout (after the previous step probably only space and newline characters) and possibly enhanced by markup symbols. Unfortunately, the graphic tokens, usually defined as anything between two layout symbols, need not coincide with the *linguistic tokens*.

The most obvious exception is formed by the concatenation of words and punctuation marks. Other cases where graphic tokens may be split are enclitic forms, such as "he's", and contractions, such as "'twas". The opposite is also possible: two or more graphic tokens may have to be combined into one linguistic token. Examples are proper names, such as "John Jones", and compounds, such as "in spite of". Usually the decision about what constitutes a token will have been made during the definition of the tagset (cf. 4.3.1). There is, however, one special case where two graphic tokens can almost always be combined, viz. *hyphenated words* generated by line breaks. We will examine this case in detail below.

If markup is present in the input, it may or may not be useful. Most of the markup can be ignored and there are even cases where it is necessary (for some operations) to ignore the markup, e.g. markup within a word. Some markup, however, can be extremely useful during tokenization.[3] Font information can help decide whether or not to combine tokens (e.g. a company name in bold font) or split them (e.g. a footnote reference in smaller font and superscripted), layout information can be a great help in pinpointing sentence boundaries and there are many more examples. However, no systematic investigation of such exploitation of markup has been made so far, mostly due to the already mentioned variety in typesetting conventions. As a result, we too will neglect markup in the rest of this chapter.[4]

The tokenization process depends strongly on the type of text which is being processed, so that an analysis of the tokenization problems in the specific type of text must be done (with ways of checking the separator/non separator status of characters). Learning from the corpus being tokenized is also very important as we demonstrate in the sections below.

---

[3] And during later stages of tagging as well, e.g. the convention to italicize foreign words and technical terms in English text can help identify them.

[4] An internet search for *dehtml* will lead to pages such as http://www.math.utah.edu/~beebe/dehtml which provide code for eliminating HTML markup from files.

For a tokenizer to be effective, it must not only be able to recognize tokens but also signal in its output where tokens begin and end. This means adding in additional marks into the input string in order to isolate tokens. It is possible to use symbols in a special markup formalism (e.g. SGML) for this but sometimes simpler measures suffice. Two very common token separation marks are the space character and the newline character. With the space character, each token is separated from the next by one or more spaces. With the newline character, each token appears on a separate line. The advantage of the newline character is that the token itself may contain a blank, for example, a morphological analyser may expect to see "hot dog" is a single token which happens to contain a space. Given the following text,

```
``Stop!'' Dr. John shouted.
```

the space-introducing tokenizer would output:

```
``  Stop  !  ''  Dr.  John  shouted  .
```

and the newline-introducing tokenizer would give:

```
``
Stop
!
''
Dr.
John
shouted
.
```

9.2 REGULAR EXPRESSIONS

In order to perform tokenization it is necessary to describe which patterns of characters are considered to be a token and to perform some action (such as break them off or join them together) each time they appear in an input file. The types of patterns that are generally needed to recognize tokens fall into a class of formal languages called *regular expressions*. This section will review this concept.

9.2.1 Definition

Regular expressions are expressions which can be built up using concatenation, repetition and alternatives. The set of all words matched by a regular expression is called the language recognized, or accepted, by that expression. Regular expressions are defined in the following way:

- Any character A is a regular expression, denoting the language of all words consisting of just that character A.

- If B is a regular expression and C is a regular expression, then BC is a regular expression, denoting the language of any word accepted by B followed by a word accepted by C.
- If B is a regular expression and C is a regular expression, then B | C is a regular expression, denoting the language of any word accepted by B, by C or by both.
- If B is a regular expression, then (B) is regular expression denoting the same language as B.
- If B is a regular expression then B* corresponds to any sequence of zero or more words belonging to B.
- If B is a regular expression then B+ corresponds to any sequence of one or more words belonging to B.
- If B is a regular expression then B? corresponds to zero or one word belonging to B.

We can use regular expressions to describe the structure of most tokens that have to be recognized by the tokenizer. In order to treat the full alphabet of a language, it is useful to have two extensions to the above description of regular expressions:

- A wild-card character that matches any character in the alphabet. This character is often defined in a given regular expression language as either "." or "?".
- A way of describing any character in an interval of characters. For example, Unix regular expressions allow you to use the expression "[A-Z]" to stand for any upper-case unaccented character. This enters into descriptions of upper-case abbreviations, like "U.S.A." which is recognized by the regular expression "([A-Z][.])+".[5]

Regular expressions can be recognized by a *finite state machine* (cf. Hopcroft and Ullman 1979:13–54 or Aho and Ullman 1992:Ch.10), an automaton consisting of states and labelled arcs. The machine begins in a start state and each input character allows the machine to follow an arc to another state. When the machine arrives in a final state, the string is said to be recognized, or to be accepted by the finite state machine. An extension of finite state machines are *finite state transducers* which consume an input symbol and produce output on each state transition (cf. Chapter 12). A tokenizer is really a transducer since it introduces extra characters, such as spaces or newlines, when it accepts a complete token.

[5] In the Unix implementation of regular expressions, the period is the wild-card character that matches any single character. In order to use it as the simple character period, the period has to be protected by an escape character such as "\", or included in brackets "[.]" which makes the system read it as a normal non-wild-card character.

9.2.2 *Regular expression tools* LEX *and* AWK

Computer compilers perform tokenization of a computer program, in order to recognize keywords and variables, as the first step in the compilation process. As a result of the need for this step, there are many tokenization tools available on computer systems. In the Unix-based world, there are two general tools which allow a user to write a natural language tokenizer: LEX (chap. 3 of Aho *et al.* 1986) and AWK (Aho 1988). Both tools are based on regular expressions and C-language commands to be executed when the regular expressions are found in the input text. There are two major differences between LEX and AWK. First, LEX compiles into C-code whereas AWK is interpreted. Secondly, AWK works input line by input line which means that you can write expressions assuming that they are limited to one line, whereas LEX considers the input file as one long string and your token expressions must deal explicitly with the newline character which ends each line of the input text.

Let's see how to write a simple English tokenizer using AWK. The main action that we want a tokenizer to perform is to isolate the recognized token by inserting spaces or newlines. In what follows, we will assume that we are building a newline-inserting tokenizer. AWK allows us to specify a regular expression within an input line and to perform certain actions, for example, inserting token boundaries (in our case, newlines), around matched patterns. Here we begin with the assumption that tokens are not split over two different lines.

9.2.3 *An example of a tokenizer*

A tokenizer can be designed by specifying

- separators,
- abbreviations,
- exceptions.

Below we present AWK code which shows how these three types of information can be used to make a deterministic tokenizer.

Separators are characters which always separate a token from other tokens. In many document retrieval systems, a space and a newline are the only separators but, as we described above, we sometimes want a different tokenization than the graphic one and hence use a number of other separators as well. Some of these are separators wherever they are found, e.g. the comma in English. Others are only separators when they occur in specific positions. We can distinguish left and right separators, i.e. separators that separate when they appear at the beginning of a word or at the end of a word. An example of a left separator in English is the single quote, an example of a right separator is a colon which should be detached from its preceding word when followed by a known separator, like a space, but retained within a word when followed by some other character, as in a World Wide Web address such as "http://www.whitehouse.gov".

The first step in tokenizing is to isolate all strings by replacing all known separators by token boundaries. AWK performs a first separation automatically for us, by dividing each input line into a number of fields (called \$1, \$2, ...) using the space, tab and newline characters and/or any other characters we wish to define as field separators. We could use this to put token boundaries around punctuation which always acts as a separator. In the program given below we use only the default field separators: the space and the tab. Then we insert token boundaries around each left separator preceded by a boundary and after each right separator followed by a boundary.[6] At this point the tokenizer has dealt with all unambiguous separators.

Abbreviations, in most European languages at least, pose serious problems for tokenization with their string-ending periods. The period is the prime example of an ambiguous separator, which may or may not be a token ending separator. For example, in "etc.", the abbreviation for "et cetera", the period is generally accepted as part of the token. The period ending a sentence is generally considered as a separate token, yet in certain cases, a period acts simultaneously as a sentence terminator and as part of an abbreviation, as in:

```
I live in Baker St.
```

As we shall see below, deciding whether a period is a sentence terminator or part of an abbreviation can not be done with 100% accuracy over open-ended text. When we make a deterministic tokenizer, we want to make choices that are right, though, as often as possible. Exceptions can be included as a short list of corrections to the preceding choices and we can thus have special cases taken into account.

The AWK code implementing all this, then, reads:

```
# The following is an GAWK program for tokenizing English

# Comments, like this line begin with a sharp sign (#)

# 'GAWK' is a GNU version of Unix's awk, freely obtainable
# on the Internet, for example at ftp://prep.ai.mit.edu/pub/gnu

# What follows BEGIN is executed before any lines are read.

BEGIN {

# Here we initialize some regular expression expressed as strings.
```

[6] The AWK program given below is only a simple version, which assumes that there will never be two adjacent left or right separators. To correctly handle cases like "they'll:" a more extensive program is needed.

```
    Letter =
      "[A-ZÀÁÂÃÄÅÆÈÉÊËÌÍÎÏÒÓÔÕÖØÙÚÛÜÝa-zàáâãäåæèéêëìíîïòóôõöøùúûüç]";
    NotLet =
      "[^A-ZÀÁÂÃÄÅÆÈÉÊËÌÍÎÏÒÓÔÕÖØÙÚÛÜÝa-zàáâãäåæèéêëìíîïòóôõöøùúûüç0-9]";
    AlwaysSep = "?!()\";/\\|,`" ;
    BeginSep = "('|&)";
    EndSep = "('|:|-|'S|'D|'M|'LL|'RE|'VE|N'T|'s|'d|'m|'ll|'re|'ve|n't)";

# GAWK supports associative arrays (hash tables).
# Here we give a non-zero value to all strings that we
# explicitly consider as abbreviations.

    Abbr["Co."]=1;
    Abbr["Corp."]=1;
    Abbr["vs."]=1;
    Abbr["e.g."]=1;
    Abbr["etc."]=1;
    Abbr["ex."]=1;
    Abbr["cf."]=1;
    Abbr["eg."]=1;

    Abbr["Jan."]=1;
    Abbr["Feb."]=1;
    Abbr["Mar."]=1;
    Abbr["Apr."]=1;
    Abbr["Jun."]=1;
    Abbr["Jul."]=1;
    Abbr["Aug."]=1;
    Abbr["Sep."]=1;
    Abbr["Sept."]=1;
    Abbr["Oct."]=1;
    Abbr["Nov."]=1;
    Abbr["Dec."]=1;
    Abbr["jan."]=1;
    Abbr["feb."]=1;
    Abbr["mar."]=1;
    Abbr["apr."]=1;
    Abbr["jun."]=1;
    Abbr["jul."]=1;
    Abbr["aug."]=1;
    Abbr["sep."]=1;
    Abbr["sept."]=1;
    Abbr["oct."]=1;
    Abbr["nov."]=1;
    Abbr["dec."]=1;

    Abbr["ed."]=1;
    Abbr["eds."]=1;
    Abbr["repr."]=1;
    Abbr["trans."]=1;
    Abbr["vol."]=1;
    Abbr["vols."]=1;
```

```
    Abbr["rev."]=1;
    Abbr["est."]=1;
    Abbr["b."]=1;
    Abbr["m."]=1;
    Abbr["bur."]=1;
    Abbr["d."]=1;
    Abbr["r."]=1;
    Abbr["M."]=1;
    Abbr["Dept."]=1;
    Abbr["MM."]=1;
    Abbr["U."]=1;
    Abbr["Mr."]=1;
    Abbr["Jr."]=1;
    Abbr["Ms."]=1;
    Abbr["Mme."]=1;
    Abbr["Mrs."]=1;
    Abbr["Dr."]=1;
  }

# The following commands are applied to all input lines.
# The GAWK default field separators are the space and the tab.

# Change tabs into spaces.

{ gsub("\t"," "); }

# Put blanks around characters that are unambiguous separators.

{ gsub("["AlwaysSep"]"," & "); }

# If a word is a separator at the beginning of a token separate
# it there.

{ gsub("^"BeginSep,"& "); }

{ gsub(NotLet""BeginSep,substr("&",1,1)" "substr("&",2)) };

# Do the same for final separators.

{ gsub(EndSep"$"," &"); }

{ gsub(EndSep""NotLet,
    substr("&",1,length("&")-1)" "substr("&",length("&"),1));
}

# GAWK has divided the input line into fields using
# the tab and the space character as separators.
# NF is a GAWK variable automatically set to the number of fields
# GAWK also creates variables $1, $2, $3, ..., $NF containing
# the field strings.

# Loop over the fields and if a field contains a letter followed by
```

```
# a period, check if it is an abbreviation.

{ for(i=1;i<=NF;i++)
    if($i ~
     /[A-ZÀÁÂÃÄÅÆÈÉÊËÌÍÎÏÒÓÔÕÖØÙÚÛÜÝa-zàáâãäåæèéêëìíîïòóôõöøùúûüç][.]$/)
    {

# If the field is explicitly found in the abbreviation list (Abbr)
# or matches the regular expression below, keep the period attached.

        if($i in Abbr)
          continue;
        if($i ~
          /^([A-Za-z]\.([A-Za-z]\.)+|[A-Z]\.|[A-Z][bcdfghj-np-tvxz]+\.)$/)
           continue;

# If not, insert a space before the period.

        gsub("[.]$"," .",$i);
    }
}

# Finally, change all spaces to newlines and print the tokenized
# line.

{ gsub("[ \t]+","\n"); print }
```

9.3 AMBIGUITY IN TOKENIZATION

Just as all other processes in NLP, a tokenizer is faced with ambiguity. And just as all other processes, knowledge about the language (and about the world) is needed to choose the correct interpretation at least often enough to produce an output that is useful. The main knowledge sources for a tokenizer are lexicons and corpora that have been manually tokenized. The corpora can serve a second purpose, namely that of a testbed for the tokenizer. Any tokenizer can be tested by comparing its output to a hand-prepared corpus and to count the differences.

In this section we will examine several frequent sources of ambiguity. For each source we will describe one or more strategies of handling them and test how successful these strategies are in processing actual corpus text.

9.3.1 Splitting graphic tokens

As with any step in the natural language processing of text, tokenization can be done with more or less language knowledge. The more knowledge of the language implemented in your system, the better the tokenization. The simplest technique, separating tokens using whitespaces (blanks, tabs and new-lines), requires no lexical knowledge.

But this simplest technique has the disadvantage of not correctly recognizing sentence boundaries as well as poorly abstracting away surface differences between individual words as shown in, for example, the following distinct tokenizations of "dog":

```
dog    dog'    dog,'    dog"    dog.    dog).
```

A slightly more evolved technique is to recognize certain characters as punctuation and others as word-forming parts. This step involves a tiny bit more lexical knowledge, i.e. knowing something about how words are formed. More lexical knowledge can be introduced into the tokenizing system up to the point of detailing the entire regular lexicon of the language. In the next few sections we describe the results to be expected from a tokenizers using different levels of lexical knowledge. In order to validate and quantify the results, we use the 1-million-word Brown corpus, which has been hand-tokenized, as providing the correct result. We use as a metric how well the tokenizing method at a given level of lexical knowledge is able to divide the original corpus into the same sentences as the hand-tokenized Brown corpus.

Sentences end with punctuation. The exclamation point and the question mark are almost always unambiguous examples of such punctuation. The semicolon is sometimes a separator of list elements and sometimes a sentence separator. But the most prevalent of ambiguous separators is the period which is extremely ambiguous. It is not at all trivial to decide when it is a full-stop, a part of an abbreviation or both. In the Brown corpus, there are 52511 sentences ended by a full stop (period or question mark) and 3569 (about 1 in 15) contain at least one non-terminal period. If one were to consider every period as a full stop, then 93.20% of the original 52511 sentences would be correctly recognized. In some cases, one might consider this most simple of heuristics as sufficient. In the following sections, we will see how this sentence recognition count can be improved by adding increasing levels of linguistic sophistication.

What you can do without a lexicon. Isolating word and sentence boundaries involves resolving the use of ambiguous punctuation. Some structurally recognizable tokens contain ambiguous punctuation, such as numbers, alphanumeric references (e.g. "T-1-AB.1.2"), dates (e.g. "02/02/94"), acronyms (e.g. "AT&T"), punctuations and abbreviations (e.g. "m.p.h."). Some of these classes can be recognized via regular expression grammars which predict the structure of the tokens as is illustrated below. Once these units are recognized, the only uses of separators are non-ambiguous and they can thus be used surely to delimit words and sentences.

Numbers are the least ambiguous of the structural types. Still, the structure of numbers are language specific constructions, for example the English number "123,456.78" will be written as "123 456,78" in French newspaper text.

A regular expression which recognizes the English version of numbers is

```
([0-9]+[,])*[0-9]([.][0-9]+)?
```

while a regular expression accepting the French version is

```
([0-9]+[ ])*[0-9]([,][0-9]+)?
```

These expressions would overgenerate strings, outside the class of numbers, but used as recognizers they are sufficient. One rarely sees strings such as "12,45.678" in ordinary text and even if one did one would probably want it considered as a number.

The table below gives some regular expressions for English numbers, dollar values and date-like constructions that can be incorporated into a tokenizer. Recognizing these strings eliminates some of the ambiguity of the comma and the period, since these characters are comprised in the token and are thus no longer considered as separators.

```
[0-9]+(\/[0-9]+)+                Fractions, Dates
([+\-])?[0-9]+(\.)?[0-9]*%       Percent
([0-9]+,)*[0-9]+(\.[0-9]+)*      Decimal Numbers (e.g.  1,234.56)
```

Once we recognize numbers using the above expressions, not considering their included periods as full stops, only 3340 sentences are now incorrectly recognized, adding 229 sentences to the count of correct Brown sentences, still using the simple heuristic "remaining period equals full stop". This improves sentence recognition from 93.20% to 93.64%.

Besides numbers, the other, most important, class of tokens incorporating the period as an element is the class of abbreviations. Lists of abbreviations can be long and, like lists of proper names, incomplete, since creation of abbreviations is a productive process. Let's assume now that we have no such lists and see how far we can get using regular expressions to recognize abbreviations, so that their periods will not be considered as full stops. Let's assume first that any period not followed by a blank is not a full stop. This simple heuristic adds 73 more correctly recognized sentences, so that, also using the number recognizers given above, 49244 of the original 52511 sentences are now correctly recognized, raising the percentage to 93.78%.

Even without a lexicon, we can predict the structure of some abbreviations. Let us consider three classes of abbreviations:

- a single capital followed by a period, such as "A.", "B." and "C.";
- a sequence of letter–period–letter–period's, such as "U.S.", "i.e." and "m.p.h.";
- a capital letter followed by a sequence of consonants followed by a period, such as "Mr.", "St." and "Assn.".

If we insert blanks around parentheses, commas, colons and questions marks, then 4037 such sequences are found in the Brown corpus. If we automatically consider each of these sequences as a non-sentence-ending abbreviation and not as an unabbreviated word followed by a final stop, we will be right 3835 out of 4037 times. The details are given in the table below. For example, the third class of regular-expression defined abbreviations (a word beginning in upper-case without any following vowels, such as "Mr.") matches actual abbreviations 1938 times, commits 44 errors recognizing strings as abbreviations that should not be (e.g. "Ash.") and recognizes a real abbreviation 26 times that is also a sentence terminator. In this case, the sentence ending period is absorbed in the abbreviation-ending period (cf. Nunberg 1990).

Regular Expression	Correct	Errors	Full Stop
`[A-Za-z]\.`	1327	52	14
`[A-Za-z]\.([A-Za-z0-9]\.)+`	570	0	66
`[A-Z][bcdfghj-np-tvxz]+\.`	1938	44	26
Totals	3835	96	106

This means that, without consulting a lexicon, but only by using the structure of the words we will correctly recognize 3935 of the non-numeric token-ending periods as part of an abbreviation (out of 4951 (330 unique) true Brown abbreviations). We will introduce 96 errors by recognizing true full stops as false abbreviations and another 106 by correctly recognizing abbreviations but not realizing that they should also be full stops. The number of original Brown sentences that will be correctly recognized using the number recognizers above and this abbreviation recognition scheme is now 51282, or 97.66%. 825 sentences still contain some type of abbreviation not recognized by the above expressions and 404 sentences will have been incorrectly joined since the final stop is not recognized as such in 202 cases.

The abbreviations in Brown that do not match the above regular expressions are the following, listed in order of decreasing frequency:

> Month-Names Sen. Gen. Rev. Gov. U.S.-State-Abbreviations fig. Rep. Ave. Corp. figs. Figs. 24-hr. lbs. Capt. yrs. dia. Stat. Ref. Prof. Atty. 6-hr. sec. eqn. chap. Messrs. Dist. Dept. ex-Mrs. Vol. Tech. Supt. Rte. Reps. Prop. Mmes. 8-oz. viz. var. seq. prop. pro-U.N.F.P. nos. mos. min. mil. mEq. ex-Gov. eqns. dept. Yok. USN. Ter. Shak. Sha. Sens. SS. Ry. Rul. Presbyterian-St. P.-T.A. Msec. McN. Maj. Lond. Jas. Grev. Gre. Cir. Cal. Brig. Aubr. 42-degrees-F. 400-lb. 400-kc. 36-in. 3-hp. 3-by-6-ft. 29-Oct. 27-in. 25-ft. 24-in. 160-ml. 15,500-lb. 12-oz. 100-million-lb. 10-yr. 1.0-mg. 0.5-mv./m. 0.1-mv./m. 0.080-in. 0.025-in.

What you can do with a corpus. In order to reduce this list of non-recognized abbreviations without referencing a lexicon, you can use the corpus itself as a filter for identifying abbreviations. Let us define as a *likely abbreviation* any string of letters terminated by a period and followed by either a comma or semi-colon, a question mark, a lower-case letter, or a number, or followed by a word beginning with a capital letter and ending in a period.

Using this definition of likely abbreviations matches 239 of the 330 unique abbreviations in the Brown corpus, but introduces a large number of false positives such as

based become behavior better board box break bull's-eye ...

which are words that happen to end sentences that are followed by another sentence beginning with a number.

We can apply the corpus itself as a filter by eliminating from the list of likely abbreviations those strings that appear without terminal periods in the corpus. This drastically reduces the collection of likely abbreviations to 231, but still misrecognizes strings such as "furlongs", "light-hearted" and "rev'rend" as abbreviations.

Likely Abbreviation	Total (unique)	Correct (unique)	Incorrect (unique)
`[A-Za-z][^ ]*\.([,?;]| [a-z0-9])`	947	239	718
not appearing without period	231	197	34

When we use the corpus as a filter for accepting the 231 candidates as likely abbreviations and accepting all structures with internal periods (i.e. not ending in a period) or ending in a period and of the form "[A-Z]\." or "[A-Za-z]\.([A-Za-z0-9]\.)+" as non-terminal abbreviations, then 337 sentences are incorrectly divided. Another 266 sentence ends are incorrectly identified as sentence internal abbreviations, mistakenly joining 532 sentences, meaning that 51642 of the 52511 original Brown sentences are now correctly recognized. This gives us a 98.35% recognition rate, after using the corpus as a filter but without any lexical access.

The abbreviations which are still uncaptured by this technique are the following

No. Sept. Rev. Jan. fig. Mass. Corp. no. Pa. La. 24-hr. cf. Tex. Mt. Miss. in. Wash. Hon. 6-hr. eqn. chap. a. Ore. Mar. sp. oz. hp. ex-Mrs. Tech. Supt. Mmes. Minn. Eq. Ed. Colo. 8-oz. u. seq. prop. nos. mos. min. mil. fed. ex-Gov. eqns. ed. al. Yok. Vs. Tenn. Sha. Sens. SS. Presbyterian-St. Pfc. OK. McN. Maj. Kas. Eng. Del. Cmdr. Cal. App. 42-degrees-F. 400-lb. 400-kc. 36-in. 3-hp. 3-by-6-ft. 29-Oct. 27-in. 25-ft. 24-in. 160-ml. 15,500-lb. 12-oz. 100-million-lb. 10-yr. 1.0-mg. 0.5-mv./m. 0.1-mv./m. 0.080-in. 0.025-in.

What you can do with a corpus and lexicon without abbreviations. The observations above suppose that the abbreviation recognition process has no access to a lexicon. Let us examine what can be gained by using a lexicon to look up the litigious cases. Suppose now that, instead of trying to solve all the ambiguities during this tokenization phase, tokenization is reduced to number recognition and splitting words on spaces and unambiguous separators. Then every word ending a sentence as well as real abbreviations ending with the period will be sent to the morphological analyser with a trailing period. It will then be the role of the morphological analyser to decide if the trailing period should be isolated as a separate, sentence-ending, character. Under this supposition, the Brown corpus produces 52430 letter-initial tokens ending in a period that must be sorted. Suppose that we have a complete lexicon, containing at least all the words in the Brown corpus, except abbreviations and proper names. Can we discover abbreviations using this method?

Consider this ordered filter on all strings terminated by a period:

1. If it is followed by a lower-case letter, comma or semi-colon, it becomes a *known abbreviation.*

2. If it is a lower-case string, is not a known abbreviation and exists as a word in the lexicon without a final period, it is not an abbreviation, otherwise it is an abbreviation.

3. If it begins with an upper-case letter, is not a known abbreviation and appears elsewhere in the corpus without a trailing period or only appears once or twice in the corpus, it is not an abbreviation (probably a proper name).

4. Otherwise, it is an abbreviation.

The list of known abbreviations defined under (1) contains 183 unique upper- and lower-case abbreviations (occurring a total of 1003 times in Brown). A sample of such such known abbreviations, given in order of decreasing frequency, follows:

U.S. Jr. Mr. U.N. i.e. Co. p.m. e.g. S. a.m. etc. Inc. St. D.C. B.C. A.L.A.M. vs. Calif. lb. cm. ...

The list derived from (2) captures most of the cases in the corpus. There are 42865 lower-case initial strings appearing with a final period. 458 of these instances correspond to known abbreviations and 42344 others correspond to words without final periods appearing in the lexicon. In some instances these words are really abbreviations. This happens when some string appears as both as an entire word and in an abbreviation, such as "fig" also appearing as "fig." for "figure." If we consider all these 42344 cases as sentence-ending non-abbreviated words, then we misrecognize 29 sentences which contain

chap. fed. fig. no. nos. u.

since these words are not considered as sentence-internal abbreviations (which they really are in these sentences) but as full stops.

63 other instances (19 words) are recognized by step (2) as abbreviations:

ca. cf. ed. eqn. eqns. ex-Gov. ex-Mrs. figs. hp. mil. min. mos. oz. pp. r.p.m. seq. sp. v. yrs.

By the time we reach step (3), we have decided in 46474 of the 52430 period-terminated string cases. Step (3) has to decide the case of the remaining 6056 upper-case initial possible abbreviations. It finds that 4628 of these correspond to upper-case initial words somewhere else in the corpus without a final period. Of the remaining 1428, 583 appear only once or twice, so they are not considered abbreviations either.

This heuristic incorrectly identifies all occurrences of the following strings as sentence-ending non-abbreviations since they appear elsewhere without a period:

App. Cal. Del. E. Ed. G. Jan. L. Mar. No. P. Rev. SS. Sept. Tech. V. W.

or only one or two times in the corpus:

Aubr. Brig. Cf. Cmdr. D.J. D.W. E.O. E.T. Eng. Eq. F.S.C. H.L. H.M. H.P.R. H.W. I.L. J.D.H. J.H. Jas. K.G. K.J.P. Kas. Maj. McN. Mfg. Mmes. N.A. N.D. N.L. P.L. P.S. P.m. Pfc. Presbyterian-St. Pt. R.H. R.L. Reps. Rte. Rul. Ry. S.S. Sens. Sha. Spec. Supt. T.W. U.S.C. U.s. Vol. Vs. W.G. W.H. W.M. W.R. Wm. Yok.

Step (4) identifies all the remaining candidates as abbreviations:

Atty. Aug. Capt. Ch. Christendom. Col. Dec. Feb. Fig. Figs. H.M.S. Hon. Lt. Martinez. Mrs. Mt. Nov. Oct. Op. Pp. Prof. Ref. Rep. Schaack. Sec. Sen. Stat.

Combining the abbreviations recognized by all four steps, only 205 sentences are erroneously split because they contain as yet unrecognized abbreviations, but 351 sentences end in strings thought to be sentence internal abbreviations. So we will incorrectly join 702 sentences. In other words, the above method of using a lexicon without abbreviations and the corpus as a filter to tokenize recognizes 51604 sentences out of 52511 original Brown sentences, or 98.27%. This slight degradation comes from the fact that "in." is recognized as a known abbreviation by Step (1) and so the 79 sentences ending in "in." are incorrectly joined to the sentence following them.

What you can do with a lexicon and a list of abbreviations. Consider now a lexicon that has not only all the lower-case words in the corpus, but also contains frequent abbreviations, here meaning titles ("Mr.", "Mrs.", "Dr.", "Sen."), month name abbreviations ("Jan.", "Feb.", "Mar."), U. S. state abbreviations ("Ala.", "Calif.", "Penna.") and some common abbreviations ("etc.", "fig.", "no.", "Co.", "Ltd.", "Corp.") but not abbreviations like ("in.").

Now we can implement the following procedure, given a sequence of letters terminated by a period: 1) if it is followed by a lower-case letter, comma or semi-colon, then it is an abbreviation; 2) if it exists as an abbreviation in the lexicon, consider it as such; 3) otherwise, consider the word as a sentence terminator. Using the following list as a list of abbreviations in the lexicon provides us with only 74 sentences contain unrecognized non-terminal abbreviations candidates in the Brown corpus. And we still have the original 207 sentences which end in an abbreviation that cannot be recognized correctly by any of the above techniques, giving a success rate of 52023 correctly recognized sentence boundaries out of 52511, or 99.07%.

The abbreviations recognized here are:

Strings containing internal periods, Single-Letters, State-Names, Titles and the following: Assn. Av. Ave. Bldg. Blvd. Cf. Co. Corp. Ct. Dept. Dist. Eq. Fig. Figs. Inc. Jas. Jr. Ltd. Mfg. Msec. Mt. Mts. No. Op. Rd. Rte. Sr. St. Stat.

Tech. USN. Vol. Vs. Yo. a. al. ca. cc. cf. cm. cu. dia. ed. eqn. eqns. etc. fig. figs. ft. gm. hp. hr. kc. l. lb. lbs. mEq. mc. mg. mil. min. ml. mm. mos. nw. oz. p. pl. pp. prop. sec. sq. v. var. viz. vs. yd. yrs.

The strategies compared. For a comparison of the relative success of the different strategies we have described, we now conclude this section with a table summarizing all the results:

Tokenization Strategy	Number of Brown Corpus sentences correctly recognized	percentage
Every sentence a full stop	48942	93.20
Regular expressions to recognize numbers	49171	93.64
Regular expressions to recognize numbers and internal periods	49244	93.78
Regular expressions to recognize numbers and abbreviations	51282	97.66
Corpus filtering with regular expressions to recognize numbers and abbreviations	51642	98.35
Lexicon and corpus filtering with regular expressions to recognize numbers and abbreviations	51604	98.27
Lexicon with common abbreviations and regular expressions to recognize numbers and abbreviations	52023	99.07

9.3.2 Combining graphic tokens

As we already mentioned above, the tokenizer must not only split graphic tokens into parts, but must sometimes also combine everal graphic tokens into a single linguistic token. One situation where this is necessary is that of line-ending hyphenations that split a word into two parts. In this case, the tokenizer must reconstruct a single unit from the two parts. Removing end-of-line hyphens seems to be a simple matter, e.g. we can use the simple strategy of deleting any end-of-line hyphen preceded by a lower-case letter and joining the two separated strings together as one word.

To test this simple dehyphenation strategy, we again took the Brown corpus and applied to the hand-tokenized version an operation that introduced end-of-line hyphens. Many word-processing programs exist which can perform this operation. We used an old Unix program called *nroff*, which formats a document to be right and left-justified text with words hyphenated at line ends in order to have an equal number of characters in each line. Now we had two versions of the original corpus: one without end-of-line hyphens and one with them. Then, we applied the dehyphenation program and verified that we got the same text as in the beginning, tabulating any differences introduced by the program.

We found that the type-setting program introduced 12473 end-of-line hyphens. Joining these lines using the simple idea expressed above produced 11858 correct dehyphenations and 615 errors (4.9%), i.e. words which did not appear in the original text. Examples of erroneously joined words are "ring-aroundthe-rosie", "rockcarved", "rocketbombs", "rockribbed", "roleexperimentation", "rookie-of-theyear", "satincovered" and "sciencefiction". This experiment gives a taste of the type of choices that must be made during tokenization. Here, if one had access to a dictionary and morphological package at this stage, one could test each of the 12473 cases by analysing the constituent parts and making more informed decisions, but such a mechanism is already rather sophisticated and its construction is rarely considered for such a preliminary stage of linguistic treatment. One may consider the 615 errors (out of 1 million words) as so many unknown words to be treated at some later stage or just accept them as noise in the system.

There are, however, also situations where token combinations are not so conveniently signalled as in line-ending hyphenations. In many linguistic applications, units of more than one token are sometimes considered as whole. Some linguistic treatment might want to consider all the tokens of a proper name as one unit, or certain multi-word expressions like "hot dog," or even whole idioms like "cold as hell."

Apart from special cases, e.g. where multi-token units are signalled by markup, recognition of such units is beyond the capabilities of the simple tokenization methods presented here. Although some of these methods, such as filtering against a corpus, can be useful for this task as well, we will not go into a discussion of techniques for recognizing and joining multi-token units here.

10 LEXICONS FOR TAGGING*

Anne Schiller and Lauri Karttunen

10.1 INTRODUCTION

In this chapter we will discuss the issue of constructing a lexicon which can be used in tagging.

The purpose of a tagger lexicon is to provide for each *token* of a text its potential properties. What this means, exactly, depends on the tagging system of which the lexicon is part. At the input side of the lexicon component, the tokens which correspond to lexicon entries must be compatible with the tokenizer output (cf. Chapter 9). This also means, e.g., that if the tokenizer recognizes multi-word units, the lexicon should cover them as well.

At the output side, there is more variability. The simplest type of tagger lexicon assigns to each wordform one or more alternative wordclass tags. The choice of the tagset determines most of the lexicon content, viz. the tags (cf. Chapters 4 and 5). The tagger lexicon may also include additional information about the inflection and the lemma or lemmas that word is an instance of. Other information may be necessary because of the tagger method, e.g. frequency information: while the Xerox HMM tagger can work perfectly well with equiprobable tags, Brill's rule-based tagger (cf. Chapter 15) needs the tags to be ordered, the most frequent reading coming first.

The structure or encoding of the lexicon depend of course on the individual tagger methods and programs. For the examples in this chapter we will concentrate on finite-state methods as an efficient means to represent and operate on lexicons.

*The following work was accomplished at Xerox Limited.

H. van Halteren (ed.), Syntactic Wordclass Tagging, 135-147.

Examining the possible content of tagger lexicons and the structures used to represent that content, we come to the following broad classification:

(a) Full-form lexicons

Inflected forms paired with one or more wordclass tags. Entries are of the form

```
<wordform> [ <wordclass-tag$>$ ]+
```

This type of lexicon may be encoded as a simple finite-state automaton.

(b) Morphological lexicons

Lexicon with additional information such as lemma and morphological features for each wordform. Entries have the form

```
<wordform> [ (<lemma>) ([<morph-tag>]+) <wordclass-tag> ]+
```

A morphological lexicon may be encoded as a finite-state transducer that maps each inflected forms into one or more lemmas followed by other morphological information and a wordclass tag.

(c) Weighted lexicons

Lexicon with statistical information <P> associated with each tag:

```
<wordform> [ <wordclass-tag> ( <P> ) ]+
```

This type of lexicon may be encoded as either a simple finite-state automaton or as a *weighted* automaton (cf. Pereira *et al.* 1994).

A tagger lexicon can be constructed in various ways. The best starting point is a morphological analyser (10.2) that has enough coverage to deal with nearly all the forms that occur in any random sample of the language, excepting of course foreign words and misspellings. In this case it is very easy to include the morphological information listed under (b) above. However, for reasons we will discuss shortly, even the best morphological analyser may require adaptation to make it optimal for tagging.

If there is no morphological analyser yet available, a lexicon can be derived automatically from a sufficiently large corpus that has been analysed manually (10.3). The result is of course limited in that it does not include any forms that do not occur in the corpus. Nevertheless, if the corpus is large enough the derived lexicon may be adequate for languages like English that have little inflection. Also, the corpus can provide the statistical data mentioned in (c) and necessary for specific types of taggers (cf. Smith 1997:141-144).

The methods that we present in this chapter are based on the assumption that the tagger lexicon, of whatever type and origin, is implemented as a finite-state transducer.

A simple, acyclic transducer may be constructed from a finite list of word–tag pairs found in an analysed corpus. Nevertheless, in general such a transducer will contain cyclic paths because of numerals, compounds and iterated derivations.

10.2 MORPHOLOGY-BASED LEXICONS

We will first examine lexicon construction with finite-state techniques based on finite-state morphological analysers (cf. Chapter 12).

It has been known for a long time (Appelt and Jacobson 1988) that large wordlists can be encoded surprisingly efficiently by finite-state networks. Each word is represented in the network by a path, a sequence of labelled transitions (= arcs), that leads from the initial state to a final state. In a determinized, minimized network all paths beginning and ending with a given sequence of letters share some of the structure. For example, the paths for "pat", "pet", "pit", "pot" and "put" start and end with the same arc:

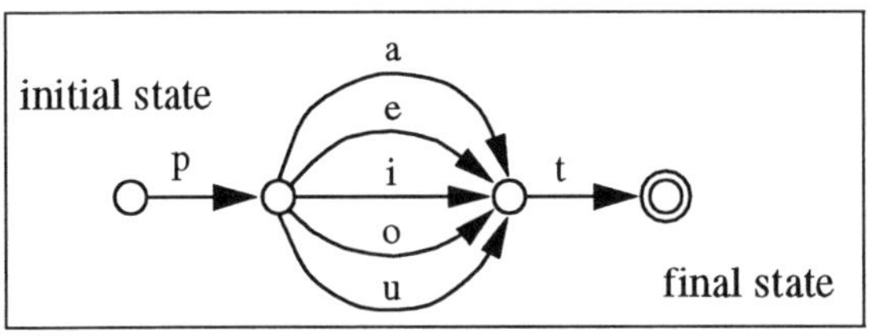

A network with cyclic paths encodes an infinite number of words. This is very important for languages like German and Finnish that have productive compounding and derivational processes that are in principle unbounded.

A network representing a simple wordlist can be upgraded to a primitive morphological analyser by adding to the end of each path one or more arcs to give some additional information about the wordform such as part-of-speech, number, case, etc.

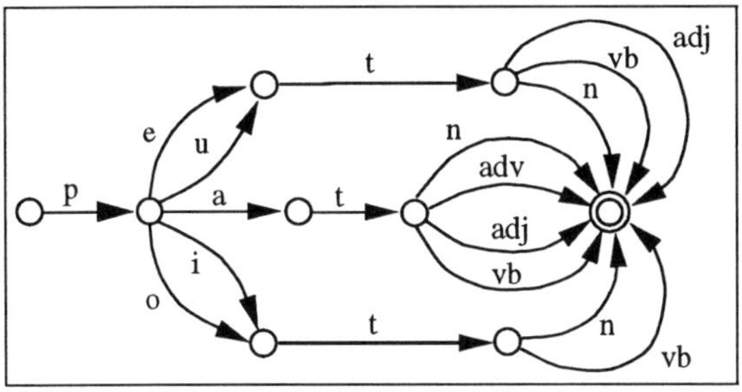

A file containing entries of the form

```
<wordform> [ <wordclass tag> ]+
```

can be compiled very efficiently into this format. The three states that contain arcs with wordclass tags each represent a particular ambiguity class: `n|vb|adj` ("pet", "put"), n|adv|adj|vb ("pat") and n|vb ("pit", "pot"). Because the network is determinized and minimized, the paths for all the words that belong to the same ambiguity class pass through the same state to a final state. For greater efficiency one can merge the alternate tag arcs into a single arc with a label that spells out the choices.

In addition to wordclass and morphological tags, one may also include a code for a function that maps the wordform to its citation form (Revuz 1991). For example, the function that maps "left" to "leave" could be encoded as a tag such as 2ave (delete the last two letters and add "ave").

The relation between a word and its analysis can be represented even more efficiently by another type of network, a finite-state transducer. A transducer is a network in which each path encodes two strings: on one side of the path the actual surface form; on the other side of the path the citation form (lemma) and one or more tags representing the morphological and syntactic characteristics of the form. Each arc is labelled by a pair of symbols, x:y, where x is a part of the lemma or a tag and y belongs to the inflected form. The graph below shows two such paths. They map the form "left" to "leave+VPast" and "leave+VPart". For the sake of clarity, we write the members of the x:y pairs on the opposite sides of the arc here and omit the two null symbols in the pairs a:0 and e:0.

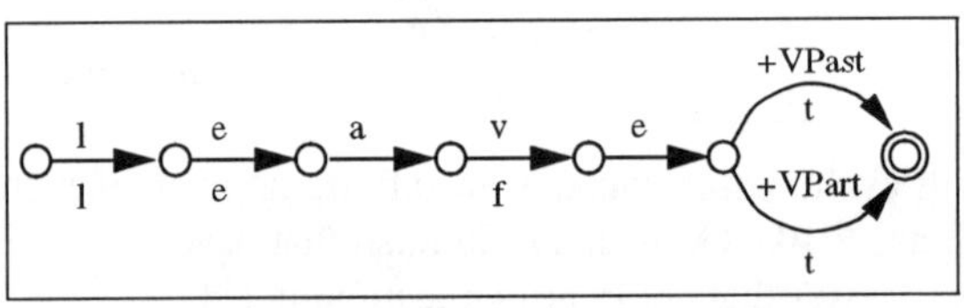

A network consisting of such paths is called a *lexical transducer* (Karttunen 1994). One can compile a lexical transducer directly from a source file containing entries of the form

```
<wordform> [(<lemma>) ([<morph-tag>]+) <wordclass tag>]+
```

but in general such transducers are made by composing a system of morphological rules (e.g. two-level rules) with a source lexicon that contains the lemmas and tags.[1] Irregular mappings, for example "leave:left", that are too idiosyncratic to be handled by general rules, can be encoded separately. Because of the productive compounding and unbounded derivational processes, in many languages the initial lexicon has a very complex structure, even before taking into account the phenomena that are handled by the rules.

Because lexical transducers are inherently bidirectional they can be used for generation as well as analysis. But for that very reason they are not necessarily as efficient

[1] For some languages there are machine readable versions of standard dictionaries. In principle, these could be transformed into such a source lexicon. In practice, however, it turns out that their adaptation is problematic and only worth the effort if they contain important additional information.

as a transducer that is tuned to be applied in a particular direction. For optimal tagging performance it is advantageous to try to make the network *sequential* (Mohri 1997) in the analysis mode. A sequential analyser maps the surface form in a deterministic fashion to a set of outputs without ever exploring any wrong paths. Even in cases where full sequentialization is not possible, a partial sequentialization may yield a great increase in speed. In practice, the analysis speed of finite-state transducers is hundreds of kilobytes of text per second.

Basically, the tags used for disambiguation correspond to the major categories given by a morphological analyser. But in general, we have to reduce the rich set of morphological categories (e.g. omitting tense information) or we have to introduce new distinctions (e.g. auxiliary vs. function verb) for tagging purposes.

- *Mapping morphological classes* is necessary whenever the whole set of morphological features cannot be handled by the tagger, e.g. if the tagger accepts only single symbols for disambiguation or if the tagset is too large (especially for statistical taggers) to get reliable results. This includes elimination of morphological features which are irrelevant for disambiguation.

- *Merging morphological classes* is recommended if the tagging system as a whole cannot reliably disambiguate the subclass information.

- *Refining morphological classes* on the other hand may increase tagger accuracy, e.g. in cases where the morphological class is not specific enough to disambiguate the context words (see examples below).

In the following subsections, we describe how such modifications can be handled within the finite-state approach. Given a finite-state morphological analyser (Karttunen 1994) the tagger lexicon is derived from the morphological lexicon by means of finite-state mapping rules (cf. Chanod and Tapanainen 1995b). These rules are compiled into a mapping transducer which is composed with the morphological lexicon transducer to obtain the final tagger lexicon. Thus, the lexical information for an inflected wordform depends on the basic morphological lexicon and on the mapping rules.

Instead of using composition of transducers one could achieve the same effect by modifying the sources from which the morphological analyser is derived and recompiling a new analyser, e.g. by applying AWK or PERL rewrite rules to each line of the source file.[2]

[2] Such a method is less efficient than applying a mapping transducer to a network which modifies a possibly infinite set of forms in one operation.

10.2.1 Direct mapping

The following example shows how a morphological lexicon is converted to a tagger lexicon by replacing the original fine-grained morphological tags by simple wordclass tags. The conversion is accomplished by composing the original lexicon with a transducer that replaces the original tags by the appropriate single wordclass tags. The mapping transducer is compiled from a regular expression that has the general form

$$\mathtt{T}_1 \texttt{ <- } \mathtt{S}_1\texttt{, } \mathtt{T}_2 \texttt{ <- } \mathtt{S}_2 \ \ldots \ \mathtt{T}_n \texttt{ <- } \mathtt{S}_n \ \texttt{||} \ \texttt{L \_ R} \quad .$$

In effect it says "replace $\mathtt{S}_1$ by $\mathtt{T}_1$ etc. when preceded by `L` and followed by `R`" where the `S`s describe particular fine-grained sequences of morphological tags, `T`s are the corresponding simple wordclass tags and the left and the right context is represented by `.#.`, the beginning- and end-of-string marker. It is a type of replace expression in the Xerox regular expression calculus (cf. Karttunen *et al.* 1997). The expression `$[+Noun ?* +Sg]` represents all strings that contain the morphological markers `+Noun` and `+Sg`, perhaps separated by other symbols. Here `$` is the "contains" operator and `?*` represents any string of symbols.

When the transducer derived from the replacement expression below is composed with the network for the morphological source lexicon in (a), the resulting network encodes the minimal tagger lexicon in (b). Because the replacement erases the lemma as well as the unwanted markers, the tagger lexicon is a transducer that simply maps surface forms to one or more alternative tags.

Example (1): Spanish

- **Mapping rules** replace lexical information by a wordclass tag

```
[ +NOUN_SG <-  $[+Noun ?* +Sg],
  +NOUN_PL <-  $[+Noun ?* +Pl],
  +VFIN    <-  $[+Verb +PInd],
  +VIMP    <-  $[+Verb +PImp],
  +VINF    <-  $[+Verb +Inf],
  +VINF_CL <-  $[+Verb +Inf ?* +Pron] || .#. _ .#. ]
```

- (a) Morphological Source Lexicon

```
casa        casa+Noun+Fem+Sg
            casar+Verb+PInd+3P+Sg
            casar+Verb+Imp+2P+Sg
casarme     casar+Verb+Inf|yo+Pron+Acc+1P+MF+Sg
            casar+Verb+Inf|yo+Pron+Dat+1P+MF+Sg
```

- (b) Minimal Tagger Lexicon

```
casa          +NOUN_SG
              +VFIN
              +VIMP
casarme       +VINF_CL
```

The advantage of using a finite-state transducer for such modifications is that it is very efficient and it can be applied to a cyclic lexicon containing an infinite number of forms.

10.2.2 Merging morphological classes

Using finite-state techniques, several distinct tags (or sequences of tags) can be mapped onto a single wordclass tag, thus merging morphosyntactic subclasses.

Example (2): French

- **Mapping rules:**

```
[ +V_P3SG   <-  $[+P3 +Verb],
  +V_P1P2   <-  $[[+P1|+P2] +Verb] || .#. _ .#. ]
```

- (a) Morphological Source Lexicon

```
chante     chanter+SubjP+SG+P3+Verb
           chanter+SubjP+SG+P1+Verb
           chanter+IndP+SG+P3+Verb
           chanter+IndP+SG+P1+Verb
           chanter+Imp+SG+P2+Verb
```

- (b) minimal tagger lexicon

```
chante     +VERB_P3SG
           +VERB_P1P2
```

10.2.3 Refining morphological classes

In some cases the morphological analyser may not provide a subclassification which is part of the wordclass tagset.

For the classification of verbs, the EAGLES proposal for encoding morphosyntactic phenomena in lexicons and corpora (cf. Chapter 11; also Monachini and Calzolari 1996) recommends the distinction of *verb finiteness* (i.e. finite verb forms, infinitives and participles) and *verb type* (i.e. auxiliaries, modals and main verbs). The German tagset described by Thielen and Schiller (1996) distinguishes three verb types: VV-, VM- and VA-, where VA- is used for auxiliaries ("haben", "sein", "werden"), VM- for modals ("dürfen", "können", "sollen", ...) and VV- for all other verbs. But in fact this classification is based only on the lemmas, not on the actual function in a sentence. Thus, this distinction within the tagset does not introduce any ambiguity for the verb forms themselves, but it can help disambiguate the *finiteness* of a given verb and of other verb forms in its context.

As in the previous examples, we derive the tagger lexicon by composing the original morphological lexicon with a transducer that replaces the morphological tags with a single wordclass tag. However, in this case we need to do this transformation in two steps with two mapping transducers. For the sake of brevity, we first define `Modal` as the union of the modal auxiliaries and use that defined term to express the special mapping rule that introduces the special `+VMFIN` and `+VMINF` tags and eliminates all other lexical information from these forms. This transformation is applied first. The second general mapping transducer retags all the remaining forms in the lexicon.

Example (3): German

- **Defining Modal**

```
define Modal [ w o l l e n | m ü s s e n | d ü r f e n |
               k ö n n e n | s o l l e n ]
```

- **Modal Mapping rules**

```
[ +VMFIN   <- $[Modal +Verb +Inf],
  +VMINF   <- $[Modal +Verb [+Indc|+Subj|+Imp]] || .#. _ .#. ]
```

- **General Mapping rules**

```
[ +VVINF   <- $[+Verb +Inf],
  +VVFIN   <- $[+Verb [+Indc|+Subj|+Imp]] || .#.  _ .#.  ]
```

- (a) Morphological Lexicon

```
wollen  wollen+Verb+Inf
        wollen+Verb+Indc+1P+Pl+Pres
        wollen+Verb+Indc+3P+Pl+Pres
        wollen+Verb+Subj+1P+Pl+Pres
        wollen+Verb+Subj+3P+Pl+Pres
gehen   gehen+Verb+Inf
        gehen+Verb+Indc+1P+Pl+Pres
        gehen+Verb+Indc+3P+Pl+Pres
        gehen+Verb+Subj+1P+Pl+Pres
        gehen+Verb+Subj+3P+Pl+Pres
```

- (b) Tagger Lexicon

```
wollen  +VMINF
        +VMFIN
gehen   +VVINF
        +VVFIN
```

The following example shows two sentences containing a sequence of a finite modal and an infinitive main verb. The tags in (1a,b) reflect only verb finiteness, whereas the tags in (2a,b) include also the verb type distinction.

(1a) Sie wollen/VFIN gehen/VINF .
They want (to) go

(1b) Er sagt, daß sie gehen/VINF wollen/VFIN .
He says that they (to) go want

(2a) Sie wollen/VMFIN gehen/VVINF .
They want (to) go

(2b) Er sagt, daß sie gehen/VVINF wollen/VMFIN .
He says that they (to) go want

An experiment with a statistical (bigram HMM) tagger proved that omitting the verb type distinction lowers the disambiguation rate for ambiguous finite/infinitive verb forms. A test on a small special corpus of 22 sentences (117 tokens; see example in table 10.1) which all contain at least one modal shows:

- Without verb subclassification for modals and auxiliaries the HMM tagging result contains 13 errors – all of them VINF/VFIN confusions.
- Using subclasses VMINF, VMFIN, VMPP for modals and VAINF, VAFIN, VAPP for the verbs "haben", "sein", "werden" reduces the number of errors on the same corpus to 5.

10.2.4 Adding residual wordclasses

Finally, some entries can be added to the tagger lexicon, especially tokens which are in general not part of morphological analysis, such as

- punctuation marks (e.g. ".", "!", "?", ":");
- numeric expressions (e.g. "12,567");
- abbreviations, acronyms (e.g. "AAAI", "A.X.B.").

Table 10.1 Part of a test corpus for modal verb experiments.

Word	Tagset 1	Correct tag	Tagset 2	Correct tag
Sie	PERSPRO	=	PERSPRO	=
müssen	VFIN	=	VMFIN	=
ausgeben	VINF	=	VVINF	=
,	CM	=	CM	=
was	RELPRO	=	RELPRO	=
sie	PERSPRO	=	PERSPRO	=
ausgeben	VFIN	VINF	VVINF	=
wollen	VINF	VFIN	VMINF	VMFIN
.	SENT	=	SENT	=
Er	PERSPRO	=	PERSPRO	=
sagt	VFIN	=	VVFIN	=
,	CM	=	CM	=
daß	COSUB	=	COSUB	=
sie	PERSPRO	=	PERSPRO	=
alles	INDPRO	=	INDPRO	=
ausgeben	VINF	=	VVINF	=
werden	VFIN	VINF	VAINF	=
müssen	VINF	VFIN	VMFIN	=

Tagset 1: .., VFIN, VINF, ..
Tagset 2: .., VAFIN, VAINF, VMFIN, VMINF, VVFIN, VVINF, ...

The infinite sublexicons for numeric expressions and acronyms can be compiled from very simple regular expressions and these cyclic lexicons can then be unioned with the main lexicon into a single transducer. In this way, the tagger does not need any special mechanism for dealing with abbreviations and numbers.

10.3 CORPUS-BASED LEXICONS

In the previous section we described a way to derive a tagger lexicon from a general-purpose morphological lexicon. Generally, such a lexicon does not contain any statistical information such as frequency of wordforms or probabilities for the associated tags.

This type of lexicon can be used by non-statistical taggers, like constraint-based taggers (e.g. Karlsson 1990; Voutilainen *et al.* 1992; Chanod and Tapanainen 1995b) or by the Xerox HMM tagger (Cutting *et al.* 1992) which associates probabilities with ambiguity classes and not with words.

Other statistical methods (e.g. Church 1988; Brill 1992) rely on word–tag frequencies and therefore require lexicons which contain probabilities associated with wordforms. These probabilities are computed from large (manually) tagged corpora and included in the lexicon.

The way statistical information is represented in the lexicon depends on the tagging method:

- In Church's (1988) approach, the lexicon contains for every surface wordform a list of tags. Each of these tags is associated with a *lexical probability*. The lexical probability of a TAG for a given WORD is the ratio between the frequency of the pair (WORD,TAG) and the overall frequency of WORD in a large tagged corpus.
- In Brill's (1992) approach, the lexicon does not contain explicit numbers. Each word in the lexicon is associated with a list of tags of which the first one has the highest probability and all the others may appear in no particular order.

Unless manual modifications are applied, the coverage and quality of a corpus-based lexicon depends entirely on the training corpus, The size and variety of the corpus determines the coverage of the lexicon: if the corpus is too small or too specific, the lexicon may lack words or readings and it may not contain reliable statistical information (cf. 6.3.5).

Errors in the corpus imply errors in the lexicon: corpus typos or erroneous tag assignments will result in erroneous lexicon entries. If the training corpus is large enough and the error rate low, these errors will not significantly influence the overall statistical information. But some cases may require corrective actions, e.g. rare words which occur with a wrong tag in the corpus should appear correctly in the lexicon.

The following examples illustrate some of the problems that may occur when deriving a lexicon from a corpus. The sample corpus is extracted from the Brown corpus and contains 40,000 tokens. In general, we can distinguish four types of problems:

1. **lexical gaps**:
   (a) **missing wordforms**: Wordforms of the language which do not occur in the corpus.
   Example: The training corpus does not contain the word "computer" (only the plural form "computers"). Therefore no lexicon entry will be created for this word.
   (b) **incomplete entries**: Wordforms for which not all readings occur in the corpus.
   Example: The training corpus contains a single instance of the word

"abuse" which is tagged as a singular noun (NN). The derived lexicon will thus lack the verb reading of "(to) abuse".

2. **lexical errors**:

   (a) **wrong entries**: Misspelled or generally mistagged wordforms.
   Example: The word "thru" occurs only once in the whole corpus and is tagged as adjective (JJ) instead of preposition (IN). So the lexicon will contain an entry for the word "thru" but associate a wrong tag with it.

   (b) **partial errors**: Occasionally mistagged wordforms.
   Example:: The word "a" appears 796 times in the corpus. 795 times it is tagged as article (AT), but in one case it is (incorrectly) tagged as proper name (NP).[3]

How can these problems be resolved? The first solution is to use a larger tagged corpus as a basis for the lexicon construction. If no such tagged corpus is available this requires resources for manual tagging. If a corpus is available, but annotated with a different tagset, a conversion or mapping tool is necessary to produce the appropriate tags. A second solution is to use other lexical resources (like a morphological analyser) to extend the actual corpus lexicon.

10.3.1 Enlarged Training Corpus

A larger training corpus will contain more wordforms, and thus *lexical gaps* (1a and 1b) will be less frequent than with our small sample corpus. But even if a very large corpus is available, it will never cover all the potential words of a language which may occur in a text to be tagged (cf. Chapter 13) – especially if the language is highly inflectional or very productive for compounding (e.g. German, Finnish, Turkish).

With a large training corpus *lexical errors* (2a and 2b) may be reduced if lexicon entries are only accepted beyond a given threshold (e.g. if a word–tag pair occurs at least twice) in order to prevent inclusion of typos or accidentally wrong tags. If the lexicon entries are associated with probabilities, filtering may not be necessary, provided that the frequency of erroneous wordforms in the corpus is low.

10.3.2 External Lexical Resources

If other lexical resources (e.g. morphological analysers) are available, these can be used to bridge gaps in the corpus lexicon or to check for lexical errors.

[3]It may well be that a tagging judged to be a partial error is correct after all, but only in a very specific context. In such cases, we may still want to remove it from the lexicon because it interferes with the normal operation of the tagger, e.g. the word "over" can be a noun, but only in a text about cricket, and should rather not be included as such in a general lexicon.

In the case of *missing wordforms* (1a) a morphological analyser (or unknown word handler; cf. Chapter 13) can be used at run-time whenever the tagger encounters a word which cannot be found in the lexicon. If the tagging method requires lexical probabilities this must be associated with a strategy to assign default probabilities for unseen words.

While non-lexicalized words forms are easy to detect in a new text, it is more difficult to check if the lexicon contains the appropriate reading (tag) for each word of an unseen text. Therefore, the lexicon should be checked for *incomplete entries* (1b) before it is used, e.g. by applying a morphological analyser in order to make sure that the corpus lexicon contains all potential tags for each entry.

Similarly, the corpus lexicon can be checked for *lexical errors* by comparing it with a morphological analyser. Completely *wrong entries* (2a) should be corrected, *partial errors* (2b) may be neglected if they are associated with very low probabilities.

To summarize, a morphological lexicon or analyser is very useful for fixing gaps and errors in a corpus-based lexicon. But as it is difficult to identify all such cases, the reverse procedure appears to be preferable, viz. to derive the core tagger lexicon from a morphological lexicon (as described in 10.2), and later add corpus specific lexicon entries when necessary. Furthermore, if the tagger method requires it, one must then add statistical information computed from a tagged corpus and, possibly, default probabilities for unseen words.

11 STANDARDIZATION IN THE LEXICON

Monica Monachini and Nicoletta Calzolari

11.1 THE INITIATIVE FOR STANDARDIZATION

Lexicons, as described in the previous chapter, are a valuable resource, not only for wordclass tagging but also for many other applications in the broad area of language engineering (LE), which encompasses fields such as computational linguistics and Natural Language Processing (NLP). Furthermore, the last decade in particular has seen an increasing use of corpora for computational lexicography, other corpus-based research and development of applications, all of which has led to the general recognition of the value of 'authentic' data.

However, awareness has also grown that information derived from 'raw' material presents obvious limitations and usually some linguistic processing is needed. Moreover, the creation of a clean and useful lexicon resource can never be fully automatic and will have to entail a substantial amount of human effort. This also means that such an activity is quite expensive and sponsoring is unlikely when the resource is designed for only one specific application or tool.

In order to be worth the effort, resources should be *reusable*, *interchangeable*, *shareable* and, preferably, also compatible with a multi-lingual environment. In effect, this means that all such resources should be designed according to a common set of consensual and broadly agreed specifications, i.e. *standards* (Calzolari 1994; Calzolari and Zampolli 1994).

H. van Halteren (ed.), Syntactic Wordclass Tagging, 149-174.

On the side of standards for the *physical* representation of language data, widely used norms exist, constituted by *ISO* standards. Moreover, on the *formal* level, the huge work of the *Text Encoding Initiative (TEI)* (Sperberg McQueen and Burnard 1994) produced norms for the representation of textual data, covering multi-lingual character sets up to meta-languages for representing the structure of documents based on SGML.

There has also been strong encouragement by the CEC to the development of standards in the area of reusable linguistic resources. After a cycle of feasibility projects such as ET-7 in the area of lexicons (Heid and McNaught (eds.) 1991) and NERC in the area of corpora (Calzolari, Baker and Kruyt (eds.) 1995 and Zampolli 1995), the *Expert Advisory Group for Language Engineering Standards (EAGLES)*[1] was launched in 1993, within EC DG-XIII's Linguistic Research and Engineering programme, with the specific aim of defining standards, when feasible, and preparing the ground for future standards for those areas which are not yet mature enough, in the field of language resources for language engineering.

The areas of concern to EAGLES are not only computational lexicons and text corpora (cf. Chapter 5), but also grammar formalisms, evaluation and spoken language. The basic idea behind the EAGLES work is for the group to act as a catalyst for pooling together concrete results coming from current major European projects. The general strategy has been to use accepted standards (e.g. TEI) and build onto them, thus getting the opportunity to concentrate more on the *content* side, i.e. the interpretative level, which is the aspect more difficult to agree on.

A general description of EAGLES work is given by Calzolari and McNaught (1996) and results of the first phase are publicly available on the Web:

```
http://www.ilc.pi.cnr.it/EAGLES96/home.html
```

Shareable, pre-competitive, publicly available resources, encoded according to EAGLES common specifications and criteria, and the tools for manipulating them, are already being partially developed in a number of EC projects. Furthermore, larger initiatives for developing more comprehensive resources are foreseen at the European level.

In this chapter, we tackle the issue of lexicon standardization for the morphosyntactic level of description, after providing some historical background, from different but interrelated perspectives: the relationship between lexicon and corpus, the EAGLES proposal for morphosyntactic encoding with a discussion on methodological aspects, some languages and projects where the recommendations are actually used, and validation activities.

---

[1] The NERC study and the EAGLES project are both coordinated by A. Zampolli.

11.2 INTERDEPENDENCE BETWEEN LEXICON AND CORPUS

It is generally accepted that there are concrete direct links between corpus and lexicon which make the two inseparable at many levels and for different purposes in the field of Linguistic Engineering (LE). This link extends to morphosyntactic annotation, where on the one hand information from a lexicon is needed in the tagging process and, on the other, information in (especially annotated) corpora is useful for the creation of a lexicon.

As a result, the two EAGLES subgroups working on morphosyntax (the Lexicon and the Corpus Working Groups) have always worked in close cooperation. The choices and approaches that were undertaken were guided by awareness of the necessity of having harmonized lexicon and corpus information at the morphosyntactic level and of defining *compatible sets of attributes and values* for the two.

The recommendations produced by the two subgroups (for lexicon specifications, Monachini and Calzolari 1996; for corpus tagsets, Leech and Wilson 1996) should be seen as two complementary sets. They examine similar sets of phenomena from different viewpoints: the more general one, the lexicon encoding, capable of being directly mapped onto an application-oriented one, the corpus annotation.

11.2.1 Lexical encoding vs. corpus tagsets

The need for two different perspectives was put into evidence during standardization work within EAGLES and related actions which joined forces with EAGLES, e.g. the MULTEXT project (cf. the introduction in Monachini and Calzolari 1996; Bel, Calzolari and Monachini (coords.) 1995). The differences are already reflected at the level of terminology: the terms *feature* and *specifications* are preferred when talking about lexicon descriptions, while the terms *tag* and *tagset* are used for the information associated with words in context, i.e. in corpus annotation. As we will see in more detail below, morphosyntax in a lexicon shows a *fine granularity*, while corpus tags tend to reflect *broader categories* and to be more language dependent.

In principle, it ought to be possible to avoid such differences, since, for the sake of reusability, both lexical specifications and corpus tagsets should

- be, as far as possible, independent from specific applications;
- aim at a general description of language use;
- be as detailed as possible: fine-granularity of the information allows users, when necessary, to rearrange categories without much difficulty.

However, unless we assume a fully manual annotation of corpora, the corpus tagset must be related to at least one specific application: an automatic tagger. This means that the choice of tagset must take into consideration the restrictions on state-of-the-art taggers.

11.2.2 Tagsets as collapsed feature specifications

The first question is whether a tagger can be expected to disambiguate between different lexicon descriptions or different types of typical homographies present in different language types. Experience shows that some distinctions are difficult to capture automatically with a high enough rate of accuracy. In some languages, for example, the disambiguation of *indicative present* and *subjunctive present* in a corpus is extremely difficult to achieve by automatic means. Some verbs have different forms for the indicative and the subjunctive (e.g. Fr. "venir": indic. "viens", subj. "vienne"; It. "venire": indic. "vieni", subj. "venga"). Others, however, do not (e.g. Fr. "manger": indic., subj. and imper. "mange"; It. "mangiare": indic. and subj. "mangi").[2] In the latter case, disambiguation can only be achieved with very complex parsing of sentences or alternatively manually, in a post-tagging phase. As a result, such distinctions are often not represented in corpus tagsets, leading to 'collapsed' tags representing several sets of feature bundles.

Another reason not to use too fine-grained a tagset is that, in order to train some types of taggers, statistical tables based on co-occurrences of tags are needed. A choice for a large tagset implies the use of a very large corpus to train the disambiguator. In a smaller corpus rare co-occurrences would not be observed often enough or even missed altogether (cf. 16.4.1). In many large scale projects (e.g. Penn Treebank, BNC, IBM France), the need for fully automatic tagging has led to very coarse tagsets, usually under 100 tags. In the MULTEXT project as well (Bel, Calzolari and Monachini (coords.) 1995), specifications proposed for French foresee 249 different lexical descriptions, but these are subsequently collapsed to only 74 corpus tags.

On the other hand, one should be very careful in being led purely by the capabilities of taggers in one's choice of tagset. If the underspecification is pushed too far, the tagset may become useless for specific applications.

11.2.3 Multi-linguality

As already mentioned above, resources should also be usable in multi-lingual environments and standardization should therefore preferably extend across several languages. However, it may be inappropriate or undesirable to have completely identical tagsets across languages. Indeed, it would be useful to preserve certain language-specific peculiarities and distinctions. As an example of this, consider the peculiarities related to person differentiations in verbal morphology.

[2] The situation is even a bit more complicated, as the different forms also differ in their possible values for the feature *person*. Fr. "mange" may be 1st or 3rd person (singular) when indicative or subjunctive, but only 3rd person (singular) when imperative. It. "mangi" may be any person (singular) when subjunctive, but only 2nd person (singular) when indicative. See also below.

In Italian, first, second and third person of the *present subjunctive* are homographic: "che io/tu/egli ami". In Spanish, the same is true for first and third person of different tenses: "Yo/El cantaba" (impf.); "Yo/El cantaría" (cond.); "Yo/El cante" (subj.). French also has homographs for different verbal persons, but these are the first and the second person of some tenses: "Je/Tu viens"; "Je/Tu étais".

Taking into account that the subject in Italian and in Spanish is not obligatory, and that the tagger cannot know if the preceding noun phrase is in fact the subject of the verb, there is no easy way to discriminate between the homograph forms. Hence, conflating tags can be recommended, containing different information for the different languages: for instance, for Spanish, a tag marked as *non-second-singular* or as *first-third singular* form; for Italian as *first-second-third*. Finally, for French, the collapsed tag could be *non-third-singular* or *first-second-singular*, but, due to the obligatory presence of the lexical subject in the context, the tagger may also be able to disambiguate between two different tags for each verbal form.

Experience gained while working jointly between EAGLES and MULTEXT demonstrated that it is indeed not possible, nor desirable, to arrive at identical tagsets across languages, even for those within the same language family. The need for idiosyncratic tagsets for each language has also been confirmed within PAROLE-LE (PAROLE Tech. Ann. 1994).

11.2.4 Lexicon specifications as an interface between tagsets

We have seen that, even after standardization, there will be several tagsets in use, showing differences inspired by language peculiarities or by the needs for specific purposes. The question is then how to *relate different physical tagsets*. This problem has been tackled and addressed within EAGLES and MULTEXT.

In the EAGLES Corpus subgroup, in particular, it has been stated that each physical tagset should be "mappable (automatically, if possible) on to a set of attribute-value pairs in conformity with the guidelines". In order for such a mapping to be feasible, it has been proposed that each tagset have an internal language-neutral representation, defined as Intermediate Tagset (cf. 5.3). This internal representation can

- act as a means to map local tagsets onto the EAGLES tagset;
- facilitate the comparison between different local tagsets;
- act as intermediate mapping between the tags of the corpus and the labels of the lexicon used to tag the corpus.

The concept expressed in the last point constitutes the basis of the whole EAGLES work on morphosyntax. The comparability and the harmonization of the linguistic properties represented in different tagsets can be obtained only by relating the tagset to the lexicon, i.e. by defining each physical tag according to the common specifications contained in the lexicon. In fact, these specifications, if stable enough and agreed and harmonized across languages, make different physical tagsets compatible and mappable on each other. In other words, lexical specifications are used as a common platform across languages, as a sort of *interface*, which permits different tagsets to 'communicate'.

Within the EAGLES Tagset Mapping task, the mapping of two different tagsets via the lexicon has been concretely implemented by automatic means (Teufel 1996b). The experiment was conducted on two English tagsets (UPenn and BNC) by means of a powerful computational tool available as a research prototype. The lexicon specifications used as interface between the two tagsets have been formally modelled in typed hierarchies, expressing all constraints and dependencies between values and each physical tag of the two tagsets has been defined according to these lexicon specifications. The tool, using as input the semantics of each tag and comparing it with the lexicon hierarchies by means of PROLOG rules, performs the mapping automatically, allowing for queries on the two tagsets through a single homogeneous query format.

The same philosophy has been adopted within the ELSNET experiment (ELSNET Reusable Resources Task Group), a pilot project launched to prove the viability of the EAGLES proposal. Within this framework, an 'integrated package' was produced jointly for German and Italian consisting of: a sample corpus of about 50,000 occurrences annotated on the basis of the EAGLES specifications and manually checked for Italian (cf. Corazzari and Monachini 1995),[3] the two tagsets defined according to the EAGLES Italian and German Guidelines for the Lexicon, the Guidelines themselves and, optionally, tagging tools trained on the EAGLES lexicon specifications. The choice of similar text types and the definition of the two tagsets according to the EAGLES lexicon specifications makes the two text samples and the two tagsets comparable.

We will conclude this section with a detailed example, taken from the MULTEXT work (cf. 11.6.1), of how lexical lexical descriptions can be mapped onto corpus tags. The French MULTEXT group (Veronis, Khouri and Meunier 1994) proposed that such a mapping should be applicative, i.e. "each lexical description should map to one and only one corpus tag, while it is not possible to do the reverse" due to the limitations of current tagging techniques.[4] Under these circumstances, the mapping can be done with *conversion tables* which relate two different items.

[3] For the Italian Reference Corpus, see Bindi, Monachini and Orsolini (1991).

[4] They also stated that a situation where corpus tags are more precise than a lexical description (i.e. one lexical tag corresponds to more than one corpus tag) should be, in principle, avoided.

Given the tags SMS and NCMS,[5] defined as[6]

TAG	Lex.Spec.	Definition
SMS	Ncms-	Noun, common, masc., sing., no-case
NCMS	Ncms-	Noun, common, masc., sing., no-case

there are no problems in comparison as their definition maps both to the same lexical specification. However, in other cases, the mapping between tag and lexical specifications can create problems when under-informative corpus tags are to be directly mapped onto the lexical descriptions each one subsumes.

An example taken from the Spanish MULTEXT proposal for the wordform “me” is:

TAG	Lex.Spec.	Definition
P1S	Pp1msa-	Personal pronoun, 1st pers., masc., sing., acc.
P1S	Px1msa-	Reflexive pronoun, 1st pers., masc., sing., acc.
P1S	Pp1fsa-	Personal pronoun, 1st pers., fem., sing., acc.
P1S	Px1fsa-	Reflexive pronoun, 1st pers., fem., sing., acc.
P1S	Pp1msd-	Personal pronoun, 1st pers., masc., sing., dat.
P1S	Px1msd-	Reflexive pronoun, 1st pers., masc., sing,, dat.
P1S	Pp1fsd-	Personal pronoun, 1st pers., fem., sing., dat.
P1S	Px1fsd-	Reflexive pronoun, 1st pers., fem., sing., dat.

For this wordform it has been decided to use a single tag P1S, conflating all the possible lexical descriptions. This was done on the basis of the assumption that an automatic tool would have disambiguation problems in assigning the correct analysis among all the lexical descriptions since such disambiguation would require syntactic analysis.

In order to avoid redundancy in the conversion tables and to make tag optimization work easier, it was proposed to study the possibility of having *intermediate representations* which prepare the conflation of information and which facilitate automatic mapping from lexical descriptions onto tags. This intermediate internal notation (which follows the EAGLES philosophy of the Intermediate Tagset; cf. 5.3) was proposed in the form of *regular expressions*. These are built using *operators* in order to sum up the information referred to by different lexical descriptions and conflated in a given tag. Operators can be:

- *Match any* (.), i.e. wild-card character

[5]The tags used to exemplify issues and problems dealt with in this section come from the tagsets proposed in the language-specific applications of four of the MULTEXT partners.

[6]Within MULTEXT, due to its strong task-oriented definition, the lexicons to be run under automatic tagging tools have been produced in the format of wordform lexical lists with lexicon specifications and corpus tags associated to each wordform.

- *List* ([...])
- *Disjunction* (..|..)
- *Negation* (^)

Expressed in the regular expression language, the above verbose notations for P1S would appear in the following, more elegant format:

```
P[px]1.s[ad]-      P1S
```

It should be noted that regular expressions can be regarded as a convenient way to map the lexical descriptions onto the corpus tags since, in many cases, the information in the lexicon is more precise than the information we can/want to have in the corpus tagset. In addition, regular expressions could make translations between the lexical description and corpus tags easier and enable the automatic generation of conversion tables.

11.3 THE EAGLES PROPOSAL FOR MORPHOSYNTACTIC ENCODING

The general approach underlying EAGLES follows the ET-7 proposal (Heid and McNaught (eds.) 1991) of looking for the basic phenomena at the morphosyntactic level of linguistic description, at the more granular level, and providing the most detailed set of features in order to encode the phenomena. This approach is applied over the full set of European languages, trying to reach the same level of granularity for the description of each of them. The description is to be *multi-lingual,* collecting features relevant for the core of all the EC languages, as well as *multi-functional,* avoiding bias towards a particular application.

The obvious consequence of the adoption of both *high granularity* and *coverage of many languages* is that a large repository of potentially useful lexicon specifications is formed. In this way, all the features which are necessary for the description of the basic phenomena in the many practices and languages considered are juxtaposed. This large collection of features does not, and cannot, constitute a consistent system to be implemented as such, but is a *redundant inventory of all the possible features relevant for that category across the different languages* (an ET-7 conformant big repository, according to the *data pool* model). At the general level of the EAGLES lexicon specifications proposed for each category, there is, by definition, i.e. by the very way in which they are constructed, no property of *monotonicity* (and no necessity of it). In contrast, *redundancy* and *conflicting values* may be found. Within the EAGLES approach, it is assumed that not all the values presented are relevant for all the languages. There are cases in which some do not apply and this has to be made clear at the language-specific level.

It is only at the language-specific level – due to the peculiarities of the language-specific systems – that the problem of the representation formalism arises and monotonicity can be looked for. Each language-specific system can afterwards be implemented as a specific application of the general redundant set by picking up the features and values appropriate for its system (see e.g. the work undertaken within the EAGLES Validation phase (Monachini 1996) and the lexicon instantiations of the PAROLE project (Monachini (coord.) 1996)).

11.3.1 Methodology of standardization

In order to create such an inventory and proposed standard, a bottom-up approach was adopted, in agreement with the general EAGLES methodology, aiming for the reasonable goal of standardization by consensus. The step-wise process towards consensus building can be summarized in the following phases: survey of the field, proposal of common specifications, discussion among specialists, application to languages and in projects, testing and validation, and feedback and revision. These phases have been implemented in a cyclical way, giving rise to a process of adjustments and re-applications and continuous exchanges between the partners and external collaborators. The procedure has ensured a *broad language coverage, cross-sectorial actions* and *compatible specifications across languages and applications,* as will be shown below.

The first step has been the survey and comparison of the main existing encoding practices for morphosyntactic description in both lexicons and corpora, with the aim of deriving a consensual proposal from their comparison. The morphosyntactic descriptions and encoding schemes in lexicons that were involved in the comparison phase are the following:

- The MULTILEX model as presented in the the Final Report on Morphology (MULTILEX 1993).
- The GENELEX model for the encoding of the morphological and syntactic levels in a lexicon, originally studied for French (GENELEX Apr. 1993; GENELEX Sept. 1993).
- The specific AlethDic application of the GENELEX model (GSI-Erli 1993).

Information on corpus annotation practices, as compiled in comparable surveys, was also included:

- The proposal of a consensual nucleus of morphosyntactic information encoded by the most common existing tagging practices presented in the framework of the NERC Project (Monachini and Östling 1992a, 1992b), which took into account the following tagsets: Penn, Gothenburg and Brown for American English, BNC, LOB, Lancaster and ENGTWOL for British English, ILC-DMI and EUROTRA for Italian, INaLF for French, Uit den Boogaart for Dutch. It should be noted that the NERC proposal also incorporated the list of common morphological features proposed within the TEI by the Linguistic Analysis Committee (TEI AI1W2 1991).

- The scheme proposed by Leech and Wilson as a first outcome of the EAGLES Corpus morphosyntactic subgroup (Leech and Wilson 1993).

The survey and comparison phase allowed us to highlight the areas of common ground and some aspects of discrepancy between the different systems for classifying morphosyntactic phenomena and as well to obtain a first *inventory of consensual descriptive labels*.

A particular problem in this activity, especially since many languages are involved, is to decide when two categories can be mapped onto each other. For each scheme/language, a different choice may have been made from the many possible definitions of *grammatical category*. Similarly, there are many definitions for *common categories*, e.g.:

- In a distributional definition, a category has the same meaning in different languages (i.e. is a common category to different languages) if it enters in the same combinations.

- Following a translational definition, a category is common "if it yields isomorphic partitions of word-lists or lexicons under translation (modulus exceptions)". A category system is common to languages L0 and L1 if members of class C in L0 translate as members of C in L1.

- From a semantic point of view, a category is common to different languages if it "conveys the same linguistic information (i.e. stands for the same linguistic phenomenon) in all the languages in which it is used".

- A wordclass can be common across the languages if it reflects a linguistic category/phenomenon which is either morphologically or lexically expressed in at least two of these languages, even though there may not be a one-to-one lexical relationship for this in translation.

In short, a category can be defined as common if "its members satisfy the same criteria and tests". However, this crucially implies a clear definition of the criteria for the recognition of the members of a category (cf. 11.5). In absence of such explicit criteria, common categories in EAGLES have been empirically recognized as those which are relevant in the morphosyntactic descriptions of a number of European languages. These common categories are usable and actually used in the largest lexicons and corpora and have in general the same 'meaning' in the different languages, even though the property of *commonality* holds mainly for open classes and poses more problems for function words or closed classes.

Taking into consideration this simple equation of *common categories* and *actually used categories*, the *adequacy* of categorization, as far as user requirements are concerned, is achieved. It is empirically obtained through the bottom-up process of looking at the largest and most used lexicons and tagging schemata. Behind these lexicons and annotation schemata there have been many different types of users (of lexicons and of annotated corpora) whose needs are implicitly reflected.

From the analysis of the schemes, it was found that a lot of commonalities proposed in EAGLES are relevant for many languages. Different schools and traditions of languages can agree on such a simple set of features. Furthermore, the setting of a sort of *de-facto* standards, justified by linguistic traditions and current practices, makes sufficiently sure that these sets are adopted, as there are no strong reasons for departing from them.

11.3.2 The proposal

After analysis and evaluation of the results of the comparison we formulated a common set of specifications, which constitute the first EAGLES proposal (Monachini and Calzolari 1996) and which we will now examine in detail. The typical objects described in EAGLES are lemmas (even though lexical decisions as to what is to be to considered as a lemma are not dealt with) and wordforms at the morphosyntactic level.[7] This essentially includes information on the grammatical category or part-of-speech, their subtypes as found in lexicons and inflectional phenomena to be encoded in attributes such as gender, number, tense, etc.

The proposal corresponds, as far as possible, to the union of the analysed practices. The set of the EAGLES recommendations is presented category by category and consists of a list of labels with examples of their use. Following the current practices for NLP, a feature-based notation is used, in the form of attribute and value pairs displayed in a tabular way, referred to as *synoptical tables*. The features are articulated on different levels, corresponding to different degrees of *obligatoriness*. As already pointed out (Leech 1993; Monachini and Östling 1992b; Leech and Wilson 1993), different levels of constraints can be isolated in the morphosyntactic encoding of a category and, therefore, different levels of consensus can be suggested.

---

[7] There is always the question of how to deal with the two complementary phenomena of (a) grammatical categories split into more than one graphical unit (multi-words or discontinuous words) and (b) graphical units composed of more than one grammatical category (e.g. contractions). In general, EAGLES recommends handling multi-words as belonging to a single grammatical category and handling contractions as two separate grammatical categories, but the option of a different treatment is left open.

Table 11.1 Layout of a synoptical table.

PoS	Attribute	Attribute	Attribute	Attribute	...
MULTILEX	value	value	value	value	...
GENELEX	value		value	value	...
AlethDic	value	value	value		...
NERC	value	value		value	...
EAGLeech	value		value	value	...
EAG-L0	obligatory: PoS				
EAG-L1	recommended: minimal common core set of features				
EAG-L2a	optional: info common to languages, either not usually encoded or not purely morphosyntactic				
EAG-L2b	language-specific: language-specific info				

Each synoptical table is divided into two zones (figure 11.1). The first zone displays the information gleaned from the survey, representing current practice and allowing the immediate *comparison*. The second zone is devoted to the attributes and values constituting the EAGLES *recommendations*.

The comparison zone comprises the top half of a synoptical table. The top horizontal row displays the morphosyntactic part-of-speech (PoS) considered (first cell on the left) and the relevant morphosyntactic information, presented as attribute names. The first vertical column on the left contains the names of the encoding systems under analysis. The other columns each represent a specific attribute. The column is headed by the attribute name and the cells list the relevant values that are used within each system. If a cell is left empty, this means that the system does not mark the information.

The bottom half of the table is the proposal zone. The specifications are articulated on three levels, L0 to L2. As already stated, it is not possible to create a single standard which is usable throughout the multi-lingual and multi-purpose environment we are confronted with. In order to give flexibility to the proposal, the multi-layered or hierarchical structure has been preferred to a flat one. This also provides an easy framework for extensions and updating. The idea is that, going from Level 0 to Level 2, the amount of information is increased and more granularity is achieved.

Level 0 (L0) contains the category (or PoS), i.e. Noun, Verb, Adjective, etc. The category is also presented in the proposal by Leech and Wilson (1996; see also Chapter 5) as the only *obligatory feature*.

Level 1 (L1) presents the grammatical features, the agreement features such as Gender, Number, Person, etc., which are usually encoded in lexicons and corpora: these are considered as *recommended features* constituting, together with the information of the upper level, the *minimal common core set of features* for the PoS.

Level 2 presents further extensions, representing information less regularly encoded. It is split into two sublevels corresponding to the reason for it not being in the common core. Level 2a (L2a) contains information which is pertinent to all or many languages and is considered useful and easy to standardize but is either not yet usually encoded by current practices or not purely morphosyntactic (e.g. countability for nouns): these are to be considered as *optional features*. As a rule of thumb, a value is put here if it is relevant to more than three languages. Level 2b (L2b) presents *language-specific features* not belonging to the set of recognized common features. The values presented in this row are labelled with the abbreviation of the language to which they are pertinent.

Within this flexible framework, the user is allowed to choose the most appropriate level of encoding, which may vary according to different languages, applications and purposes. The possibility of incorporating idiosyncratic classes and distinctions after the common core features makes the proposal adaptable to different user needs and extensible enough without interfering with the compatibility. In such a way, the risks linked to a rigid standardization (already mentioned in Chapter 5) are avoided. At the same time, the structure is rigorous enough to permit comparison among different practices which do not reach the same granularity of encoding (at least up to the level of the minimal core) and among different languages.

As an example, let us examine the EAGLES synoptical table for Nouns (figure 11.2). The names of the attributes and values are presented in the tables in an abbreviated form, due to the obvious limits allowed by the page. Almost always, the abbreviations are self-explanatory, e.g. "Gen" stands for Gender and "Num" for Number. In the actual EAGLES document, the full names of the attributes with the relevant values are listed, in the *Comment* section which follows each of the tables.

The abbreviations are not intended to represent by themselves a proposal of how to encode these attributes concretely in lexicons. Instead they are only labels to represent a distinction that can be marked in the various European concrete practices using different symbols, e.g. for the value `singular` of the attribute Number, *singular, sing, sg, s, 1* or whatever label can be used. What is important is the mappability from a concrete system of labels to the set of proposed labels, so that the table can serve as an interface or interchange proposal.

11.4 INSTANTIATION IN DIFFERENT LANGUAGES

Within the EAGLES project, the first set of recommendations has been concretely tested and quite extensively evaluated on real corpora and lexicons, by applying it to the European languages (Danish, Dutch, English, French, German, Greek, Italian, Portuguese, Spanish) in a cyclical process of applications, adjustments and revisions, which leads to the final and stable set of recommendations.

Table 11.2 Synoptical table for nouns.

N	Typ	Gen	Num	Cas	Count	Def	Infl	Decl
Mult	com prop	m f n	sg pl	nom gen dat acc voc	cou mass			
Gen	com prop	m f	sg pl					
AlD		m f n	sg pl					
NERC	com prop	m f m+f	sg pl s+p	nom gen dat acc bas				
Leech	com prop	m f n c	sg pl	nom gen dat acc voc bas	cou mass			
L0	NOUN							
L1	com prop	m f n	sg pl					
L2a				nom gen dat acc	cou mass			
L2b		It c Du f(m) Du cont Sp trns Sp notr	It n	Gr voc Gr ind		Da def Da indf Da unmk	Ge weak Ge strg Ge mix	Ge no Ge adject

Further confrontation of the EAGLES specifications with specific languages has taken place in the framework of two large EC projects. Within the MLAP-PAROLE project (coordinated by A. Zampolli), the application and evaluation has been extended to three other languages: Irish, Catalan and Swedish (Monachini (coord.) 1996). Within the MULTEXT-EAST COPERNICUS Project, the EAGLES minimal common set of specifications were proved to fit the requirements of six Eastern European languages: Bulgarian, Czech, Estonian, Hungarian, Romanian, Slovene (Monachini (coord.) 1995). Some distinctions for the peculiarities of these languages have been added at the language-specific level and concrete lexicons have been built.

The implementation of each language specific system as a specific instantiation of the general set by picking up the appropriate features and values is a completely new fact in this kind of work. Indeed, the application work serves both as a refinement and as a validation of EAGLES itself, thus increasing the strength of the proposal and covering the need for a *broad language perspective*.

During the instantiation exercises, in particular when the specifications have been detailed for real lexicon building, the range of pertinent values for each feature and any further constraints have been made explicit as far as possible. Aspects of the hierarchical organization of the features have also been solved.

An agreed format has been decided so that a common structure and comparability among languages is ensured:

- The morphosyntactic feature pertinent to a category in a specific language is represented in the form of a table. Examples from the specific language are given for each attribute value and, when necessary, criteria for the application of the feature are described.

- The attribute name is presented in bold characters, the values are listed in italics and examples are provided, together with the information on the corresponding tag in a language-specific lexicon encoding scheme or corpus tagset (if available).

- If a language does not have any values for a given attribute, the cell of the example (and of the tag) is left empty.

- If the attribute has, in a particular language, more values than in the proposed common system, the extra value is inserted in the second part of the table (with the note *l-spec* to the left).

- If, in a particular language, some features are not applicable at all, they are dealt with in the Comments section and the feature is not represented as a table.

Here we show several examples of such a tables, extracted from Monachini and Calzolari (1994).

The following tables display features and values pertinent to Italian nouns:

Attribute	*value*	It. example	It. tag
Type	*common*	insegnante	**S**
	proper	Mario	**P**

Attribute	*value*	It. example	It. tag
Gender	*masculine*	(il) libro	**S/m**
	feminine	(la) casa	**S/f**
	neuter		
l-spec	*common*	(l')insegnante	**S/n**

Attribute	*value*	It. example	It. tag
Number	*singular*	(il) libro	S/ms
	plural	(i) libri	S/mp
l-spec	*invariant*	(la/le) attivita'	S/fn

The next table offers an example of constraints in the application of features in the presence of a given attribute and/or value in the Italian simple-form verbal paradigm.

SIMPLE FORMS		
Finiteness	**Verb-Form/Mood**	**gramm. features**
finite	*ind*	**Tense Person Number**
	sub	**Tense Person Number**
	impr	**Tense Person Number**
	cond	**Tense Person Number**
non-finite	*inf*	**Tense**
	ger	**Tense**
	part	**Tense Number Gender**

The specifications emerging after the language specific exercises, were included in the set of recommendations as language specific extensions and circulated again among the partners for new cycles of revisions, until the set was considered acceptable and stable enough.

Experience shows that the adopted representation, as described above, with the concrete applications which display and exemplify the attributes and values and in addition providing their internal constraints and relationships, makes the proposal self-explanatory. Other groups can easily test the specifications on their language, simply by following the method of the applications.

The language specific instantiations are presented in the EAGLES document as an integral part of the proposal in the sense that they constitute the required implementation of the proposal itself.

11.5 GUIDELINES FOR THE VALIDATION PHASE

After the definition of the set of specifications for the lexicon encoding at morphosyntactic level, and the experience gained in the initial instantiations described in the previous section, a clear need for formalizing the inventory of labels and making the specifications operational through guidelines has emerged.

The production of complete *full-fledged guidelines*, i.e. clear criteria and tests for the application of the abovementioned specifications and their *formalization*, has been carried out in parallel on four European languages, giving rise to four documents: French (Von Rekovsky 1996), German (Teufel 1996a), English (Teufel 1996b) and Italian (Monachini 1996). For further details about this work, see the introduction by Heid (1996).

The methodology agreed on and adopted by the partners working on the different languages foresees that each of the EAGLES main categories is dealt with on the basis of the principles described below. As an example of the whole procedure, tables taken from the Italian Guidelines (Monachini 1996) will be provided in the following sections.

11.5.1 Values pertinent to a given language

Starting from the specifications proposed in the multi-layered EAGLES proposal (that – it is worthwhile reminding here – have been collected in a very broad multi-lingual and multi-purpose approach), the first operation towards the production of guidelines for the practical encoding of a lexicon foresees the *recognition of the set of values* pertinent to the language in question. The selected values are marked in the table by using bold type (cf. the table below for Nouns). In such a way, the level reached in the encoding is immediately visualized.

11.5.2 Logic relationships between values

Another requisite of the provision of guidelines is the detailed *description of the logic relationships between the values* of the system pertinent to the language in question, in the cases in which the values belong to different levels of obligatoriness.

As already explained elsewhere in this chapter, the EAGLES lexicon specifications constitute a redundant inventory of the information which, in a bottom-up approach, have been found actually encoded in existing lexicons of different languages for different purposes. This set cannot be implemented as such, but each language has to pick up the relevant values, specifying its system and internal constraints. In the Validation task, for each language, a readily interpretable *formalized system* has been provided: the aim is to obtain a structured inventory, specifying the logic relationships between the values, in order to make others able to consistently interpret them. Following a suggestion first carefully explained in a report by Teufel (1995), an example of how the tables display these relationships will be given.

We first examine the Case system displayed in the EAGLES proposal and the relationships between the values in that system. The Case system proposed for Nouns at the level EAG-L2a contains the values `nominative, genitive, dative, accusative`. For any other value presented at a different level, the relationships to this set should be specified, e.g.:

- The value `vocative`, presented for Greek at the language specific level, constitutes an *addition* to the system.
- The value `indeclinable`, also for Greek, represents a case of non-applicativity of the attribute, or *neutralization*.
- The value `oblique`, proposed for the Case system for Pronouns (not shown here), is a *meta-value* which has the function of *replacement* of the values `dative, accusative`.

Now consider now another typical relationship. At the L1 of the EAGLES proposal, there are three values for Gender: `masculine, feminine, neuter` and two for Number: `singural, plural`.

- The value `common` for Gender, presented for Italian, does not constitute a simple addition to the system. Rather, it represents a *disjunction* relation of the values `masculine` and `feminine`. The same is true for `invariant` for Number, again for Italian, which is the disjunction of `singular` and `plural`. These values are, in other words, *meta-values*, multilabels with the semantics of disjunction.

In the tables, the interrelationships can be expressed for each language by using *indices* added to the values in question. Practically, the value set `masculine, feminine, common`, highlighted as the Italian Gender system,[8] will receive the following indices: $masc_1$, fem_1 and $comm_{D1}$.
The values `feminine` and `masculine` are given a numerical index: `1`. `common`, the value added to the EAGLES L1 at the language specific level L2b for Italian, receives the same numerical index pointing to the values it disjoins and it is further labelled with the label `D` for the disjunction relation. The same happens to the values of the feature Number.

[8] `neuter` is not relevant for Italian.

N	Typ	Gen	Num	Cas	Count	Def	Infl	Decl
L0	NOUN							
L1	**com** **prop**	$\mathbf{m}_1$ $\mathbf{f}_1$ n	$\mathbf{sg}_1$ $\mathbf{pl}_1$					
L2a				nom gen dat acc	cou mass			
L2b		It c_{D1} Du f(m) Du cont Sp trns Sp notr	It $\mathbf{n}_{D1}$	Gr voc Gr ind		Da def Da indf Da unmk	Ge weak Ge strg Ge mix	Ge no Ge adject

Similarly to the label *D* for the disjunction relation, *R* is used for the replacement relation and *A* for the addition relation.[9].

11.5.3 Constraints in the application of attributes and values

The *hierarchical relationships between subcategories* and the morphosyntactic features which are applicable to them are displayed in another table, devoted to the type hierarchy of the category (see the examples for the noun and the main verb below). The table appears subdivided into two zones. There is no hierarchical relation between features of the second zone; they *cross-classify* the feature space. In such a table, *constraints in the application of an attribute or a value* – or a set of attributes or values – in the presence of a given attribute or value are specified: for example, `common` Gender does not apply to Italian `proper` nouns; `Person` with all its values does not apply to `non-finite` forms in the Italian verbal system; only the `common` Gender applies to present participles.

NOUN		
Feature	**Values**	
PoS	noun	
Typ	com	prop
Num Gen	sg, pl, n m, f, c	sg, pl m, f
Example:	casa, libro, citta' case, libri, citta'	Monica, Mario Americhe, Appennini

[9]The *A* relation does not need numerical indices, since the values are simply added to the system.

V E R B 1 (finite forms)				
Feature	**Values**			
PoS	verb			
Typ	mai			
Fin	fin			
Mood	ind	subj	cond	imp
Tense Pers Num Gen	pres, impf, fut, past 1 2 3 sg pl –	pres, impf 1 2 3 sg pl –	pres 1 2 3 sg pl –	pres 2 sg pl –
Ex:	amo, amavo, amero', amai	ami, amassi	amerei	ama

V E R B 1 (non-finite forms)				
Feature	**Values**			
PoS	verb			
Typ	mai			
Fin	no-fin			
Mood	inf	ger	part	
Tense Pers Num Gen	pres – – –	pres – – –	pres – sg pl c	past – sg pl m f
Ex:	amare	amando	amante	amato

11.5.4 Semantics of the PoS

Another important requirement of the development of guidelines for lexicon encoding is the clear definition of the central *semantics* of the PoS. It should contain clear tests and criteria for the inclusion of an item in a given PoS.

The following list contains an example of concrete criteria and tests proposed for Italian nouns:

- Morphological criteria: the noun carries the termination mark of gender ("amico/amica") and number ("amico/amici").
- Syntactic criteria: the noun combines with other PoS (determiner, adjective) giving rise to Noun Phrases of which the noun constitutes, in general, the head. The head regulates the agreement mechanism: "Il mio caro amico parte/I miei cari amici partono". The presence/absence of the article is useful in order to distinguish between common and proper nouns. Anthroponyms, names of cities, etc. are, in general, incompatible with articles, but a lot of exceptions can be found. The criterion is, hence, not sufficient.
- Presentational/graphical criteria: proper nouns appear capitalized, e.g. "Mario", "Rossi", "i Borboni", "le Alpi". Capitalization constitutes a first test for the

assignment of the values common and proper, but, given that not all capitalized nouns are proper nouns, this criterion is necessary but not sufficient (see the delimitation table between common and proper nouns below).

Furthermore, typical examples of items to be included in a given class are provided.

pos = noun	
Description	**Examples**
Common nouns	**uomo**/noun, **amico**/noun, **febbraio**/noun, **martedi**/noun
	insegnante/noun, **rosa**/noun, **computer**/noun
Proper nouns	**Monica**/noun, **Pisa**/noun, **Rossi**/noun, **Galileo**/noun
	Alpi/noun, **Tevere**/noun, **Fiat**/noun, **Sardegna**/noun

pos=noun		**pos=adj**	
Description	**Example**	**Description**	**Example**
lexicalized nominalized adjectives nominal head of NP	il **bello**/noun della vita	adjectival forms used in an attributive way	un uomo **bello**/adj
	un **antibiotico**/noun potente		un medicinale **antibiotico**/adj
inhabitants of places	Gli **italiani**/noun	inhabitants in attributive adjectival distribution	Gli uomini **italiani**/adj
nominalized present participles	i **credenti**/noun	present participles used in an adjectival context	cattolici molto **credenti**/adj
	ho preso un **calmante**/noun		una tisana **calmante**/adj
	i **militanti**/noun del partito		sono comunisti **militanti**/adj
nominalized past participles	il **passato**/noun	adjectivally used past participles	l'anno **passato**/adj
colour	il **rosa**/noun e' un colore pastello	colour as qualification	una camicetta **rosa**/adj
noun	il giocatore e' un **esterno**/noun destro	adjective	misura il perimetro **esterno**/adj della casa
noun (comparative adj.)	i **migliori**/noun sono quelli che se ne vanno	comparative adjective	gli anni **migliori**/adj della nostra vita
noun (superlative adj.)	rappresenta l' **ottimo**/noun delle condizioni	superlative adjective	e' un **ottimo**/adj tecnico

Especially interesting in the respect of inclusion/exclusion in a PoS are transcategorization phenomena or possible confusions of a class with other classes. Is an item, e.g. a past participle, to be encoded as a verb or as a noun or as both? These questions have been addressed in the form of delimitation tables,[10] which will be more useful for a human annotator, but can also serve to deepen the semantics of a PoS. References to corpus tagging issues are sometimes also present.

11.5.5 Semantics of the features

Similarly to PoS, a description of all values is provided for each feature. Tables explain the central semantics of each value by listing subclasses of the class and giving examples. The following tables provide examples of the typical classes of the feature **Type** for Italian nouns.

type = prop	
Description	**Examples**
anthroponyms (first, last)	**Monica**/prop, **Giovanni**/prop
toponyms	**Pisa**/prop, **Parigi**/prop
names of planets	**Marte**/prop, **Venere**/prop
names of companies	**Enimont**/prop, **Mitsubishi**/prop
names of organizations, parties	**BNL**/prop, **PSI**/prop
names of fictive/mythological figures	**Zeus**/prop
names of countries	**America**/prop
names of rivers, mountains	**Tevere**/prop, **Alpi**/prop

type = com	
Description	**Examples**
concrete and abstract nouns	**casa**/com, **pazienza**/com
measurements	**litro**/com, **metro**/com, **chilo**/com
currencies	**dollaro**/com, **lira**/com
titles or forms of address	**Direttore**/com, **Professore**/com
professions	un **ingegnere**/com, un **avvocato**/com
names of food and drinks	**caffe'**/com, **whisky**/com
names of animals and plants	**cane**/com, **edera**/com
inhabitants of countries, cities	gli **italiani**/com, i **romani**/com
names of months	**gennaio**/com, **febbraio**/com
cardinal numbers	un **nove**/com, un **tre**/com
names of weekdays	**martedi**/com, **sabato**/com
nominalized adjectives	il **bello**/com, i **ricchi**/com
nominalized participles	i **possidenti**/com, il **passato**/com
nominalized infinitives	il **dare**/com, l' **avere**/com

Confusion classes are treated pairwise for all combinations where confusion is possible. Examples very close to each other have been deliberately chosen, in order to shed light on the exact borderline between two confusing contexts. See below, for the distinction between the values com and prop.

[10] In the guideline documents, the delimitation tables are further supported by an index containing all the wordforms described.

type=com		type=prop	
Description	**Example**	**Description**	**Example**
names of products	berro' una **coca-cola**/`com`	names of products	una **Coca-Cola**/`prop`
common nouns spelled as proper nouns	un **marco**/`com`	proper names	**Marco**/`prop`
currencies	un **marco**/`com`	currencies abbrev.	un **DM**/`prop`
geographical nouns	le **isole**/`com`	names of regions	**Isole**/`prop` **Vergini**/`prop`
common nouns	la **terra**/`com`	names of planets	la **Terra**/`prop`

The same procedure is applied to all the attributes and relevant values pertinent to a given class.

11.6 APPLICATION IN EU PROJECTS

The EAGLES specifications were adopted in a number of European projects, such as LRE-DELIS, RENOS, CRATER, MECOLB, MULTEXT, MLAP- and LE-PAROLE, LE-SPARKLE, ELSNET and the French project GRACE, for different types of purposes. These vary from lexicon building to information retrieval, creation of taggers and annotation schemes, tagging and finally to the development of resources and tools for NLP and MT corpus-based research. In this way, the proposal has been tested, validated and refined as well from the point of view of *continually evolving application requirements*.

The experience gained by such applications of the EAGLES proposal had a strong influence on the set of the recommendations. The specifications, in fact, were cyclically revised also according to the requirements that emerged while working in the framework of these projects. This helped to obtain harmonized recommendations for language resources which can run under the same tools and make it possible to perform comparisons across languages.

In this section, the experience of MULTEXT and PAROLE are summarized and the interaction between these two projects and EAGLES is reported. It is worth stating again that within these two frameworks, very useful *feedback* has been collected and the impact of the EAGLES proposal has been concretely measured.

11.6.1 MULTEXT

The main objectives of the MULTEXT project (MULTEXT Tech.Ann. 1993) were the definition and the implementation of a set of tools for corpus-based research and applications and the production of resources to run under these tools in a multi-lingual framework.

The interrelations, interdependencies and reciprocal influence between EAGLES and MULTEXT have been already described extensively by Bel, Calzolari and Monachini (1995) and elsewhere in this chapter. The MULTEXT experience served many viewpoints: from a theoretical point of view, it was crucial in order to clarify and fix some important aspects such as the interdependencies and/or differences between lexicon and corpus (cf. 11.2) and from a practical point of view, it is a concrete example of how the EAGLES recommendations have been concretely used to implement electronic morphological lexica. What is worth noting here is that the premises and the aims of MULTEXT, being application-oriented, i.e. automatic corpus annotation strongly influenced both the *format* and the *notation formalism* of the lexicons.

As to the format, the MULTEXT lexicons have been produced as wordform lexical lists with lexicon specifications (collected mostly from the EAGLES common core with very few language-specific extensions) and tags associated to each wordform. This has been considered a format more suited to corpus annotation and as the best and quickest way to obtain fast electronic lexica to run under the tools for experimentations.

Obviously, the formalism for the notation has been influenced by the purpose as well. The EAGLES philosophy of features represented by attributes and values has been adopted, concretely implemented in the formalism of linear strings. The string is constructed following the Intermediate Tagset of the EAGLES Corpus subgroup (cf. 5.3). Some minor changes have been introduced to the EAGLES notation, in the sense that alphabetic characters have been preferred instead of numbers, as they are considered more mnemonic. In the MULTEXT document (Bel, Calzolari and Monachini (coords.) 1995), it has been carefully demonstrated how this notation can be considered synonymous with attribute/value pairs used in current unification formalisms.

Another important MULTEXT contribution to the efforts of standardization in lexicons and corpora is the mechanism for the mapping of the lexicon specifications onto the tags and the formalism of *regular expressions* to optimize the conversion tables for the mapping in cases in which the information in the lexicon is more precise than that in a tag.

11.6.2 PAROLE

The aim of the MLAP-PAROLE project was the definition of the minimal initial technical conditions for starting the activity of actually providing Written Language Resources (PAROLE Tech.Ann. 1994)[11]. The EAGLES recommendations for morphosyntactic encoding in lexicons were taken as the starting point for finding and establishing a framework in which either to reuse existing lexical data or to start creating new data according to harmonized criteria, thus ensuring reusability.

The major contribution of the PAROLE project to the standardization efforts is the translation of the EAGLES recommendations into *operational standards*. In the setting

[11] The PAROLE project is being coordinated by A. Zampolli for the Consorzio Pisa Ricerche.

of these operational standards, a particular emphasis has been given to the practical criterion of *feasibility* in terms of money and time constraints and lexicon resources already available and reusable. Indeed, following the criterion of feasibility, the set of values pertinent to each language and relevant for a large range of NLP applications was isolated and the most appropriate level of encoding for lexicon resources was indicated from the multi-level and multi-layered EAGLES proposal.

Once the pertinent attribute-value set was circumscribed, PAROLE went a step further with respect to EAGLES and, for each language, provided a readily interpretable formalized system with the aim to obtain a structured inventory (see 11.5 for more details). Furthermore, the dependencies and the constraints in the application of an attribute or a value in the presence of a given attribute or value were also formally specified. This phase appears particularly crucial. In fact, the definition of a formalized and structured set with the logic relationships between specified values makes others able to consistently interpret the correct semantics of values, without misinterpretations, and to handle them with computational techniques.

11.6.3 Coverage with respect to languages, users and applications

The methodology adopted for the EAGLES standardization work, and the framework prepared by the Lexicon group, was designed to provide a robust coverage to the proposal with respect to various aspects.

First of all, the proposed inventory of labels implicitly mirrors the *requirements of different languages and many different types of users* of lexicons and annotated corpora and applications, viz. all those using the surveyed practices (such as MULTILEX, GENELEX, NERC survey, TEI, etc.) some of which are themselves already the result of a comparison of many different systems for many languages.

Secondly, the cycle of concrete applications and testing on real corpora and lexicons of all the European languages, and continuous revisions and adjustments of the specifications, ensures the evaluation of the recommendations in a multi-lingual perspective and increases the coverage of the proposal with respect to *different language requirements, different schools and grammatical traditions*. Moreover, since the proposal has been evaluated as compatible within consolidated practices for various kind of applications (such as BNC, IBM, Eurotra, Danish Dictionary, CELEX), it arguably also covers the requirements, needs and purposes of a *very large range of users*.

Finally, the interaction and the reciprocal influence between EAGLES and other European projects, as described in detail above, having various aims and purposes, validates and reinforces the proposal as well from the point of view of *continually evolving application requirements*. The large number of examples of instantiations and applications of the proposal also clearly demonstrates the effective *dissemination* and substantial *impact* of the proposal within the scientific community and of the *real-life applicability* of the EAGLES results.

In conclusion, the EAGLES experience shows that the road to standardization is a long but fruitful one. Standards must respond to commonly perceived needs or help in resolving common problems. In order to offer a workable and acceptable solution, they must be based on some solid platform of accepted practices. Furthermore, the process of consensus-building implies cycles of testing and refining of the specifications. It is therefore essential that the interaction between EAGLES and other projects flows in both directions (as it has been happening until now). In particular, EAGLES has benefitted and is continuing to benefit from the feedback coming from concrete applications of its proposal and, as it is continuously refined and improved, its strength and coverage is increasing.

12 MORPHOLOGICAL ANALYSIS

Kemal Oflazer

12.1 INTRODUCTION

In the previous chapters, we have seen that a lot of information about the potential tags of tokens in a text is found by lexicon lookup. Another, often complementary source of information is *morphological analysis*, i.e. the process of decomposing words into their constituents. The information about the individual constituents can be used to determine the necessary information about the word as a whole. Such information may range from basic wordclass information assigned from a fixed inventory of tags to structural information consisting of the relationships between components of the word further annotated with various features and their values (cf. Chapter 10). The English word "redness" could thus either be analysed as having the tag NN (singular noun) hiding its internal details, or be analysed by a suitable word grammar to have the structure `Adj(red) + N(+ness)` where the internal structure of the word has been made explicit.

This chapter will present issues in implementing morphological analysers to be used in wordclass tagging or other natural language processing activities, such as syntactic parsing, speech recognition, text–to–speech, spelling checking and correction, document indexing and retrieval. The purpose of this chapter, however, is not to provide a detailed coverage of various aspects of computational morphology; the reader is referred to several recent books covering this topic (see e.g. Sproat (1992) for a quite comprehensive treatment of computational morphology and Ritchie *et al.* (1992) for a description of a morphological analyser and lexicon for English). Instead, after

H. van Halteren (ed.), Syntactic Wordclass Tagging, 175-205.

a short overview of the relevant concepts involved, highlighted with some examples from different languages, this chapter will present issues involved in implementing an industrial strength high-coverage morphological analyser using the *two-level morphology* approach using tools that are either publicly or commercially available.[1] The presentation will not only focus on the usual topics of implementing *morphophonemic/morphographemic* phenomena and *morphotactics* (word grammar) but also more mundane issues such as foreign words, acronyms and abbreviations, numerical tokens, etc., which turn out to be quite important when one has to process real text. This part of the presentation will be based on Turkish, a Ural-Altaic language with agglutinative wordforms. Apart from being the native tongue of the author, and a language that has not until recently been computationally investigated, Turkish is quite interesting for an exposition of this nature for a number of reasons: Turkish (along with languages like Finnish and Hungarian) exhibits phenomena such as vowel harmony which do not show up in Western European languages. Turkish also has very productive inflectional and derivational morphological phenomena. The latter may pose challenging issues in developing a tagset, as the number of forms one can derive from a root form may be in the thousands (some researchers actually give a much higher figure in the millions; cf. Hankamer 1989). Owing to this productivity, Turkish exhibits a quite complex morphotactics, an issue typically not found or not addressed in morphological analysers for many European languages. To illustrate this, we can provide the following rather exaggerated example of a Turkish word: "uygarlaştıramayabilecek-lerimizdenmişsinizcesine"[2] which has the structure:

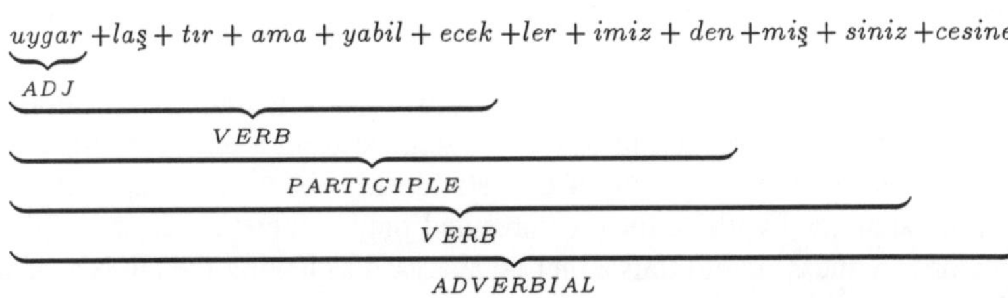

Despite this complexity, the rules governing Turkish morphology are for most part quite regular and hopefully easily understandable. Understanding issues in developing a morphological analyser for Turkish may actually be quite helpful in dealing with

[1] We will not touch upon quite number of issues such as rule compilation, rule conflicts and their resolution, non-concatenative morphological combinations or the details of specific systems, such as PC-KIMMO or the Xerox Tools, and refer the interested reader to more technical sources with ample coverage of these topics, such as Antworth (1990), Karttunen and Beesley (1992) and Karttunen (1993).

[2] Meaning "behaving as if s/he was one of those whom we could not civilize." Obviously this is not a word that one would use everyday. Turkish words found in typical text average about 10 letters.

many languages that have, for a number of reasons, received less attention from a computational viewpoint.

The chapter starts with a brief overview of morphology and computational morphology and then presents an overview of two-level morphology as a mature state-of-the-art paradigm to implement wide-coverage morphological analysers. It then discusses two general systems for implementing two-level morphological analysers, PC-KIMMO and the Xerox Finite State Tools, covering and contrasting issues such as ease of development, rule compilation, tracing and debugging facilities, speed, memory requirements, etc. This section will mainly look at Turkish as a source of quite interesting problems in implementing an analyser, some of which were alluded to above.

12.2 MORPHOLOGY

Morphology is the study of the structure of the words and how words are formed by combining smaller units of linguistic information called *morphemes*. We will briefly summarize some preliminary notions on morphology, taken from the book by Sproat (1992).

Morphemes can be classified into two groups depending on how they can occur: *free morphemes* can occur by themselves as a word while *bound morphemes* are not words in their own right but have to be attached in some way to a free morpheme. The way in which morphemes are combined and the information conveyed by the morphemes and by their combination differs from language to language. Languages can be loosely classified with the following characterizations: *Isolating languages* are languages which do not allow any bound morphemes to attach to a word. Mandarin Chinese with some minor exceptions is a close example of such a language. *Agglutinative languages* are languages in which bound morphemes are attached to a free morpheme like beads on a string. Turkish, Finnish, Hungarian and Swahili are examples of such languages. In Turkish, e.g., each morpheme usually conveys one piece of morphological information such as tense, agreement, case, etc. *Inflectional languages* are languages where a single bound morpheme (or closely united free and bound forms) simultaneously conveys multiple pieces of information. Latin is a classical example. In the Latin word "amō" (I love), the suffix +ō expresses 1st person singular agreement, present tense, active voice and indicative mood. *Polysynthetic languages* are languages which use morphology to express certain elements (such as verbs and their complements) that often appear as separate words in other languages. Sproat (1992) cites certain Eskimo languages as examples of this kind of a language.

12.2.1 *Types of morphology*

There are three main types of morphological processes involving morphemes.

Derivational morphology produces a new word usually of a different part-of-speech category by combining morphemes. The new word is said to be *derived* from the original word. For example, the noun "happiness" is a word derived from the adjective "happy". A derivational process is never demanded by the syntactic context the word is to be used in.

Inflectional morphology introduces relevant information to a word so that it can be used in the syntactic context properly. Such processes do not change the part-of-speech, but add information like person and number agreement, case, definiteness, tense, aspect, etc. For instance in order to use a verb with a third person singular subject in present tense, English syntax demands that the agreement morpheme +s be added, e.g. "comes". Turkish will indicate possible functions for a noun phrase, but requiring that a case morpheme be attached to the head of the phrase, e.g. "ev+i" (the accusative form of "ev" ("house") which can only serve the function of a direct object).

Compounding (cf. 4.3.1) is the concatenation of two or more free morphemes (usually nouns) to form a new word (usually with no or very minor changes in the words involved). Compounding may occur in different ways in different languages. The boundary between compound words and normal words is not very clear in languages like English where such forms can be written separately though conceptually they are considered as one unit, e.g. "firefighter" or "fire-fighter" is a compound word in English while the noun phrase "coffee pot" is an example where components are written separately. German is the prime example of productive use of compounding to create new words on the fly, a textbook example being "Lebensversicherungsgesellschaftsangesteller" consisting of the words "Leben" ("life"), "Versicherung" ("insurance"), "Gesellschaft" ("company") and "Angesteller" ("employee") with some glue in between.

12.2.2 Types of morphological combination

Morphemes can be combined together in a number of ways. In *purely concatenative* combination, the free and bound morphemes are just concatenated. *Prefixation* refers to a concatenative combination where the bound morpheme is affixed to the beginning of the free morpheme or a stem, while *suffixation* refers to a concatenative combination where the bound morpheme is affixed to the end of the free morpheme or a stem. Turkish uses purely concatenative morphological combination with only suffixes attaching to a free morpheme.

In *infixation*, the bound morpheme is inserted to the stem it is attached to. An example is the derivation of "fumikas" ("to be strong") from "fikas" ("strong") in the Bontoc language (Sproat 1992). In *circumfixation*, part of the attached morpheme comes before the stem while another part goes after the stem. In German, e.g., the past participle of a verb such as "tauschen" ("to deceive") is indicated by "getauscht".

Arabic, a language that has long been of interest and challenge to computational morphology uses *templatic* combination where a root word consisting of just consonants is modulated with a template of consonant and vowel alternations. For instance the root "ktb" (meaning the general concept of writing) can be combined with the template CVCCVC to derive new words such as "kattab" ("to cause to write") or "kuttib" ("to be caused to write").

Reduplication refers to duplicating (some part of) a word to convey morphological information. In Indonesian, e.g., total reduplication is used to mark plurals: "orang" ("man"), "orangorang" ("men") (Sproat 1992). Turkish uses partial reduplication for a limited number of adjectives to derive some emphatic adjectives: "sarı" ("yellow"), "sapsarı" ("very yellow").

In *zero morphology*, derivation/inflection takes place without any additional morpheme. In English the verb "to second (a motion)" is derived from the ordinal "second".

In *subtractive morphology*, part of the wordform is removed to indicate a morphological feature. Sproat (1992) gives the Muskogean language Koasati as an example of such a language, where a part of the form is removed to mark plural agreement.

12.2.3 Computational morphology

Computational morphology studies the computational analysis and synthesis of wordforms for eventual use in natural language processing applications. Almost all applications of computational analysis of wordforms have been on written or orthographic forms of words where tokens are neatly delineated. Since the main theme in this book is the processing of written language, we will from now on assume that we are dealing with written forms of words.

Morphological analysis breaks down a given wordform into its morphological constituents, assigning suitable labels or tags to these constituents. Morphological analysis has analogous problems to all those in full-blown parsing albeit usually at a smaller scale. Words may be ambiguous in their wordclass, e.g. in French, a form such as "danse" has six interpretations:[3]

1)	danse	V(danse)+MOOD/Subj+AGR/3SG	lest s/he dance
2)	danse	V(danse)+MOOD/Subj+AGR/1SG	lest I dance
3)	danse	V(danse)+MOOD/Imp+AGR/2SG	(you) dance!
4)	danse	V(danse)+MOOD/Ind+AGR/3SG	(s/he) dances
5)	danse	V(danse)+MOOD/Ind+AGR/1SG	(I) dance
6)	danse	N(danse)+GEN/Fem+AGR/3SG	dance

In a language like Turkish, whose morphology is more extensive, words may be divided up in a number of ways, e.g. a simple word like "oyun" may be decomposed into constituents in four ways:

1)	oyun	N(oyun)+AGR/3SG+POSS/none+CASE/nom	game
2)	oy+un	N(oy)+AGR/3SG+POSS/2SG+CASE/nom	your vote
3)	oy+[n]un	N(oy)+AGR/3SG+POSS/none+CASE/gen	of the vote
4)	oy+un	V(oy)+MOOD/imp+AGR/2SG	carve !

[3]Unless the output of a specific system is being presented, we will display morphological parses by a sequence of feature/value pairs. [..] indicates elided material.

A number of systems have been developed for computational morphology. These have been mainly intended for a specific language (e.g. DECOMP for English, cf. Allen *et al.* 1987; keçi, for Turkish, cf. Hankamer 1986) though the underlying ideas can be extended to certain other languages. Computational morphology has gained a substantial boost after Koskenniemmi's work which introduced the *two-level morphology* approach (Koskenniemmi 1983). This work was immediately followed by substantial activity to apply the approach to many different languages (Alam 1983; Lun 1983; Karttunen 1983; Karttunen and Wittenburg 1983; Khan 1983) and eventually lead to language independent software tools such as PC-KIMMO (Antworth 1990) and the Xerox Finite State Tools (Karttunen 1993; Karttunen and Beesley 1992).

12.3 TWO-LEVEL MORPHOLOGY

Two level morphology posits two distinct levels of representations for a wordform: the *lexical level* refers to the abstract internal structure of the word consisting of the morphemes making up the word and the *surface level* refers to the orthographic representation of a wordform as it appears in text. The morphemes in the lexical level representation are combined together according to language-specific combination rules possibly undergoing changes along the way, resulting in the surface level representation. The changes that take place during this combination process are defined or constrained by language-specific rules. Such rules can be considered to define the correspondence between the string of symbols making up the lexical level representation and the string of symbols making up the surface level representation. For instance, in English, the lexical form of the word "blemishes" can be represented as `blemish+s` indicating that the root word is `blemish` and the plural marker is the bound morpheme `+s` combined by concatenation indicated by the `+`. The English spelling rule of *epenthesis* requires that an `e` has to be inserted after a root ending with `sh` and before the morpheme `s`, resulting in `blemishes`. We textually represent this correspondence by aligning the lexical and surface characters that map to each other as shown below. In this example and in the examples to follow later the symbol `0` stands for the *null symbol* of zero length which never appears in any surface form when printed.

Lexical: `blemish+0s`
Surface: `blemish0es` `blemishes`

A two-level description for a language requires that two components be specified. The *morphographemic component* describes the orthographic changes between lexical and surface levels. The *morphotactics component* describes how the morphemes from the inventory of root words and affixes in the language make up the words. Current implementations of two-level morphology usually assume that morphemes are combined by concatenation which seems to be sufficient at least for languages on which NLP applications are developed. The two components are then used by a *morphological analysis engine* (either directly at run time or after a compilation process) to perform

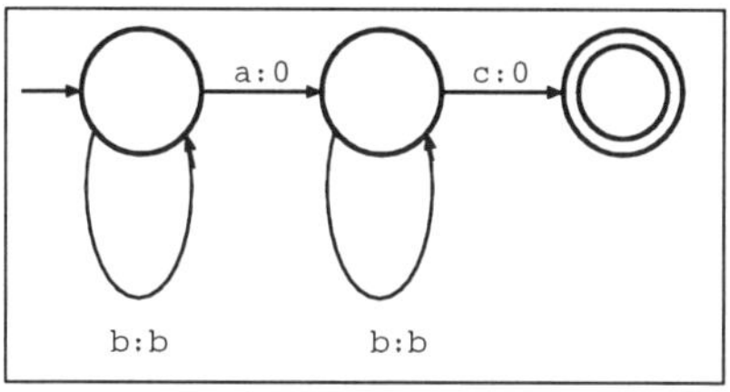

Figure 12.1 The finite-state recognizer for $(b : b)^*(a : 0)(b : b)^*(c : 0)$.

morphological analysis. They could also be used by a *morphological generation engine* which can be used in applications like language generation.

12.3.1 The morphographemic component

The morphographemic component describes the spelling changes that take place between the lexical and surface levels when morphemes are combined to make new wordforms. The changes are expressed by a set of *two-level rules* each of which describes one specific phenomenon (such as epenthesis above), along with the contexts the phenomenon occurs in and whether it is obligatory or optional.

Before we proceed further, some automata-theoretic background would be helpful. Let us consider a finite alphabet whose symbols are actually pairs of atomic symbols `l:s`, where `l` is a lexical symbol and `s` is a surface symbol. One can define regular languages over such pairs of symbols using regular expressions. For instance given the alphabet $A = \{\mathtt{a:0}, \mathtt{a:a}, \mathtt{b:b}, \mathtt{c:0}, \mathtt{c:c}\}$, the regular expression

$$R = (b : b)^*(a : 0)(b : b)^*(c : 0)$$

describes a regular language containing examples like `b:b b:b b:b a:0 b:b b:b c:0`, where the first three `b:b` pairs match $(b : b)^*$ in the regular expression, `a:0` pair matches the $(a : 0)$, the next two `b:b` pairs match the $(b : b)^*$ and finally the `c:0` pair matches $(c : 0)$. We can also view this string of pairs of lexical–surface symbols as a correspondence, showing the sequence of lexical and surface symbols separately:

Lexical: `bbbabbc`
Surface: `bbb0bb0` `bbbbb`

Such a regular expression can be converted into a finite-state recognizer over the same alphabet using standard techniques, as shown in figure 12.1 (cf. e.g. Hopcroft and Ullman 1979).

Another way to view this recognizer is as a *transducer* that maps between strings consisting of the lexical symbols and strings consisting of the surface symbols.[4] Thus, for the example above, the lexical string `bbbabbc` would be transduced to the surface string `bbbbb`, if the lexical level is treated as the input string and the surface level is treated as the output string. The transduction would be in the reverse direction if the roles of the levels are interchanged. On the other hand, the lexical string `bbabbbb` cannot be transduced because it is missing a `c` at the end and hence cannot lead the transducer to its final state.

In general, regular expressions are too low a notation to describe morphographemic changes or correspondences. Two-level morphology provides higher-level notational mechanisms for describing constraints on strings over an alphabet, called the *set of feasible pairs* in two-level terminology. The set of feasible pairs is the set of all possible lexical–surface pairs. Morphographemic changes are expressed by four kinds of rules that specify in which context and how morphographemic changes take place. The contexts are expressed by regular expressions (over the set of feasible pairs) and describe what comes on the left (`LC`, for left context) and on the right (`RC`, for right context), of a morphographemic change.

The *context restriction rule* `a:b => LC _ RC` states that a lexical `a` may be paired with a surface `b` only in the given context, i.e. `a:b` may only occur in this context (if it ever occurs in a string). In this case *the correspondence implies the context.* For instance in English, the `y:i` correspondence (in a word like *happiness* is only allowed between a consonant (possibly followed by an optional morpheme boundary) and a morpheme boundary. This is expressed by a rule like `y:i => C (+:0) _ +:0` where `C` denotes a consonant.

The *surface coercion rule* `a:b <= LC _ RC` states that a lexical `a` *must be* paired with a surface `b` in the given context, i.e. no other pairs with `a` as its lexical symbol can appear in this context. In this case *the context implies the correspondence*. Note that `a:b` is not prohibited from occurring in other contexts. For instance in English, the `s` in a genitive suffix has to be deleted on the surface if the previous consonant is an `s` that belongs to the plural morpheme. One would express this by a rule of the sort `s:0 <= +:0 (0:e) s +:0 ' __`. Note that there are other contexts where an `s` may be dropped, but no obligatorily.

The *composite rule* `a:b <=> LC _ RC` states that a lexical `a` must be paired with a surface `b` in the given context and this correspondence is valid only in the given context. This rule is the combination of the previous two rules. For instance, in English the `i:y` correspondence (as in *tie+ing* being *tying*), is valid only before an `e:0` correspondence followed by a morpheme boundary followed by an `i`, and furthermore, in this context,

---

[4] Such transducers are slightly different from the classical finite state transducers in that (i) they have final states just like finite state recognizers and (ii) a transduction is valid only when the input leads the transducer into one of the final states.

a lexical `i` *has* to be paired with a surface `y`. This is expressed by the composite rule `i:y <=> __ e:0 +:0 i`.

The *exclusion rule* `a:b /<= LC _ RC` states that lexical symbol `a` may not be paired with a surface symbol `b`, i.e. `a:b` cannot occur in this context. For instance, the `y:i` correspondence in the context restriction rule above cannot occur if the morpheme on the right hand side starts with an `i` or a `'` (the genitive marker). Thus a rule like `y:i /<= C (+:0) _ +:0 [ i | ' ]` prevents the context restriction rule from applying in situations like *try+ing* or *spy+'s*.

The constraints expressed by these rules are compiled into finite-state recognizers which operate in parallel on the lexical and surface symbol pairs. A given string of lexical–surface pairs is accepted by a collection of such recognizers if none of the individual recognizers ends up in a rejecting state.

We will illustrate the possibilities of this system with some examples for two-level rules and the corresponding recognizers.

Turkish Vowel Harmony. Turkish has a phenomenon called *vowel harmony* where with some exceptions, the vowels in suffix morphemes have to agree in certain phonetic features with the most recent vowel in the stem the morpheme is attached to. For instance, in its (considerably) simplified form, if the surface representation of the last vowel in the stem is a front vowel (one of "e", "i", "ö" or "ü" in Turkish), then an unrounded back vowel (which will be represented in the lexical representation by the symbol as `A`) in a morpheme is resolved as "e" on the surface. Otherwise, if the last vowel is a back vowel (one of "a", "ı", "o" or "u"), then it is resolved as "a". The following data exemplifies this phenomenon:

Lexical: `masa+lAr` `N(table)+AGR/3pl`
Surface: `masa0lar` `masalar`

Lexical: `okul+lAr` `N(school)+AGR/3pl`
Surface: `okul0lar` `okullar`

Lexical: `ev+lAr` `N(house)+AGR/3pl`
Surface: `ev0ler` `evler`

Lexical: `gül+lAr` `N(rose)+AGR/3pl`
Surface: `gül0ler` `güller`

Thus, we have two feasible pairs in our set of feasible pairs with `A` as its lexical symbol: `A:a` and `A:e`. Let us also assume we have the additional feasible pairs `a:a, b:b , ..., z:z`, (called the default pairs), and `+:0` for morpheme boundaries. The rule

```
A:a  <=>  [ A:a | a:a | ı:ı | u:u | o:o ]
          [ b:b | c:c | ...   | z:z ]* +:0
          [ b:b | c:c | ...   | z:z ]* __
```

indicates the A should be paired only with an a in the left context comprising 1) a surface back vowel (indicated by the first set of alternatives following <=>), 2) followed by any number of feasible pairs of consonants pairing with themselves (indicated by the second set of alternatives), 3) followed by a morpheme boundary (+ : 0) and 4) followed by again any number of consonant pairs. The right context is irrelevant, which is shown by there being is nothing after the __.

As such, the rule looks quite verbose and clumsy, but a little bit of additional notational convention leads to quite succinct rule descriptions. We define the following shortcuts:

- @ acts as a wildcard, matching any symbol
- Vback indicates any surface back vowel
- C is the set of all surface consonants

So @:Vback would denote the set of feasible pairs whose surface symbol is a back vowel. With these conventions, we can write the rule above in a much shorter form:

```
A:a <=> @:Vback @:C* +:0 @:C* __
```

Although the compilation of two-level rules to finite state transducers is beyond the scope of this chapter,[5] let us investigate what the transducer for this rule would look like. We want the transducer for this rule to reject a string of feasible pairs either if A is paired with some surface symbol other than an a in the specified context or if A is paired with a without the requisite context. Otherwise it can accept the string. The transducer in this case is quite simple, as depicted in figure 12.2.[6]

The transducer will go into the *sink state* (state 0) if it encounters an A:a before it encounters the requisite left context or if encounters an A:e (the other feasible pair with A as its lexical symbol) in the context specified.

Two-level transducers are traditionally described using state tables such as:

	@	@	+	A	A	@
	Vback	C	0	a	e	@
1:	2	1	1	0	1	1
2:	2	2	2	2	0	1

The accepted notation in two-level terminology is that states with a . in their label are non-final or rejecting, while states with : are final accepting states. State 1 is the start state, while state 0 is a rejecting sink state which has no transitions to any other state.

[5] See Antworth (1990) on how to manually compile two-level rules.

[6] @ : @ is a special case of wild-card use, which does not stand for all possible feasible pairs, but only those that are not covered by the other transition labels. Note also that the tranducer in figure 12.2 has been simplified for expository purposes as it allows any number of occurrences of + : 0. However, this is unproblematic as the input never contains such a repetition.

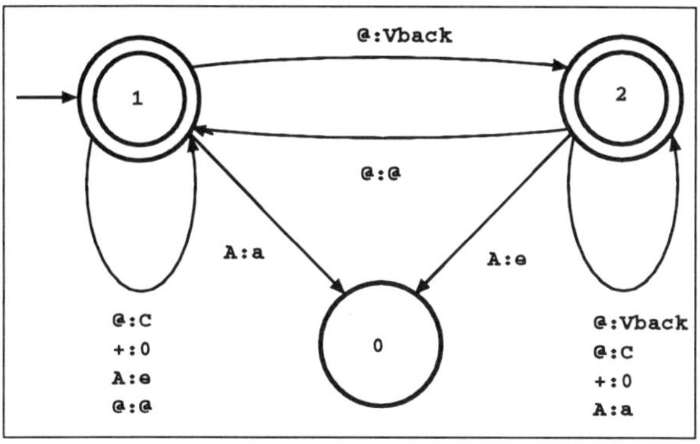

Figure 12.2 Transducer for the A:a vowel harmony rule.

For example this recognizer would accept a lexical–surface pair of strings

Lexical:	masa+lAr	N(table)+AGR/3pl
Surface:	masa0lar	masalar

by going through the following sequence of pairs:[7]

Step	State	Input (Matches)	Next State
1	1	m:m (@:C)	1
2	1	a:a (@:Vback)	2
3	2	s:s (@:C)	2
4	2	a:a (@:Vback)	2
5	2	+:0 (+:0)	2
6	2	l:l (@:C)	2
7	2	A:a (A:a)	2
8	2	r:r (@:C)	2
9	2	#:# (@:@)	2

After the input is consumed, the machine is in state 2 which is an accepting state, hence this sequence is accepted. However, for the pair of strings

Lexical:	masa+lAr
Surface:	masa0ler

[7] There are many morphographemic phenomena which are sensitive to recognition of the beginning or the end of a word. The symbol # denotes the special word boundary symbol which marks the beginning or the end of a word. In our tracing examples we will just show this symbol to mark the end of a word.

the recognizer would go to the sink state 0 at step 7 as the vowel harmony condition is violated.

This recognizer, then, handles the back vowel cases. To cover the complementary instance of this kind of vowel harmony, we have to combine it with its companion recognizer,

	@	@	+	A	A	@
	Vfr	C	0	e	a	@
1:	2	1	1	0	1	1
2:	2	2	2	2	0	1

which corresponds to the front vowel (`Vfr`) rule

```
A:e => @:Vfr @:C* +:0 @:C*_
```

Epenthesis in English. For another example we look to the phenomenon of epenthesis in English, where an `e` is inserted on the surface. The phenomenon can be exemplified by the following data:

Lexical:	`fox+s`	`kiss+s`	`church+s`	`spy+s`
Surface:	`foxes`	`kisses`	`churches`	`spies`

The two-level rule describing epenthesis could be written as:

```
+:e <=> [ Csib | s h | c h | y:i | o ] __ s [+:@ | #]
```

where `Csib` stands for the sibilant consonants {`s, x, z`}. Note that, in this example, instead of using a `+:0` pair and a `0:e` pair, a single pair `+:e` has been used. The right context is such that either a further morpheme boundary or the end of a word may follow the `s`.

Clearly English has many more phenomena and the reader is referred to more detailed sources for these such as Ritchie *et al.* (1992) or Karttunen and Wittenburg (1983). In addition, there are quite a number of other sources for information on writing two-level rules and one can refer to those for more comprehensive treatments of both general and language-specific phenomena (e.g. Antworth 1990).

12.3.2 The morphotactics component

So far we have assumed that both the lexical and surface sides were somehow available when we checked whether all the morphographemic constraints were satisfied or not. This by itself is not very interesting, since the whole point in morphological analysis is to find out if a given surface form is valid, and what the (possible) underlying representation(s) is (are).

In order to check if a given surface form corresponds to a properly constructed word in a language, one needs a model of the word structure. This model includes the root words for all parts-of-speech in the language (nouns, adjectives, verbs, adverbs, connectives, pre/postpositions, exclamations, etc.), the affixes and the paradigms of how root words and affixes combine to create words. Once one has a such a model, it is possible to use a test-and-generate approach to generate a possible lexical representation for a word and then check if this lexical form actually corresponds to the given surface form, subject to all the morphographemic constraints. This generate-and-test approach is constrained by the surface form so that the search for possible lexical representations need not be grossly inefficient in practice (see Barton (1986) for the computational complexity of two-level recognition and Koskenniemmi and Church (1988) for an empirical performance evaluation of a two-level analyser).

Most two-level systems provide simple finite state mechanisms for describing lexicons of root words and affixes and how they are combined. This approach makes the assumption that all affixations are essentially concatenative or can be 'faked' with concatenation. For instance, PC-KIMMO represents the root words and affixes and their sequencing with interlinked lexicons. A simple fragment of a lexicon for English is shown in figure 12.3.[8] Here, the `ALTERNATION` keyword defines possible continuation lexicons that should be tried after the lexicon entry naming that alternation as its successor. Each lexicon entry has the *lexical* representation of a root word or an affix as its first part. The second part names the successor alternation with `End` indicating no further successors. The third part is the *gloss* which corresponds to what the morphological analyser prints out when a match occurs.

In PC-KIMMO, the recognition engine operates in the following fashion when attempting to recognize a given surface form:

1. Starting with the initial lexicon, the whole lexicon network is traversed in a *depth first* manner.

2. During this traversal, the linked lexicon structure is used to incrementally create partial lexical forms, by concatenating the lexical symbols from the entries in the lexicon.

3. Whenever the candidate lexical string is extended by a symbol all two-level rule constraints are concurrently checked by all corresponding finite state recognizers (note that the surface form is already available).

4. If none of the recognizers raise any objections to the partial string so far, then the string is extended further.

5. If at least one of the recognizers rejects the partial string by going to a sink state, then the search backs up: the last lexical string that was concatenated is stripped off and a new candidate lexical symbol is appended and checked. If no such candidates are found, then search backs up to a prior point where alternative lexical symbol choices are available.

[8] Such a set of lexicons is sometimes also represented with a finite state machine where arcs are labelled with pairs of lexical forms and corresponding glosses.

```
ALTERNATION VERBS                        AUX IRREGULAR_VERBS  REGULAR VERBS
ALTERNATION REGULAR_VERB-SUFFIXES   V_SUFFIX1 V_SUFFIX2  V_INFL
ALTERNATION NOUN_SUFFIXES                    .....
ALTERNATION ADJ_SUFFIXES                     .....
ALTERNATION ADV_SUFFIXES                     .....
...

LEXICON REGULAR_VERBS
abandon          REGULAR_VERB-SUFFIXES        "V(abandon)"
...
zip              REGULAR_VERB-SUFFIXES        "V(zip)"
END

LEXICON V_SUFFIX1
+ant             NOUN_SUFFIXES                "+N(ant)"
...
+or              NOUN_SUFFIXES                "+N(or)"
...
+ability         NOUN_SUFFIXES                "+ADJ(able)+N(ity)"
END

LEXICON V_SUFFIX2
+able            ADJ_SUFFIXES                 "+ADJ(able)"
...
+ibly            ADV_SUFFIXES                 "+ADJ(able)+ADV(ly)"
END

LEXICON V_INFL
+ed              End                          "+PAST"
+ing             End                          "+PROG"
+s               End                          "+3SG"
0                End                          ""
END
```

Figure 12.3 Lexicons in PC-KIMMO.

6. If, when the string is completed, none of the recognizers reject the string, then the constructed lexical form corresponds to the given surface form.

Although the linked lexicon mechanism is quite useful for implementing certain paradigms, it limits the options of morpheme combinations to just concatenation. This is generally not a real problem: prefixing, suffixing and compounding phenomena can be implemented in an obvious way. Phenomena such as infixation or circumfixation can also be handled, though not as cleanly (Antworth 1990). On the other hand, when developing a wide-coverage and accurate morphological analyser, one needs to be concerned with a number of issues such as handling *exceptions* to general paradigms, handling *lexicalized derivations* and preventing *overgeneration* by *restricting affixation*.

All languages have wordforms which are *exceptions* to the general inflectional or derivational paradigm. For instance, a large number of English verbs do not follow the general paradigm for inflections. The past tense form of `eat` is not formed by the suffix `+ed`, but rather is `ate`. The way to handle this is quite obvious when the number of wordforms are small: one can manually enter all regular and exceptional forms directly into the lexicon. Thus, in the simplified lexicon above, `eat` would be associated with an alternation which only points to the derivational suffixes and the inflected forms would be entered into the same lexicon as:

```
...
eat      VB_DER_SUFFIXES   "V(eat)"
eats     End               "V(eat)+3SG"
ate      End               "V(eat)+PAST"
eaten    End               "V(eat)+PART"
eating   End               "V(eat)+PROG"
...
```

If there are a number of exceptional cases that behave similarly, it is certainly possible to create affix lexicons for those cases and have their successor point there, rather than explicitly listing all the forms. For instance the wide coverage morphological analyser developed (using PC-KIMMO) at University of Pennsylvania (Karp *et al.* 1992) has eight separate continuation classes for verbs. Verbal roots are partitioned into eight classes, depending on which subset of the three verb suffixes apply regularly to the verbs in that class. For instance, the verb `admire` is in a class which can take all the suffixes regularly, while `teach` is in a class which only takes the `+ing` and `+s` suffixes regularly. The exceptional and idiosyncratic cases are entered explicitly as described above.

Another problem area is the treatment of *lexicalized derivations*. In English, e.g., the word “application” has a lexicalized interpretation when it is used to mean “a computer program”. On the other hand, in a sentence like “The application of this theory to this case is useful.”, the same word is a noun derived from the verb “apply”. Note that in both cases the final part-of-speech of the word is noun so the difference does not matter in the case of part-of-speech tagging. But if the output of the morphological analyser is to be used in parsing, the parser may want to know if the word is derived from a verb to check any relationships with verbal complements. An obvious solution is to have an entry for the lexicalized form in the appropriate lexicon and have the derivation paradigm generate it from the original form as well.

Finally, there is the matter of *overgeneration*. There are many instances of productive derivational affixes which apply only to a restricted subset of the roots. For instance, in English the suffix `+ity`, which derives a noun from an adjective, is not applicable to all adjectives. If the analyser is to be used with the assumption that only valid input will be processed, then overgeneration is not a real problem. But usually, the same two-level morphological description is used for generation or spelling correction where the generation capabilities may be used to construct words and then one certainly would like to generate only valid words. Hence, if the aim of the morphological analyser is to accept and analyse all valid words in a language and reject all words that are not in the language, then one has to build the morphotactics component very accurately. This can be done by partitioning the root lexicons into classes depending on the subsets of suffixes that can be applied. Since the underlying mechanism, based on linked lexicons, is quite low level, this may become a non-trivial and perhaps unmanageable task. One may have to duplicate certain affix lexicons with very minor changes to account for minor variations in morphotactics. And worse, when the dependencies governing morphotactics are long distance, one may have to duplicate quite a substantial portion of the lexicons. This may easily become a maintenance and debugging nightmare. This is generally a much more serious problem for languages with rich inflectional and derivational morphology, such as Turkish and Finnish, as compared to a language like English. Other solutions are possible as well. Some systems provide special lexical symbols with `0` as their surface realization (e.g. *diacritics* in the Xerox TWOLC rule compiler), which can be used in the lexical representation of appropriate words. These symbols can be referred to by the rules to block or enable any affixation, but are irrelevant to any rule that does not refer to them.

12.3.3 Development tools

The two-level methodology is supported by several development tools. One of these is the PC-KIMMO system, to which we already have made quite a few references. The system exists in two different versions.

PC-KIMMO Version 1 is a publicly available system for developing two-level morphological analysers. It is quite a robust system and has been used to develop a number of quite substantial analysers (e.g. Oflazer 1993, 1994; Karp *et al.* 1992).

The main advantages of PC-KIMMO Version 1 are:

- It is publicly available at no cost (see `http://www.sil.org`).
- It runs on UNIX, DOS and MacOS platforms and the data developed is portable between these platforms (except perhaps for character set encoding).
- It provides quite comprehensive facilities for debugging and tracing analyser operations.
- In addition to being accessible as a stand-alone program the morphological analysis engine can be accessed from user programs.
- It has facilities for processing large batches of words for either analysis or generation, or for comparing with previous results.

- A quite comprehensive user's manual and a book specifically intended to be used with this system are available.

However, PC-KIMMO also has a number of drawbacks that need to be taken into account when developing a large system:

- The rules have to be given as finite state transducers as the software system proper does not provide a rule compiler. There is a preliminary version of a compiler called KGEN that is expressly built for PC-KIMMO rule files, but this compiler is limited in a number of respects. The compiler is a very crucial component in the development of two-level rules, especially for languages with a large number of morphographemic constraints sometimes interacting in unforeseen ways. Hand-compiling is feasible only for the simplest of rules and is very prone to errors.
- PC-KIMMO Version 1 provides linked lexicons as the only mechanism for constructing lexicons. These lexicons are stored as a trie data structure in the underlying analysis engine. There is some removal of redundancy with the trie compression, but nevertheless the necessary storage may be quite substantial. As mentioned earlier, the linked lexicon mechanism is too low a mechanism for implementing complex morphotactic paradigms.
- PC-KIMMO is quite slow in analysis especially with large lexicons. On an Ultra-SPARC 1 system, it can process about 10 words per second when loaded with Turkish description of about 30,000 root words and 35,000 proper nouns.

Version 2 of PC-KIMMO is an enriched version of the original system. The rule component now supports the use of multiple character symbols (multigraphs) and can accept recognizer/transducer descriptions in the format generated by the Xerox TWOLC compiler (see below). Another very important improvement in this version is the inclusion of a unification-based word grammar component, along the lines of the Ritchie *et al.* (1992) and based on the PATR formalism (Shieber 1986). The word parser produces a tree structure with morphemes at the leaves and nodes in the tree being annotated with any relevant features. For instance, for the word "foxes", the word grammar of the English description would produce

```
             Word
              |
             Stem
       --------|-------
     Stem            INFL
     ROOT             +s
     `fox             +PL
```

```
[cat:  Word
 clitic: -
 head: [pos: N
        number: PL]
 root_pos: N
 root:'fox]
```

Since this output makes the internal structure of the word explicit, it is very easy to integrate the output of this analyser with a syntactic parser. The word grammar is specified using a context-free backbone augmented with constraint equations on the set of morphosyntactic features. In addition to providing a richly annotated word parse, the main advantage of the word grammar is in providing a more powerful model of morphotactics. This lets the developer express morphotactic variations and long-distance dependencies directly in the grammar. The practical effect is that a more accurate and less overgenerating morphotactic component can be constructed.

There is also another rich set of software tools (developed by Xerox and known as the *Xerox Finite State Tools*), based on the theory of finite state calculus developed by Kaplan and Kay (1994), for building and manipulating finite state recognizers and transducers. Although there is support for building classical two-component two-level morphological analyser systems, their current approach is now more and more based on the notion of regular relations applied to NLP problems at different levels starting on tokenization all the way up to finite state parsing.

For developing morphological analysis systems Xerox provides TWOLC (Karttunen and Beesley 1992) for developing the morphographemic rule component and LEXC (Karttunen 1993) for developing the lexicon component. TWOLC is a very sophisticated two-level rule compiler, accepting rule definitions with a very general syntax and powerful operators, and is able to produce finite state transducers with very compact representations. It contains a facility for intersecting the automata generated for a rule set to obtain a single transducer that combines the constraints of all rules. There is also extensive support for testing the rules. LEXC is a lexicon compiler which builds a finite state transducer from a lexicon specification. The main mechanism for lexicon specification is again a linked lexicon model very much like PC-KIMMO Version 1, at least as far as the developer is concerned. There is however a very important provision for describing regular expressions as lexicon entries. This comes in very handy when building number or date parsers right into the lexicon.

The finite state transducers built by both LEXC and TWOLC are determinized and minimized as far as possible (it may not be possible to completely determinize transducers!) and then combined by composition to obtain a final transducer (cf. figure 12.5 below) which contains the complete morphological model of a language. There is no rule interpretation at run time. A generic transduction engine fed with the morphological transducer will map any surface form to a lexical form and vice versa. The system is very fast. For instance for Turkish with the lexicon sizes given earlier, it can process about 2000 words per second on an UltraSPARC 1 system. Commercial grade versions of the Xerox system have been quoted as providing a much higher performance and much better transducer compression.

The development of high-level and efficient tools for algebraic operations on finite state machines exemplified by the Xerox Finite State Tools has fostered development of morphological analysers which use the full power of finite state calculus. A recent interesting example of this is Beesley's morphological analyser for Arabic (Beesley 1996) in which the lexicon entries for roots are actually regular expressions and finite state calculus operations are used to restrict the overgenerating lexicon entries by the templates. Another example is Chanod's system for French (Chanod 1994), where linguistic generalizations shared across various inflectional paradigms are expressed as separate transducers and then composed to produce the final transducer.

12.3.4 Developing a Morphological Analyser

Developing a morphological analyser for a language involves a considerable amount of planning and preparation. The choice of the tool and/or the formalism depends on a number of issues. As discussed earlier there are public domain tools available which are quite usable but these have considerable drawbacks (such as speed or lack of a rule compiler) if one wants to develop an industrial strength system. Commercial systems such as the Xerox Finite State Tools on the other hand provide speed, good compilation and debugging facilities, and additional machinery for handling problems outside the classical two-level paradigm.

Before any attempt to code the required information in the formalism of the selected tool, the developer should compile:

1. A list of root words annotated with parts-of-speech and other information that may conceivably be needed in the analysis (e.g. any semantic information that may be required for morphotactics). It may also be necessary to include the citation forms (e.g. infinitive forms of verbs) of the roots, if these are different from the root forms.

2. A list of morphemes along with the associated morphological information they encode. Obviously this information has to be represented in a form that is in line with the tagset that will be used or that can be converted to the tagset encoding.

3. A model of the morphotactics described as paradigms or as a high-level description of the sequence of morphemes.

4. A comprehensive listing of the morphographemic phenomena, along with as many examples as possible for each.

5. A large corpus of tokens collected from various sources on which the resulting analyser will be used.

Unless the language has complex morphotactics (such as Turkish or Finnish), the most crucial and time consuming component in the process is the coding and debugging of the rules that describe the morphographemic processes. In the case of two-level morphology, this involves abstracting phenomena and coding them using the rule formalism. This is usually not a trivial process, but Anthworth (1990) provides some very good guidelines and examples for this process (see Karttunen and Beesley (1992) for further examples.) The following table from Antworth (1990) aids one in choosing the rules describing a given morphographemic phenomenon, such as the realization (inhibition) of a lexical symbol `l` as the surface symbol `s` in given left and right contexts `LC` and `RC`.

Rule to use	**Is l:s allowed in context LC _ RC ?**	**Is l:s only allowed in context LC _ RC ?**	**Must l always correspond to S in context LC _ RC ?**
`l:s => LC _ RC`	Yes	Yes	No
`l:s <= LC _ RC`	Yes	No	Yes
`l:s <=> LC _ RC`	Yes	Yes	Yes
`l:s /<= LC _ RC`	No	NA	NA

Rule ordering is not really an issue in two-level morphology, but sometimes rules may interact or conflict in unforeseen ways and these will need to be resolved by careful re-crafting and/or refining of context descriptions (see Karttunen and Beesley (1992) for details on this). Debugging the rule component involves repeated testing of the rules on the examples compiled in (4) above until all are satisfactorily handled.

Getting the morphotactics right involves careful linking of root and affix lexicons so that all valid sequences are allowed while invalid sequences are disallowed. Machinery provided in the two-level development tools is usually sufficient to implement this sequencing, but sometimes one may need to enforce long distance constraints (across multiple morphemes) and this may cause substantial complexity in the lexicon component. Certain tools such as Xerox's provide sophisticated mechanisms for automatically enforcing such constraints.

Once the lexicon and rule components are coded, real testing should be done on actual corpora to gauge and improve coverage. This is quite a time-consuming process as one continually encounters new words, which have to be retrofitted to the lexicon, and as yet uncovered phenomena, for which rules have to be updated. This process continues until a stable and satisfactory point is reached. From this point on, maintenance is a low-intensity effort where updates become progressively infrequent.

12.4 A MORPHOLOGICAL ANALYSER FOR TURKISH

We conclude this chapter with a description of the implementation of a morphological analyser for Turkish that has been built using the Xerox Finite State Tools. The analyser is two-level in spirit but uses additional levels for a variety of reasons. It uses about 30,000 Turkish root words and about 35,000 proper name roots and implements all derivational and inflectional paradigms of Turkish quite accurately. It has its roots in an earlier PC-KIMMO implementation (Oflazer 1993, 1994).

Turkish is an agglutinative language with word structures formed by productive affixations of derivational and inflectional suffixes to root words. Turkish has finite-state but nevertheless rather complex morphotactics. Morphemes added to a root word or a stem can convert the word from nominal to verbal or vice versa. The morphotactics is highly productive, leading to such adverbial constructs such as the one on page 176, and circular, at least from a competence point of view.[9]

The surface realizations of morphological constructions are constrained and modified by a number of phonetic rules. Vowels in an affixed morpheme have to agree with the preceding vowel in certain aspects to achieve *vowel harmony*, although there are a small number of exceptions. Under certain circumstances vowels in the roots and morphemes are elided. Similarly, consonants in the root words or in the affixed morphemes undergo certain modifications and may again sometimes be deleted. Nevertheless, things are quite regular, especially compared to many European languages. However, the assimilation of a large number of words into the language from various foreign languages – notably French, Arabic and Persian – have resulted in word formations which behave as exceptions to many rules. Our intention in this section is not to cover a two-level description of Turkish morphology in gruesome detail (for a detailed exposition, see Oflazer 1994) but rather to highlight aspects that are potentially of more general interest to researchers who intend to build two-level descriptions for various languages.

12.4.1 Requirements

Underlying the design of the morphological analyser, we find a number of requirements.

1. It was to be used in a number of applications: (i) in parsing, as a component of the lexicon (Güngördü and Oflazer 1995), (ii) in morphological disambiguation (Oflazer and Tür 1996), (iii) in generation (Hakkani and Oflazer 1998) and (iv) as the basis for spelling correction (Oflazer 1996).

2. The morphosyntactic representation produced had to cater to at least all these applications, abstracting from all details of the surface representation. Furthermore, full derivational history of the word form had to be represented.

[9] Turkish allows, among other examples, words of the sort *ev+(ler+de+ki)**, (those things at those things at ... at the houses). Here the morphotactics has to loop back.

Table 12.1 Inflectional morphosyntactic features for verbs.

FEATURE	VALUES	INTERPRETATION
CAT	VERB	Major Category
TYPE	TRANS, INTRANS	Minor Category
VOICE	REFLEX, RECIP CAUS, PASS	Verbal Voice
POLARITY	POS, NEG, NEGC	Sense polarity
COMP	POSSIBILITY, etc.,	Modal Compounding
TAM1	PAST, NARR, PRES, IMP, PROG OPT, NECES, FUT, DESR	Tense-Aspect-Mood Marker 1
TAM2	PAST, NARR, COND	Tense-Aspect-Mood Marker 2
AGR	1SG–3PL	Person Agreement

```
REFLEX: Reflexive voice, RECIP: Reciprocal/Collective voice
CAUS:Causative voice, PASS: Passive voice
COMP corresponds to a set of 10 modal compounding suffixes, the most
important of which is the equivalent of the possibility auxiliary
(''can''/''may'') in English.
Most of the others do not have any close equivalents in English.

PAST: Past Tense, NARR: Narrative Past Tense, PRES: Present Tense
FUT:  Future tense, PROG: Progressive
IMP: Imperative, OPT: Optative Mood, NECES: Necessitative Mood
DESR: Desire mood, COND: Conditional
```

3. For (i) and (ii) above, all analyses and interpretations of a word had to be produced even if the underlying morpheme structure is the same.[10] Furthermore, to alleviate any problems in (iii) and (iv) the morphotactics has to be very accurate with minimal overgeneration.

4. We had to be able to process free-running text such as one finds in newspapers: news items, editorials, etc. Thus it had to be wide-coverage with root lexicons augmented with words from economic, social, legal and technical jargon, in addition to the standard Turkish words one finds in published word lists. Such texts contain lots of proper names,[11] foreign and/or unknown words, abbreviations and acronyms, and numerical tokens, which, though written with numerals in orthography, nevertheless go through the same suffixations demanded by the syntactic context. All these put additional demands on the morphographemic component.

[10] Note that this does not involve any analyses due to polysemy of the root. We are only concerned with morphosyntactic ambiguity.

[11] What makes this a problem in Turkish especially for morphological disambiguation is that any Turkish word with its inflectional and derivational markers is a potential last name.

5. In addition to standard morphographemic phenomena in Turkish, one has to be concerned with the intermingling of such phenomena with words imported into the language from foreign languages. These cause many exceptions and have to be cared for in the morphographemic component. For instance, the author's name "Kemal" will violate vowel harmony when something is affixed (e.g. "Kemal'i"and not "Kemal'ı" because of the interaction between "a" and the palatal "l". A similar phenomenon is found in the French import "alkol" which has an accusative form "alkolü" as opposed to "alkolu".[12]

6. The system had to run with sufficient speed so as not to be the bottleneck in performance.

Let us briefly touch upon some of the issues above:

Morphosyntactic Features. The morphological representation that we have opted to produce is a linear *feature–value* representation of the morphological information encoded by the morphemes, accounting for the fact that certain morphemes signal multiple features, while certain features are signalled by a zero morpheme. The inflectional features for finite verbs, e.g., are presented in table 12.1.

There are also many very productive derivational morphemes – too numerous to list here – that transform words between all major parts-of-speech. The final representation that we would like to get should also represent any derivational history. The linear feature–value sequence represents derivations by the target part-of-speech category and the suffix (if any), marking the derivation, e.g. the word "geldiğimdeki" ("at the time I came"), would have the feature sequence:

```
[[CAT VERB] [ROOT gel] [SENSE POS]
                       [NOUN DIK] [AGR 3SG] [POSS 1SG] [CASE LOC]
                       [ADJ KI]]
```

The linear representation can, when necessary (e.g. for use in parsing), be transformed into a hierarchical representation representing the inflectional features of the last derived category and the derivational history (figure 12.4).

Tokens with numerical components. Tokens involving digits or numbers are frequently found in real text. This again may not be a problem in languages like English where the only tokens that may follow a number may be an "s", "'s" or an ordinal abbreviation. But in Turkish, one encounters tokens like "122.ye" ("to the 122nd") or "33'ünün" ("of 33 of them") where the suffixation proceeds depending on how the numerical component is actually pronounced. Thus the morphotactic engine has to

[12] Some sources still make a distinction in the orthography of such words by putting a little cap on the vowel, but the ISO Latin-5 standard for Turkish character set does not make such distinctions.

```
┌                                              ┐
│CAT      ADJ                                  │
│         ┌                                 ┐  │
│         │CAT      NOUN                    │  │
│         │AGR      3SG                     │  │
│         │POSS     1SG                     │  │
│         │CASE     LOC                     │  │
│STEM     │         ┌                ┐      │  │
│         │         │CAT      VERB   │      │  │
│         │STEM     │ROOT     gel    │      │  │
│         │         │SENSE    POS    │      │  │
│         │         └                ┘      │  │
│         │SUFFIX   DIK                     │  │
│         └                                 ┘  │
│SUFFIX   KI                                   │
└                                              ┘
```

Figure 12.4 Hierarchical representation of a morphological derivation.

have quite a sophisticated finite state parser in order to be able to handle numeric constructions such as these and additional ones like "16:15'te" (time designator), "3:4" (ratio), "%23'ü" (percentage), "2/3'si" and "2/3'ü" (ratios, but suffixation again proceeds depending on the pronunciation of the ratio – numerator or denominator last). The morphographemic rules for vowel harmony and vowel and consonant ellipsis also must take into account the phenomena in such numerical constructions.

Foreign words and abbreviations. Most foreign words used in a text follow their standard orthography in the original language. On the other hand, they do receive the Turkish morphosyntactic markers dictated by the context, the form of the suffixation again being based on the original pronunciation. This may make some constraints like vowel harmony inapplicable on the graphemic representation, though harmony is in effect in the pronunciation. For instance, one sees the form "Carter'a" where the last vowel in "Carter" is pronounced so that it harmonizes with `a` in Turkish, while the `e` in the written surface form does not harmonize with `a`.

A similar problem occurs with abbreviations such as "PTT" or "TV", whose orthographic form cannot supply sufficient information for morphographemic constraints. So a form like "PTT'ye", ("to the PTT") cannot be processed, since it does not have any vowels in its orthographic representation to serve as context for various morphographemic constraints.

The simplest solution to this problem is to collect as many such root forms as possible, place them into the appropriate part-of-speech lexicons and add *hidden* lexical symbols to their lexical representation which indicate how the word is really pronounced. The hidden symbols are considered when evaluating the morphographemic rules but disappear at the surface level (i.e. they are paired with `0` as their surface symbol). For instance, the lexical form for "Carter" would actually be `Carte`<u>`HidA`</u>`r`,

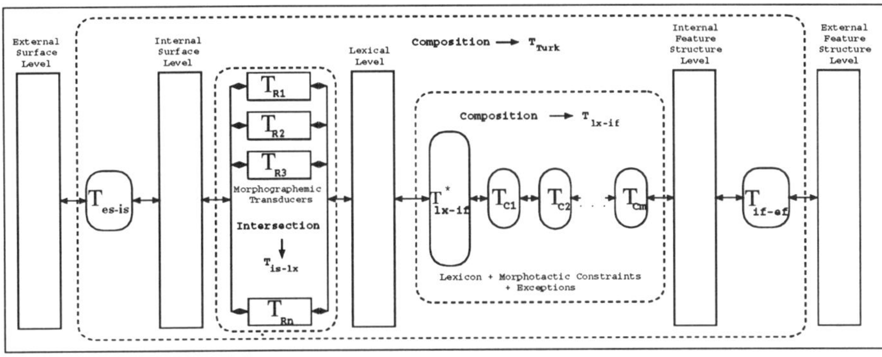

Figure 12.5 The architecture of the Turkish morphological analyser.

with `HidA:0` in the list of feasible pairs. A more principled solution should involve phonetic representations of the words.

12.4.2 System architecture

The morphological analyser has been based on the general architecture proposed by Karttunen (1993), with some additional components. The system has the five level architecture shown in figure 12.5.

The *external surface level* corresponds to a platform/OS/standard specific coding of the surface representations of Turkish words.

The *internal surface level* corresponds to the traditional surface level in two-level morphology. It uses a platform independent ASCII encoding of Turkish surface characters.

The *lexical level* corresponds to the traditional lexical level in two-level morphology, where the internal form of the word is represented as a sequence of morphemes.

The *internal feature structure level* corresponds to the information that is essentially provided by the lexicon glosses in PC-KIMMO. However, in PC-KIMMO itself these glosses are not accessible except for printing. In our system this level serves the very important level of controlling and fine-tuning morphotactics.

The *external feature structure level* corresponds to the actual output of the morphological analysis.

When given a Turkish word at the external surface level, such as: "yediriyordum" ("I was causing someone to eat (it)") the analyser produces the output at the external feature structure level:

```
[[CAT=VERB][ROOT=ye][VOICE=CAUS]
  [SENSE=POS][TAM1=PROG][TAM2=PAST][AGR=1SG]]
```

which indicates the word is a verb with root `ye`, in causative voice, with past tense, progressive aspect and first person singular agreement. The representations of the internal levels are not available directly: in fact, they do not exist as they are factored out during the construction of the analyser.

The analyser as a whole (T_{Turk}) is a large finite state transducer of about 250,000 states and 840,000 transitions, that maps between the external surface level and the external feature structure level. All morphographemic and morphotactic variations are handled at compile time, so there is no rule interpretation at run time: transduction is done directly. This machine is constructed from four finite state transducers, each of which maps between two levels.

The first transducer, T_{es-is}, maps between the external surface level and the internal surface level. On UNIX systems, e.g., there is not much of a support for special Turkish characters so we encode them by prefixing them with a "\". On PC or MacOS platforms one would choose an appropriate character encoding supported by the underlying system and use that as the external surface level encoding. T_{es-is} also deals with upper-case character folding and translates everything to lower case in the internal representation. On a Unix system this would map `\uz\ulm\u\st\um` ("I had felt sorry") to `UzUlmUStUm` while on a PC or a MacOS system with Turkish character support, it will map `üzülmüştüm` to the same internal surface string.

The second one, T_{is-lx}, maps between the internal surface level and the lexical level. It corresponds to the morphographemic mapping defined by the two-level rules in traditional two-level morphology. It is constructed by intersecting the transducers for each of the two-level morphographemic rules. For instance, it will map the internal surface string `UzUlmUStUm` into its lexical form `Uz+Hl+mHS+DH+Hm`.[13]

The third one, T_{lx-if}, maps between the lexical level and the internal feature structure level. This is the transducer that comprises the root word and suffix lexicons and the morphotactic paradigms. It is the most complicated and crucial transducer and is elaborated further below. As an example, consider the word `yaptırttığında`. This has two lexical level representations. The first, `yap+DHr+t+DHk+sH+nDA` ("at the time you caused someone to have someone else do (it)"), is mapped to the internal feature structure representation

```
[[CAT=VERB][ROOT=yap][VOICE=CAUS-DIR][VOICE=CAUS-T]
  [SENSE=POS][NOUN=DIK][AGR=3SG][POSS=3SG][CASE=LOCn]]
```

and the second, `yap+DHr+t+DHk+Hn+DA` ("at the time s/he caused someone cause someone else do (it)"), to the representation

[13] Here, `H` is a lexical symbol denoting a high-vowel whose surface realization is one of `ı, i, u, ü` and `D` is a lexical symbol whose surface realization is one of `d` or `t`.

```
[[CAT=VERB][ROOT=yap][VOICE=CAUS-DIR][VOICE=CAUS-T]
  [SENSE=POS][NOUN=DIK][AGR=3SG][POSS=2SG][CASE=LOCy]]
```

Here, the root word is represented using internal surface symbols. The voice markers and case markers bear signs of the underlying morpheme (e.g. CAUS-DIR indicates that the causative is marked by the +DHr morpheme). Although such details are not relevant at the external feature structure level, they are crucial in controlling the morphotactics as detailed below.

The fourth transducer, T_{if-ef}, cleans up the internal feature structure representation, mainly by replacing things like CAUS-DIR or LOCn by more meaningful tokens like CAUS or LOC and by changing the lexical form of the root word so that it corresponds to the external surface lemma form. So, for instance, for the first representation above one would get

```
[[CAT=VERB][ROOT=yap][VOICE=CAUS][VOICE=CAUS]
  [SENSE=POS][NOUN=DIK][AGR=3SG][POSS=3SG][CASE=LOC]]
```

at the external feature structure level.

The complete transducer for Turkish, T_{Turk}, is constructed by composing the transducers above:

$$T_{Turk} = T_{es-is} \circ T_{is-lx} \circ T_{lx-if} \circ T_{if-ef}$$

The composition operation eliminates all intermediate levels of representation and provides a direct mapping between the external surface level and the external feature structure level. Thus, e.g., when yaptırttığında is fed to T_{Turk} on the lower side, both

```
[[CAT=VERB][ROOT=yap][VOICE=CAUS][VOICE=CAUS]
  [SENSE=POS][NOUN=DIK][AGR=3SG][POSS=3SG][CASE=LOC]]
```

and

```
[[CAT=VERB][ROOT=yap][VOICE=CAUS][VOICE=CAUS]
  [SENSE=POS][NOUN=DIK][AGR=3SG][POSS=2SG][CASE=LOC]]
```

are produced at the upper side (the right side in the figure). And when one of these are provided at the upper side, the original Turkish word and its all capitalized and initial capitalized variants are produced at the lower side.

The transducers T_{es-is} and T_{if-ef} are quite simple and do not really warrant any further attention but it is instructive to look at T_{is-lx}, which is responsible for the morphographemic phenomena, and especially at T_{lx-if}, which is responsible for very accurate morphotactics.

12.4.3 The morphographemic transducer: T_{is-lx}

T_{is-lx} implements all the morphographemic phenomena in Turkish. It is actually the *intersection* of the transducers which are produced by the Xerox rule compiler TWOLC from rules using the Xerox syntax as specified in Karttunen and Beesley (1992). There are a total of 31 two-level rules covering phenomena such as: vowel harmony, vowel and consonant ellipsis, gemination in words borrowed from Arabic and Persian, devoicing, phenomena involving proper noun suffix separators and productive phenomena in words borrowed from Western languages. The resulting transducer for all these rules, obtained after intersection, has about 3600 states and about 170,000 transitions.

The two-level rules implemented at this level are essentially the ones in the original PC-KIMMO implementation (Oflazer 1994). There are, however, quite a number of additional cases covered, related to phenomena such as exceptions, numerical tokens, foreign words, abbreviations, etc.

It is quite instructive to examine the real version of the `A:e` vowel harmony rule used as an example earlier in 12.3.1:

```
A:e <=> [ VOWEL:FRONTV |
          HidE:0 |
          â:a | û:u | ô:o |
          NumECons:0 | NumEConsT:0 | NumE:0 |
          NumÜCons:0 | NumÜConsT:0 | NumÜ:0 ]
          [ ':' | CONS | =:0 | +:0 | 0:CONS ]+ __
```

The first part of the left context consists of a number of alternatives regarding the vowel that will enforce the harmony:

`VOWEL:FRONTV` is the basic case covered in the earlier example where one of the front vowels is present as the last vowel on the surface, either in the root or in a preceding morpheme.

`HidE:0` handles all those cases of acronyms or abbreviations whose orthography does not contain any vowel, but whose pronunciation has a front vowel (actually just `e`) as the last vowel (e.g. "PTT"). Such entries in the lexicon have a lexical symbol `HidE` in their lexical representation. This symbol disappears on the surface while enforcing the vowel harmony constraint in words like `PTTdeki`.

The `â:a | û:u | ô:o` alternatives deal with the cases of certain vowels which surface as back vowels but harmonize as front vowels due to a number of reasons. The lexical entries for such words contain a lexically different vowel representation.

The next set of options deal with a variety of cases of the pronunciation of numerical tokens. The parser for such constructs inserts into the lexical representation a symbol that represents how the last vowel and consonant of that number (not necessarily of the last digit) are pronounced. The vowel harmony rules then take this into account and act appropriately.

Once the vowel context is set, it can be followed by one or more of the proper noun suffix separator (`' : '`), any number of consonants, morpheme boundary, lexical symbols that disappear and consonants that are geminated (`0 : CONS`). This is intentionally kept quite over-generating as the sequence is actually enforced by the lexicon.

The proper noun suffix separator (represented by an apostrophe) has to be dealt with carefully in *all* the morphographemic rules. The reason is that all morphographemic changes on the left side of a morpheme boundary (in the stem) are blocked if the proper noun suffix separator is present, even though such a distinction is not made in pronunciation. The effect becomes clear when we compare the proper noun "Işık" and the common noun "ışık" ("light")

Lexical:	`Işık+'+nHn`	`ışık+nHn`
Surface:	`Işık0'00ın`	`ışığ00ın`
	`Işık'ın`	`ışığın`

For the proper noun, the apostrophe blocks the `k : ğ` pair in the orthography (but not in pronunciation).

12.4.4 The morphotactics transducer: T_{lx-if}

T_{lx-if} is compiled from Turkish root word and suffix lexicons using the Xerox lexicon compiler LEXC. The basic implementation of the paradigms is intentionally overgenerating. With very minor exceptions, the lexicons and their linking are specified so that they very closely follow the paradigms.

Although it would be possible to implement the paradigms with lexicons linked through continuation classes, exceptions and idiosyncrasies start causing maintenance and debugging problems after a while. Consider the following examples of morphotactic constraints:

1. Personal pronouns have the same inflectional paradigm as nouns, except that possessive suffixes (which happen to be in the middle of the paradigm) are not applicable.

2. 1SG and 2SG pronouns have irregular forms in dative case.

3. The choice of the suffixes to mark a certain morphological feature depends on the morphographemic and morphotactic context, e.g.

    i) The causative voice marker morpheme is `+t` following a previous causative marker (except `+t`) or a verbal root ending in a vowel.

ii) The case morphemes for certain cases are different depending on whether they follow nominals with 3rd person possessive suffixes or monosyllabic 3rd person personal or demonstrative pronouns.

iii) Which verbal agreement suffixes are used depends on the combination of the previous tense-aspect-mood markers.

4. Derivations should be carefully controlled to prevent duplicate applications of the same derivation or conflicting derivations. For instance, in Turkish the `+ci` suffix and its allomorphs productively derive nouns from nouns, but should not be applied repeatedly. The `+li` suffix ("with") should not be applied after the suffix `+siz` ("without") as that would produce a semantically anomalous form.

Although all these (and many more) constraints could be handled within the finite state lexicon mechanism, by duplicating portions of lexicons here and there, manually handling them is cumbersome and verification is quite a nightmare. Instead, we can use the full power of finite state constraints to incorporate such exceptions with minimal effort. We again use the notion of transducer composition. We start with T^*_{lx-if}, a (significantly overgenerating) transducer that is constructed by just implementing the regular derivational and inflectional paradigms, i.e. a given inflectional or derivational suffix is assumed to be applicable productively as long as it is in the correct morphotactic order. This results in a rather clean lexicon structure.

The morphotactic constraints beyond the basic paradigm ordering are expressed as constraints on regular relations using finite state operators over the alphabet of internal feature structure representations. Each such constraint C_i is then compiled into a finite state transducer T_{C_i} (cf. figure 12.5).

As an example, assume that constraint 1 above is expressed in C_1:

```
%[POSS=1SG%] /<= %[CAT=PRON%] %[ROOT= ?+ %] %[TYPE=PERSONAL%] ? __
```

This and five other similar rules indicate that possessive suffixes can *not* appear in a context following a personal pronoun, with any root (matching `?+`), and an agreement marker (matching the last `?`).[14] C_1 (and hence its transducer T_{C_1}) defines a regular relation that maps every string *not containing* the sequence matching the left context above to itself but rejects any string where a possessive marker follows a personal pronoun root and agreement marker. Composing $T^*_{lx-if} \circ T_{C_1}$ produces a new transducer T^1_{lx-if} which is exactly like T^*_{lx-if} except that invalid cases with regard to C_1 are weeded out during the intersection of the upper language of T^*_{lx-if} and the lower language of T_{C_1}.[15]

[14]Note that tokens such as `[POSS=1SG]` or `[CAT=PRON]` are single symbols in the internal feature structure representations. The % is the escape character in the Xerox syntax and has to be used here because [and] are also used as brackets in the rule syntax.

[15]This is slightly different than the way it is implemented in Xerox LEXC because of the direction of composition.

We then add further rules for each of the other constraints. For example 2 above,

```
%[CASE=DAT%] /<= [ %[AGR=1SG%] | %[AGR=2SG%] ] %[POSS=NONE%] __
```

disables the dative suffix for 1SG and 2SG personal pronouns while the exceptional dative entries are entered in to the lexicon directly. And for point i) of example 3,

```
%[VOICE=CAUS-T%] => [ %[VOICE=CAUS-DIR%] | %[VOICE=CAUS-AR%] |
                      %[VOICE=CAUS-HR%]  | %[ROOT= ?* VOWEL %] ] __
```

constrains the selection of the causative voice marker suffix: the first three options denote other causative markers that can precede while the last one constrains the verbal root to end with a vowel. Each such constraint rule $C_1 \ldots C_n$ leads to a new transducer $T_{C_1} \ldots T_{C_n}$ and, after composition, to new transducers $T^2_{lx-if}, T^3_{lx-if}, \ldots$, each a bit more accurate than the previous one. Eventually, we get

$$T_{lx-if} = T^*_{lx-if} \circ T_{C_1} \circ T_{C_2} \circ \cdots \circ T_{C_n}$$

which incorporates all the finite state constraints for morphotactics expressed in our rules.

There is a crucial point to pay attention to in this process. The composition operation is order dependent. However, if the rules are coded so that they specify their contexts very accurately and pass through all strings that they are not sensitive to, the final result of all these compositions is a transducer that maps the lexical level representations to an order independent intersection of the upper languages of individual transducers and that of T^*_{lx-if}.

Composition may be a time-consuming operation, especially when the transducers involved are large. Thus for the situation above it is best to perform the composition of the constraint transducers first in batches and then compose the result of this with the lexicon transducer. The number of constraints for our Turkish implementation is about 200 and all the compilations and composition operations take about 3 minutes of CPU time on an Ultra Sparc 1 system.

13 TAGGING UNKNOWN WORDS

Eric Brill

13.1 INTRODUCTION

In the previous chapters we have assumed that there was information present on each word, or at least on its underlying lemma. Most taggers rely on such lexical information, listing the allowable tags for each word, perhaps along with some probabilistic information such as the probability of the particular word and tag co-occurring. In practice, however, a wordclass tagger will also have to be able to deal with *unknown* words, for which such information is not available in the lexicon. Except in the case of a truly closed vocabulary system, such as a voice dictation system with a preset vocabulary, unknown words will always be a phenomenon that cannot be ignored. In figure 13.1, we show the number of unrecognized words in a test set from Wall Street Journal text as a function of the number of words in the training set, from which the lexicon has been built.

13.2 BEHAVIOUR OF UNKNOWN WORDS

We can study the behaviour of unknown words by examining predictive features for determining the wordclass of words that occur only once in the training corpus, the assumption being that these two sets of words should behave similarly (cf. Baayen and Sproat 1996). It is dangerous to draw conclusions about unknown words based on observations of *all* words in the training corpus. This is because unknown words have very different distributional properties than words in the language overall. For

H. van Halteren (ed.), Syntactic Wordclass Tagging, 207-216.

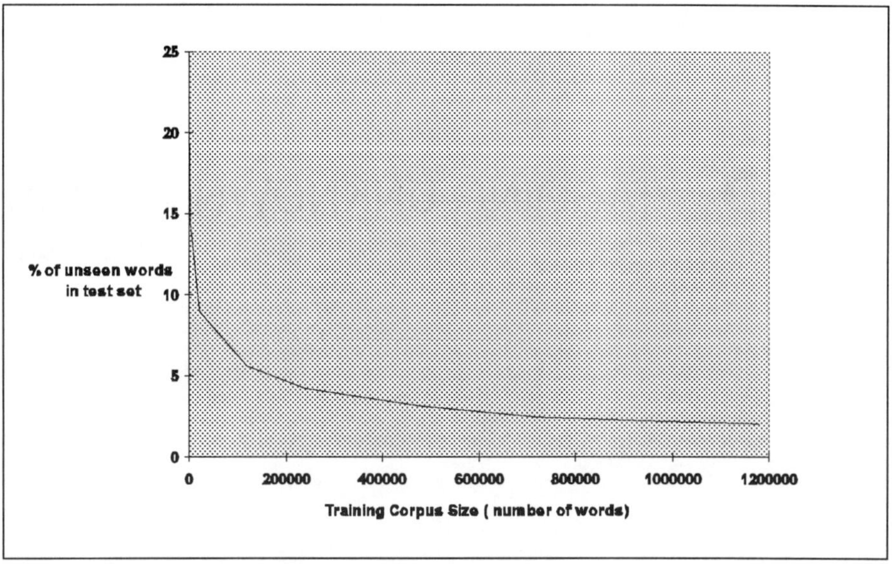

Figure 13.1 Presence of unknown words.

example, in the Penn Treebank tagged Brown corpus,[1] if we measure the entropy of the tag distribution in the entire corpus, we get $entropy = 3.83\ bits$. If we measure the entropy for words that occur only once in the training set, we get $entropy = 2.92\ bits$. Since $2^{entropy}$ is equivalent to the number of equally probable choices the model predicts, on average, we can think of this as meaning that in the corpus overall, with no lexical or contextual information, the problem of tagging is equivalent to deciding between 14.2 ($2^{3.83}$) equally likely tags, whereas for unknown words it is equivalent to deciding between 7.6 ($2^{2.92}$) equally likely tags. This makes sense, as it is very unlikely that with a sufficiently large training corpus any unseen word will be a closed-class item. Likewise, finding the most likely tag overall in the entire corpus and then tagging each word in the corpus with that word gives a tagging accuracy of 17.8%, whereas doing the same only for words occurring once in the corpus gives an accuracy of 25.0%.

Table 13.1 shows the probabilities of the ten most frequent tags in the Penn Treebank Wall Street Journal corpus (henceforth PennWSJ) both overall and for only words that occur once. It makes it quite clear that the difference between the two distributions is great.

[1] A full list of the tags used in the Penn Treebank project can be found in Appendix 17.6.

Table 13.1 Ten most frequent wordclass tags in PennWSJ.

Tag	Description	Overall Probability	Probability Among Singletons
NN	singular common noun	14.0	15.8
IN	preposition or sub. conjunction	10.5	0.1
NNP	singular proper noun	9.8	24.9
DT	determiner	8.7	0.02
JJ	adjective or ord. numeral	6.4	19.2
NNS	plural common noun	6.3	9.1
,	comma	5.2	0.0
.	sentence terminator	4.2	0.0
CD	cardinal numeral	3.7	14.8
RB	adverb	3.3	1.9

There are two fundamental problems in dealing with unknown words: the lack of lexical priors on the specific words and the fact that unknown words have significantly different distributional properties than the overall set of words observed in any training corpus. Because of this, we have no specific prior information about these unknown words and we cannot use information about the overall population of seen words to model the behaviour of unseen words.

There have been a number of approaches proposed that are able to achieve fairly high accuracy at determining the proper tag for words that did not occur in the training set. However, unknown word tagging accuracy is always significantly lower than known word tagging accuracy, even though by the entropy measure unknown word tagging is in theory a less difficult problem.

13.3 DEALING WITH UNKNOWN WORDS

When looking at specific unknown words it is often possible to predict their tag based on various cues, such as specific affixes, features such as capitalization and the context in which the word occurs. For example, look at the highlighted 'words' in the following sentences:

1. The *quabix* fell from a tree.
2. She ate *rouspitly*.
3. She ate *quasit*.
4. The *quang klampted*.
5. The *klamp quanged*.

In all of the above cases, it is fairly easy to determine the wordclass of the nonsensical words, using morphological and contextual cues. For languages with a strong morphology, such as Turkish (cf. Chapter 12), it is possible to use a morphological analyser without knowing an exhaustive list of lemmas, thus reducing the need for contextual cues. For most languages, however, morphological cues are insufficient and it is necessary to combine them with contextual ones. Although we we have decided, for this book, to separate the assignment of potential tags to words and the selection of the contextually most appropriate tag, the relative strength of the cues for unknown words makes this position rather less tenable. As a result we will often see the two processes integrated into a single whole in this chapter.

While it is not difficult to realize a set of potentially useful cues for tagging unknown words, it might be difficult using only intuition as a guide to deciding how multiple cues should interact. For instance, in the sentence "The lemming jumped", the fact that "lemming" ends in "-ing" is strong evidence that the word is not a noun, whereas the fact that it occurs between a determiner and a verb is strong evidence that it is. Let us examine how this problem can be addressed in, e.g., the framework of Markov models (cf. Chapter 16). In a trigram tagger, lexical probabilities ($P(word|tag)$) are typically estimated using the maximum likelihood estimate: $count(word\, and\, tag)/count(tag)$, as observed in a training corpus. Since unknown words are absent from the training text, this maximum likelihood estimate cannot be used. Instead, one could group all words in the training corpus that occur just once together, compute the maximum likelihood estimate for $P(one\text{-}count\ word|tag)$ and use this as the lexical probability for unknown words. However, this model does not take advantage of lexical cues such as affixes. This idea can be extended to take significant cues into account. In Weischedel *et al.* (1993) the following model was used for estimating the lexical probabilities for unknown words:

$$
\begin{aligned}
P(w_i|t_j) = &\\
&P(one-count\ word|t_j) * \\
&P(Capital\ Feature|t_j) * \\
&P(Hyphenation|t_j) * \\
&P(Endings|t_j)
\end{aligned}
$$

where Capital Feature is a binary feature which is one if the first letter of the word is a capital letter and zero otherwise, the Hyphenation feature is a binary feature indicating whether the word has a hyphen in it and Endings is a j+1-valued feature indicating which (if any) of j suffixes a word has. Note that with an HMM, lexical and contextual cues are combined automatically, as the tagger returns the tag sequence that maximizes the product of lexical and contextual probabilities for the sentence. However, the useful cues in this model had first to be determined manually. In one set of experiments done on PennWSJ, the above model achieved an accuracy of 85% at tagging unknown words.

13.4 UNKNOWN WORDS IN CASE-BASED TAGGING

In *case-based* wordclass tagging (cf. Chapter 17), all environments in which tags occur are stored, along with the correct tag that occurred in that environment. To tag new text, the environment in which an ambiguous word occurs is compared to all of the stored environments, the stored environment most similar to that of the ambiguous word is determined and the word is tagged with the most frequently occurring tag in that particular environment for that ambiguity class. Since there are no instances of unknown words in the training corpus, there are no environments that can be used for disambiguating the word. In Daelemans *et al.* (1996) the environment used to tag known words is: the two disambiguated tags to the left of the current word, the ambiguity class of the current word and the ambiguity class of the following word, along with the correct tag for the current word in this context. For unknown words, the ambiguity class of the current word is unknown, so a different environment vector is needed. One possibility is to substitute the single feature representing the ambiguity class of the word with a set of morphological features. For example, one possible context vector for unknown words could consist of:

- the disambiguated tag of the two words to the left,
- the disambiguated tag of the word to the left,
- the ambiguity class of the word to the right,
- the first letter of the current word,
- the last letter of the current word,
- the second to last letter of the current word,
- the third to last letter of the current word.

If in tagging from left to right, the word "roped" is unknown to the system in the sentence

```
The/DT horse/NN was/VBD roped/??? up/RP_RB ./.
```

the environment would be:

```
Tag 2 to left                = NN
Tag 1 to left                = VBD
Ambiguity class 1 to right   = RP_RB
First letter                 = r
Last letter                  = d
Second to last letter        = e
Third to last letter         = p
```

There would be a set of such environments extracted from the training set. In Daelemans *et al.* (1996) environment vectors for unknown words are only created for open-class words in the training set, since most unknown words will be open-class. After defining a distance measure, the above vector can then be compared to those extracted from the training set, finding the closest vector to that which was observed and assigning the unknown word the most likely tag for that particular environment. It would be learned that environments where the last letter is "d" and the second to last letter is "e", the tag of the word is most likely Verb. This approach achieved an accuracy of 90.6% when trained and tested on PennWSJ.

It is interesting to contrast this approach with the HMM approach in that the case-based approach does not need to provide the system with any prior specific knowledge about morphology, whereas the HMM approach needs a list of affixes. In the case-based approach, only the fact that letters in certain positions in a word might be significant cues need be provided, with the specifics of what letters are good cues and what these cues indicate being learned.

Note also that the case-based approach can well be used without reference to contextual cues, leading to a system which can suggest potential tags for unknown words with access to just the individual word. Such a system can then be combined with any of the systems for selecting the contextually most appropriate tag described in the remaining chapters.

13.5 UNKNOWN WORDS IN TRANSFORMATION-BASED TAGGING

Unknown word tagging can also be done within the *transformation-based* paradigm, described in more detail in Chapter 15. The transformation-based system for tagging unknown words can be implemented as follows:

Initial state tagger. For initial state annotation, some quick and dirty initial guess as to the wordclass of unknown words is used. For example, given the tag distributions shown above, we could use a simple rule tagging a word either as a noun or a proper noun, depending on whether it is capitalized. This rule achieves an accuracy of approximately 40% on PennWSJ. The initial state tagger could also be based on a more powerful approach, such as guessing the tag of a word based on a more sophisticated set of properties. There is a trade-off here between the amount of human labour needed and the accuracy of the initial state annotator. Since the goal of automated rule learning is to minimize human labour, the initial state annotators that have been explored for automatic tagging are typically very minimalist. It is quite possible that more human labour here could have significant impact in improving the eventual accuracy of the system.

Transformation templates. If we know nothing about a word, we can use distributional and morphological cues to predict its tag, or in this case to change the tags assigned to unknown words in order to improve the overall tagging accuracy. Some possible templates include:

```
Change the tagging of an unknown word from X to Y if:
    the word ends in suffix S
    the word begins with prefix P
    deleting suffix S results in a known word
    deleting prefix P results in a known word
    the word appears before word W
    the word appears after word W
```

where S and P range over all character sequences and are not limited to the known prefixes and suffixes of a language. It is important to stress that this method does not need to be provided with any explicit morphological cues, such as a set of prefixes and suffixes, but only needs to know what sorts of cues might be useful. In other words the machine is told that the language has prefixes and suffixes which might serve as good cues for determining the wordclass of a word, but the learner then discovers everything else, including what these affixes are and what wordclasses are associated with each affix. The search space of possible rules is very large. If we restrict affixes to length of less than 6, there are 26^5 ($> 10^7$) possible prefixes and suffixes that need to be considered. However, using the data-driven approach described in Chapter 15 we are able to learn rules efficiently. For a language with a different system of morphology such as Turkish (cf. Chapter 12), the template set would have to be adjusted.

Scoring function. In the case of unknown words, scoring the efficacy of rules is a little tricky. By definition, there are no unknown words in the training set. So we cannot simply measure how a particular transformation affects tagging accuracy on unknown words. There are two possibilities. We could have two training sets, where the lexicon is built from the first training set and then we consider in our second training set every word not appearing in the first set to be an unknown word and apply the learning algorithm to try to improve the tagging accuracy over these words. A problem with this is that we generally want each training set as large as possible, in order to have sufficient statistics for learning, and annotated data is still a relatively rare commodity. So it would be unfortunate to have to split the training set into two smaller sets. Alternatively, we can just use one training set, but measure the success (or failure) of different transformations on low frequency words in this set. One question is what we should consider to be low frequency words. It appears that in general, the lower the frequency of a word in a corpus, the more its form and distributional behaviour parallels that of new words to be encountered in the future. Given this information, it would make sense to consider words that occur only once in the training corpus as low frequency words for the sake of training the system to tag new words. One problem with this is that there may not be enough data to train successfully. Choosing the ideal frequency cut-off for considering a word low-frequency is an empirical decision, where increasing the cut-off frequency will result in a larger training sample, but the more

Table 13.2 First rules learned for English unknown words.

1	NN	→	NNS	Has suffix "-s"
2	NN	→	CD	Has character "."
3	NN	→	JJ	Has character "-"
4	NN	→	VBN	Has suffix "-ed"
5	NN	→	VBG	Has suffix "-ing"
6	??	→	RB	Has suffix "-ly"
7	??	→	JJ	Adding suffix "-ly" results in a word
8	NN	→	CD	The word "$" can appear to the left
9	NN	→	JJ	Has suffix "-al"
10	NN	→	VB	The word "would" can appear to the left

frequent words will be less representative in behaviour of words to be encountered for the first time in the future.

The first ten rules learned from one run of the unknown word learner on English text are listed in table 13.2. The first rule operates on the corpus immediately after it has been processed by the initial state annotator, which in this case assigned all non-capitalized words the tag NN. This rule learns that words ending in "-s" are very likely plural nouns and so it changes the tag of words with this ending accordingly. The second rule learned states that unseen words containing a period are likely to be numbers ("2.3", "1.12", etc.). The sixth rule says that regardless of what the current tag is of a word, if it ends in "-ly" then change the tag to adverb. The seventh says that regardless of the current tag, if adding "-ly" to the end of the word results in a word, then change the tag to adjective. The eighth rule states that if a word is currently tagged as a noun and a dollar sign can appear to the left of it then change the tag to a number. This distributional information can be determined either by checking all occurrences of the unknown word in the test set if using this off-line, or checking the single instance of the word currently being tagged if on-line.

When working completely within the transformation-based paradigm, the unknown word module, once trained, becomes part of the start-state annotator for the overall tagging program. As shown in figure 13.2, untagged text is first passed through the initial state annotator. If a word is unknown, it is passed through the unknown word tagger, which is itself a transformation-based system. Otherwise it is passed through the known word initial state annotator. Then the output of the initial state annotator is passed through the module which applies the set of tagging rules, the outcome being automatically annotated text. Such a transformation-based tagger achieved an accuracy of 85% on tagging unknown words when trained and tested on PennWSJ.

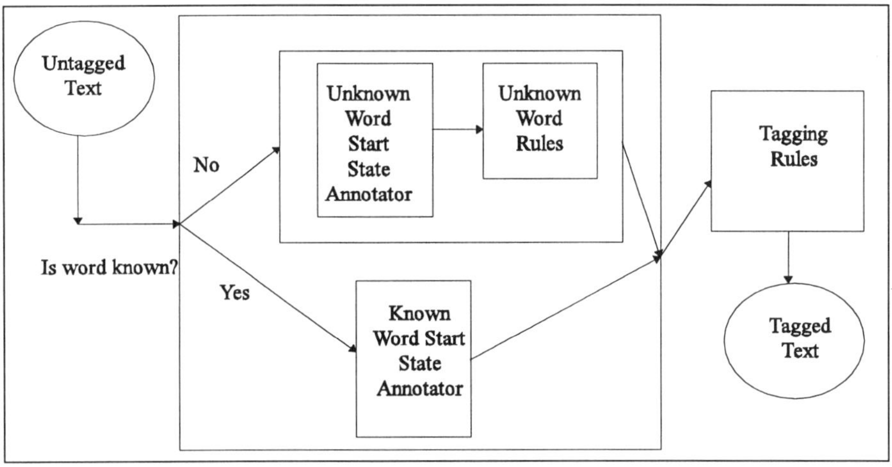

Figure 13.2 A complete transformation-based tagger.

13.6 LEXICON EXTRAPOLATION

Another type of rule-based approach has also been applied to the problem of tagging unknown words. If we have the set of allowable tags for a word in our lexicon, we can disambiguate the tags in context fairly reliably. The problem with unknown words is that we do not know a priori what the allowable tags are for that word and for this reason unknown word tagging accuracy is considerably lower than that for words appearing in the lexicon. Mikheev (1996) describes a rule-based method for determining the set of allowable tags for a word not occurring in the lexicon. The method does not rely on a manually tagged corpus, but only uses a tagging lexicon listing the set of allowable tags for words and an untagged text for training. The learning method has three steps: rule extraction, rule scoring and rule merging. Rules use as their information source the prefix and suffix of a word, as well as being able to remove a prefix and suffix from a word and obtain a word. Below are two examples of rules learned:

```
(1) [ un- (VBD VBN) ( JJ )]
(2) [ -ing (JJ NN VBG)]
```

The first rule states that if some word "X" appears in the lexicon and its allowable tags are VBD and VBN, then if the word "unX" is encountered it is an adjective. The second rule states that unknown words ending in "-ing" should be assigned the ambiguity class of JJ, NN and VBG. The first rule is a morphological rule and the second is an ending guessing rule. To extract morphological rules, all word pairs (X,Y) in the lexicon are compared and in all cases where for some string α, $X = \alpha Y$, the ambiguity class of X is Ω and the ambiguity class of Y is Π, the rule: $[\alpha - (\Omega)(\Pi)]$ is extracted. For suffix guessing rules, a word $W = c_1 c_2 ... c_n$ with ambiguity class Π will generate the set of rules: $[-c_n(\Pi)]$, $[-c_{n-1}c_n(\Pi)]$, $[-c_{n-2}c_{n-1}c_n(\Pi)]$, etc.

After the rule extraction phase, a set of rules is obtained which could be used for predicting the ambiguity class for an unseen word. However, there will be many nonsensical rules. For instance, if the word "and" appears in the lexicon and is unambiguously a coordinating conjunction (CC), this would generate the rules: $[-d(CC)]$, $[-nd(CC)]$ and $[-and(CC)]$. To overcome this problem, some rules in the set of extracted rules are purged or combined to create a more reliable rule set. First, every rule is scored to determine the likelihood that it is a viable rule and therefore should not be purged from the rule list. A guessing rule is scored by its *estimated proportion of success*: the proportion of the time when a rule can be applied that doing so results in a correct answer. Because of possible estimation error due to estimates derived from low counts, the score of a rule is a function of both estimated proportion of success and variance of the estimate. All rules are scored and only rules whose score is greater than some prespecified threshold are kept.

The final phase of training involves rule merging. In this step, rules that scored below the threshold are merged together and if the merged rule has a score greater than the threshold, it is added to the rule list. For example, given the two rules: $[-ing(AB)]$ and $[-ing(AC)]$, these rules can be merged into: $[-ing(ABC)]$.

An experiment was run where the transformation-based tagger described above was used, but for all unknown words in the test set disambiguation class guessing rules were applied and these words were added to the lexicon prior to tagging. Doing this resulted in an improvement of unknown word tagging from 84.5% to 92.2%, with an overall tagging improvement due to the improved unknown word tagging of 0.4% (from 95.9% to 96.3%).

14 HAND-CRAFTED RULES

Atro Voutilainen

14.1 INTRODUCTION

As already stated in Chapter 8, a linguistic tagger can consist of the following modules:[1]

- Tokenizer (identification of words and punctuation marks)
- Morphological analyser (introduction of ambiguities)
  - lexical analyser
  - analyser of unknown words ('guesser')
- Morphological disambiguator (ambiguity resolution)
  - grammar
  - heuristic grammar(s)

Our attention is on the last main item on the list, the morphological disambiguator. We will describe what we call the *linguistic paradigm* in tagging: disambiguation of

[1] We use a terminology here which differs from the one in that chapter. The one here is more in concordance with the literature on Constraint Grammar, a formalism playing a large role in this chapter.

H. van Halteren (ed.), Syntactic Wordclass Tagging, 217-246.

morphological ambiguities is accomplished by means of constraint rules proposed by a grammarian.[2]

The purpose of this chapter is to give practical instructions for writing a disambiguation grammar for a wordclass tagger. Most of the presentation is based on reasonably mature work rather than possible improvements not yet conclusively evaluated. Reference will often be made to Constraint Grammar, a recent successful framework for tagging and shallow parsing, as well as to the English grammar formulated in the late 1980's and early 1990's (EngCG; Karlsson 1990; Voutilainen *et al.* 1992; Karlsson *et al.* (eds.) 1995; Tapanainen 1996). An example of EngCG annotation can be found in 8.2.

After a general characterization of the statistical and the linguistic paradigms, we present an outline of a rule formalism (basically Constraint Grammar). This is followed by a routine for writing and testing a disambiguation grammar. As an illustration, a concrete working session is presented. Finally, some problems are discussed, in particular the treatment of ambiguities that are difficult for the kind of rule system outlined here.

14.2 COMPARISON OF PARADIGMS

This book presents several disambiguation methods. Why should one consider using the linguistic approach? What are the pros and cons of the statistical data-driven and the rule-based linguistic approaches to tagging?

The *statistical paradigm* has produced a large number of taggers for various languages, with the general outcome that:

- To reach state-of-the-art results, a carefully annotated training corpus of at least some tens of thousands of words, but preferably a few hundred thousand words, is necessary.
- Given that an annotated training corpus is available, a tagger can be trained with very little human effort.
- Provided with a carefully designed tagset, the tagger may be capable of fully and correctly disambiguating up to 95-97% of all words in text.
- If the input text represents a domain or style not adequately represented in the training corpus, the tagger's accuracy is likely to deteriorate (cf. 6.3.6).
- The tagger can be forced to produce multiple analyses (the 'n-best' variant), e.g. more than 97% of all words may retain the correct reading, but at a considerable

[2] In this chapter, the terms "grammar" and "grammarian" are used in a broader sense than in some other works; here grammar covers not only 'pure' syntax but also lexical generalizations.

cost in terms of remaining ambiguity (see e.g. de Marcken 1990; Kempe 1994; Brill 1994; Weischedel *et al.* 1993; Samuelsson and Voutilainen 1997).

- It is extremely difficult to improve on the 97% accuracy (recall with unambiguous output).

What about the *linguistic paradigm*? One obvious disadvantage is that so far there is rather little documentation of this approach, so it is difficult to say anything conclusive about its merits or problems. The observations presented here are largely based on the author's own work and on discussions with other people who have written disambiguation grammars for various languages. The following are likely properties of the linguistic approach:

- Contrary to common belief, a useful grammar can be written without spending years on it, given that reasonable attention has been given to the lexicon, tagset and test corpus – in fact, a few months of rule-writing and testing is likely to suffice for making a useful grammar.

- Reaching a recall of well above 99% is not particularly difficult, but reaching a precision of well above 95% at the same time may require a considerable effort.

- Correcting and improving a grammar, e.g. for customizing the system for the analysis of texts from some particular domain, is not particularly difficult.

- As in the statistical paradigm, annotated corpora are highly useful for testing and improving disambiguation grammars; the need for corpora may vary with the expertise of the grammarian and with the desired accuracy of the resulting tagger.

- If necessary, non-linguistic methods can be added on to resolve ambiguities left pending by the linguistic tagger.

14.3 RULE FORMALISM

Recent linguistic taggers have used two main techniques for reducing ambiguity:

- Alternative word (rather than sentence) readings are removed using a rule formalism that allows reference to relatively simple contextual patterns (e.g. Constraint Grammar; Oflazer and Kuruöz 1994).

- Alternative sentence readings are removed on the basis of finite state rules coded as regular expressions that can refer to complex structural patterns (various taggers based on finite state methods, e.g. Voutilainen and Tapanainen 1993; Chanod and Tapanainen 1995a).

Most recent work on rule-based disambiguation has been done with the first main technique. Though the latter technique has certain advantages in its greater expressive power, this approach is on the whole still at a somewhat intermediate stage. As a result we will focus our discussion on the former approach.

In this section we outline some central characteristics of a CG-type rule formalism. Most of the presentation is derived from a new version of Constraint Grammar known as CG-2 (Tapanainen 1996).

14.3.1 Overview

A disambiguation grammar contains constraint rules of the form "perform operation X on target Y in context Z".

The disambiguator applies all constraints, sequentially, to each utterance,[3] considering one word at a time. Since a rule sometimes is applied only after some other rules have made the context sufficiently unambiguous, the disambiguator can make several passes over the utterance, looping through the rules until no more disambiguation takes place.

In addition to a 'reliable' grammar, there may also be purportedly *heuristic* grammars. These can be applied for resolving remaining ambiguities. Generally, however, applying them leads to higher error rates. For instance, if there are three subgrammars G1, G2 and G3, it is possible to first apply G1, (low error rate), then G1+G2 for resolving more ambiguities (higher error rate) and finally G1+G2+G3 (yet higher error rate).

If it is important that all words receive at least one analysis, rules should be applied to ambiguous words only. However, it may be beneficial, e.g. for a grammar-checking application, that rules are applied also to unambiguous analyses; when even the last analysis violates a rule, the sentence should be flagged, possibly using the rule as a diagnostic. Such a feature is also useful for grammar development: likely errors in the language description could automatically be located even on the basis of raw text, by observing those words that lose every analysis.

14.3.2 Operations

The two main operations in the disambiguator are selection (SELECT) and rejection (REMOVE). In the case of selection, a rule selects one or more target readings in an ambiguous analysis when all specified context conditions are satisfied. As a consequence, all other readings are discarded. In the case of rejection, the rule specifies a context where the target readings should be discarded.

[3] For the disambiguator, an utterance is any sequence of words and tags ending in a sentence delimiter.

In effect, both these operations merely remove some of the readings already present in the lexically analysed input. It may be beneficial also to use 'addition rules' that introduce new readings on the basis of contextual information (e.g. consider analysis of names like "The The": the latter instance of "The" should receive a noun analysis, which it will probably not get from the lexicon).

14.3.3 Targets

The reading targeted by a rule is generally specified by a sequence of grammatical tags and/or base forms. We call such a sequence the *target* of the rule. For instance, the tag sequence (N NOM SG) is a legitimate target that can be REMOVEd by a constraint like

```
REMOVE (N NOM SG)
       (-1C (AUXMOD))
       (0 (INF)) ;
```

Here a noun reading is removed if there is another alternative reading for infinitives and the preceding word is unambiguously analysed as a modal auxiliary.

Though the target typically concerns only a single word analysis at a time, it may also be desirable to refer to analyses of word sequences too (especially in the context-based recognition of idioms; this feature is not available in CG-2, but cf. Oflazer and Kuruöz 1994).

14.3.4 Context conditions

The context in which an operation should be carried out can be specified as one or more statements about sequences of grammatical tags, punctuation marks and words (base forms and surface forms) around the ambiguous word. Sometimes a context condition may refer to a class of alternative, distributionally similar tags, e.g. various tags for nominals (e.g. N, A, PRON). Frequently used classes can be conveniently defined as sets, e.g. NOMINAL.[4]

Context statements in recent rule-based systems can refer to both local and 'long-range' context. For instance, the following statements are possible:

(i) -2 DET/GEN

(ii) NOT 1 (VFIN)

(iii) 0 UPPERCASE

---

[4] Such definitions are listed under the heading SETS at the start of the grammar. Usually, the definition takes the form of a list of tags, indicated by the keyword LIST. See the example grammar below.

(iv) NOT *-3 CLAUSE-BOUNDARY-MARKER

(i) requires that the second word to the left of the ambiguous word (-2) contains a feature in the set DET/GEN; (ii) requires that the first word to the right (1) does not contain the tag VFIN; (iii) requires that the morphological analysis of the ambiguous word itself (0) contains a tag that indicates that the word was written in upper case (UPPERCASE); (iv) requires that, from the third word on up to the beginning of the sentence in the left-hand context (*-3), there are no words with tags belonging to the set CLAUSE-BOUNDARY-MARKER (e.g. relative pronouns).

A problem with statements like (i) is that the word may be ambiguous and so contain some other analyses that are correct. Therefore it should also be possible to control the ambiguity of the context. This is done by using the optional *Careful mode* (cf. Karlsson 1995b), e.g. the expression

(v) -2C DET/GEN

requires that the second word to the left is an unambiguous (C) member of the set DET/GEN.

Context conditions of this kind enable the grammarian to express generalizations about word sequences in a fixed context. It may also be beneficial to make statements about word sequences in a less specified context. For instance,

(vi) *-2C REL-PRON BARRIER VFIN

states that somewhere to the left of the ambiguous word (*-2), there is an unambiguous (C) member of the set REL-PRON and there are no intervening (BARRIER) members of the set VFIN.

Context patterns can also be linked, e.g.

(vi) *-2C REL-PRON BARRIER VFIN LINK -1C NOMINAL

This expression states that somewhere to the left of the second word to the left, there is an unambiguous relative pronoun, that there are no intervening finite verbs and that the first word to the left of the relative pronoun is an unambiguous member of the set NOMINAL.

14.3.5 Sample rules

Now that we have described the components of the rules, we can show how the formalism works in practice, by presenting a number of CG rules for disambiguating English.

```
REMOVE VFININF
        (-1C (DET) OR (GEN))
        (NOT -1 (INDEP)) ;
```

This rule discards readings belonging to the set VFININF (finite and infinitive verb forms) if the first word to the left is an unambiguous determiner or genitive but it is not an independent genitive (INDEP; e.g. “theirs”).

```
"<A>" REMOVE (ABBR)
        (NOT *-1 POS)
        (1C A/NOMINAL)
        (NOT 1 PL)
        (NOT 1 PROPER) ;
```

This rule is about the wordform “A” that is ambiguous due to a determiner and abbreviation reading. The abbreviation reading of “A” is discarded if it is the first word in the sentence (the set POS contains all part-of-speech tags), if the following word is a member of the set A/NOMINAL and if the following word does not contain features for plurals or proper nouns.

```
REMOVE (<Rel>)
        (NOT *1 VFIN) ;
```

This is a simple long-range rule that discards a relative pronoun reading if there are no finite verb readings after it in the same sentence.

```
"<had>" SELECT (PAST)
        (*-1C CLAUSE-START BARRIER ("be") OR ("have")) ;
```

This rule selects the past tense reading of the wordform “had” as the correct one (thus discarding alternative readings not containing the tag PAST) if to the left there is a sentence start or a word all of whose readings are members of the set CLAUSE-START (notably relative pronouns, WH-words and subordinating conjunctions) and there are no intervening forms of “be” or “have”.

```
REMOVE (<Rel>)
        (NOT *-2 REL)
        (-1C CC) ;
```

This is a partial coordination rule. It discards relative pronoun readings if the immediately preceding word is a coordinating conjunction and there are no relative pronoun readings to the left.

Table 14.1 Frequency distribution of context positions in 3,744 EngCG-2 disambiguation constraints.

Position	*Frequency*	*Position*	*Frequency*
–9	1		
–8	3	*–8	1
–7	7	*–7	4
–6	16	*–6	3
–5	31	*–5	14
–4	94	*–4	31
–3	287	*–3	59
–2	923	*–2	213
–1	3090	*–1	338
0	2048		
+1	3426	*+1	310
+2	963	*+2	169
+3	243	*+3	37
+4	58	*+4	4
+5	14	*+5	4
+6	2		
+7	1		

14.3.6 Some facts about a large grammar

We conclude this section by showing how the formalism has been used in an actual grammar, viz. a recent version of EngCG: the EngCG-2 disambiguation grammar. In this grammar's 3,744 disambiguation rules, 12,394 context conditions are used. Table 14.1 lists the word positions used in the context conditions of the disambiguation grammar.

The right-hand columns indicates the frequency and the left-hand columns indicates the word position. '–' means 'to the left of the word being disambiguated' and '+' means 'to the right of the word being disambiguated'. The asterisk "*" indicates an unbounded context condition. For instance, '(*-8 X)' means 'Somewhere to the left, no closer than the eighth word, there is a member of the set X'.

2,048 references are made to the tags of the word being disambiguated using the context mechanism. Often these 'extra' checks make sure that some tag is not present in the analysis of the word to be disambiguated, e.g. (NOT 0 (<*>)) checks that the word does not contain the tag given to words written in the upper case. This could be used as an extra check e.g. in a constraint that would select a verb reading as the correct one (typically verbs are written in the lower case; if the word is in the upper case, it might be a nominal like a name instead).

Of the context conditions, 1,187 (almost 10%) refer to an unbounded context. These conditions often control the presence or absence of clause-level phenomena, e.g. the presence or absence of finite verbs or words occurring at sentence-internal clause boundaries (e.g. relative pronouns and WH-words).

About 21% of the context conditions (2,643 context conditions) refer to a fixed position other than the immediately neighbouring word or position zero. Often several fixed positions are used jointly to refer to some contextual pattern, e.g. (-3C (DET)) (-2C (A)) (-1C (CC)) states that the word in position zero is preceded by the sequence 'determiner – adjective – coordinating conjunction': this pattern might suggest a coordination of adjectives, so it could be used as a contextual criterion for selecting an adjective reading as the correct one in the case of words like "free" that are ambiguous due to the adjective reading and some other readings.

Distributional generalizations can be expressed at various levels of abstraction. The rules can refer to words, tags and their various combinations. Actually, quite a few constraints are devoted to disambiguating particular words. Usually these 'lexico-grammatical' constraints address frequent ambiguous closed-class words; e.g. there are 109 constraints for the wordform "that", 71 for "as", 69 for "more" and so on. This 'divide and conquer' approach tends to increase the size of the grammar (which implies a minor penalty in analysis speed), but on the other hand it often seems easier to make reliable distributional predictions about lexico-grammatical rather than purely grammatical categories.

The performance of old versions of the EngCG morphological disambiguator (1,100-1,300 rules) has been evaluated in several publications (Voutilainen *et al.* 1992; Voutilainen and Heikkilä 1994; Tapanainen and Voutilainen 1994; Voutilainen 1995b). In the system's output, 99.7–99.8% of all words generally retain the correct analysis, while the average number of alternative analyses per word has ranged from 1.05 to 1.09. The test texts represent various kinds of standard written British and American English and the size of the corpora varies from a few thousand words to about 55,000 words. The tagger's results have been automatically compared to a benchmark version prepared by two human annotators who (i) first independently disambiguate the same texts, (ii) then negotiate about the different initial analyses (usually less than a percentage point of all words), almost always reaching a consensus about the correct analysis, and (iii) finally prepare the benchmark version that reflects the outcome of the negotiations.

The new, extended version of the tagger is called EngCG-2. Its error rate is between a half and two-thirds of that of the older versions, while the amount of ambiguity it leaves is well below half of that left by the older versions. A performance test and a comparison to a state-of-the-art statistical tagger is reported by Samuelsson and Voutilainen (1997). EngCG-2 documentation and an interactive demo can be found at the following URL: `http://www.ling.helsinki.fi/~avoutila/engcg-2.html`

14.4 WRITING A DISAMBIGUATION GRAMMAR

Preferably, writing a disambiguation grammar makes use of a *grammar definition corpus* as described in 8.2. This kind of manually disambiguated corpus where all alternative analyses are present – one or more of them marked as correct – serves not only as (part of the) documentation of the adopted grammatical representation; it is also useful for testing and improving the proposed disambiguation rules.

Given that the grammarian has the following resources available:

- a reasonably large manually disambiguated benchmark corpus;
- a documentation of the grammatical representation;[5]
- lots of text;
- a mature tokenizer and morphological analyser;
- a rule formalism and a parser that not only disambiguates but also leaves a trace on the output of the applied rules,

writing a grammar can be a straightforward and, contrary to a common belief, fast process. The author uses the following method for writing constraint grammars:

1. Apply the current system (morphological analyser + grammar) to new text.
2. Formulate new rules for resolving remaining ambiguities (a few dozen at a time).
3. Apply them to the benchmark corpus.
4. Gather application statistics for the rules (how many correct applications; how many incorrect applications).
5. Observe those cases where a rule mispredicted.
6. Do one of the following:
   - Correct the rule and repeat cycle 3-6.
   - Assign the rule to a suitable heuristic subgrammar.
   - Abandon the rule.

[5] One sometimes hears arguments that documenting a grammatical representation in the form of some kind of annotator's manual is not actually of much use because (i) an exhaustive documentation will be too long or complex to learn or use and (ii) the documentation will be interpreted in several ways in any case. However, the documentation suggested here is not meant to be exhaustive in the sense that all grammatical phenomena in the language would be exhaustively discussed. In practice it seems to be the case that most types of wordclass assignment can be analysed without a manual consistently by annotators with some background in the grammar of the language; explicit documentation seems to be necessary only in a restricted number of problem cases. Experiments in annotation consistency are discussed, e.g., in Voutilainen and Järvinen (1995) and Källgren (1996).

14.4.1 A sample session

Perhaps the best way to get a feeling for the operation of the method described above is to give an actual example. Below we examine how a disambiguation grammar (using the CG formalism) is created from scratch. What we have at this stage of the grammar-writing process is the EngCG tagger minus a grammar. We will define disambiguation rules on the basis of the following two sentences:

```
A look at long-term changes in European legal customs is
pertinent here.  Medieval Europe had originally practiced
restitutive justice, a form of community customary law that
functioned through arbitration with a goal of reconciliation.
```

The analysis by the initial system is:

```
"<A>"
        "a" <*> <Indef> DET CENTRAL ART SG
        "a" <*> ABBR NOM SG
"<look>"
        "look" <SVC/N> <SVC/A> <SV> V SUBJUNCTIVE VFIN
        "look" <SVC/N> <SVC/A> <SV> V IMP VFIN
        "look" <SVC/N> <SVC/A> <SV> V INF
        "look" <SVC/N> <SVC/A> <SV> V PRES -SG3 VFIN
        "look" N NOM SG
"<at>"
        "at" PREP
"<long-term>"
        "long-term" A ABS
"<changes>"
        "change" N NOM PL
        "change" <SVO> <SV> V PRES SG3 VFIN
"<in>"
        "in" PREP
        "in" ADV ADVL
"<European>"
        "european" <*> <Nominal> A ABS
"<legal>"
        "legal" A ABS
"<customs>"
        "custom" N NOM PL
"<is>"
        "be" <SV> <SVC/N> <SVC/A> V PRES SG3 VFIN
"<pertinent>"
        "pertinent" A ABS
"<here>"
        "here" ADV
"<$.>"
"<Medieval>"
        "medieval" <*> A ABS
"<Europe>"
        "europe" <*> <Proper> N NOM SG
```

```
"<had>"
        "have" <SVO> <SVOC/A> V PAST VFIN
        "have" <SVO> <SVOC/A> PCP2
"<originally>"
        "original" <DER:ly> ADV
"<practiced>"
        "practice" <SVO> <SV> PCP2
        "practice" <SVO> <SV> V PAST VFIN
"<restitutive>"
        "restitutive" <?> <Nominal> A ABS
"<justice>"
        "justice" <-Indef> <Title> N NOM SG
"<$,>"
"<a>"
        "a" <Indef> DET CENTRAL ART SG
"<form>"
        "form" <SVO> <SV>  V SUBJUNCTIVE VFIN
        "form" <SVO> <SV>  V IMP VFIN
        "form" <SVO> <SV>  V INF
        "form" <SVO> <SV>  V PRES -SG3 VFIN
        "form" N NOM SG
"<of>"
        "of" PREP
"<community>"
        "community" N NOM SG/PL
"<customary>"
        "customary" A ABS
"<law>"
        "law" <SV> <Rare> V IMP VFIN
        "law" <SV> <Rare> V INF
        "law" N NOM SG
"<that>"
        "that" <**CLB> CS
        "that" DET CENTRAL DEM SG
        "that" ADV AD-A>
        "that" PRON DEM SG
        "that" <NonMod> <**CLB> <Rel> PRON SG/PL
"<functioned>"
        "function" <SV> PCP2
        "function" <SV> V PAST VFIN
"<through>"
        "through" PREP
        "through" ADV ADVL
"<arbitration>"
        "arbitration"  N NOM SG
"<with>"
        "with" PREP
"<a>"
        "a" <Indef> DET CENTRAL ART SG
"<goal>"
        "goal" N NOM SG
```

```
"<of>"
        "of" PREP
"<reconciliation>"
        "reconciliation" N NOM SG
"<$.>"
```

The following tentative grammar is proposed. In the comments given as italics, the reasoning underlying the formulation of the rules is recorded. Some of these constraints are actually somewhat naive; this is done intentionally for better illustrating the debugging of constraints.

```
# Experimental Constraint Grammar
# Constraints for morphological disambiguation

DELIMITERS = "<$.>" "<$!>" "<$?>" "<$HEAD>" "<$2-NL>" ;

SETS

LIST >>> = >>> ; # beginning of utterance
LIST <<< = <<< ; # end of utterance

LIST PUNCT = "<$">" "<$.>" "<$!>" "<$?>" "<$HEAD>"
        "<$2-NL>" "<$'>" "<$`>" "<$,>" "<$;>"
        "<$:>" "<$->" "<$)>" "<$(>" "<$/>" "<$[>"
        "<$]>" "<${>" "<$}>" ;

LIST CS-that = ("that" CS) ;
LIST INFMARKER = AUXMOD "do" INFMARK> ;
LIST NOM = NOM ;
LIST PLURAL = PL SG/PL PL1 PL2 PL3 ;
LIST NSG = (N SG) ;
LIST NOMINAL = A N PRON NUM PCP1 PCP2 ABBR ;
LIST INDEF = <Indef> ;
LIST V-cum-IN/ADV = "take" "bring" ;

LIST NON-ADV = A N NUM PRON DET ABBR V PCP1 PCP2 CC CS I
        INFMARK> PREP ;
LIST NON-ABBR/NUM/ADV = A N PRON DET V PCP1 PCP2 CC CS I
        INFMARK> PREP NEG-PART ;

LIST NPHEAD-ALL = ACC NUM NOM INDEP PCP1 A PCP2 ;


CONSTRAINTS
```

"A" looks difficult. Let's try "look" instead. The subjunctive reading is out because subjunctives occur in "that"-clauses only. I.e. if there is no "that" as CS, discard subjunctives.

```
REMOVE (SUBJUNCTIVE)
        (NOT *-2 CS-that) ;
```

Infinitives are preceded by an infinitive marker: "to", modal auxiliary or "do". If these do not occur in the left-hand context, discard the infinitive.

```
REMOVE (INF)
        (NOT *-1 INFMARKER) ;
```

Consider the present tense reading. This looks more difficult – a present tense verb is often preceded by a nominal. There seems to be an agreement violation, though: if "look" is a present tense verb, then "A" has to be an abbreviation acting as subject. But the subject should be a plural to agree with -SG3. So, if the word is not a plural, discard PRES -SG3.

```
REMOVE (PRES -SG3)
        (-2 >>>)
        (-1 NOM)
        (NOT -1 PLURAL) ;
```

An imperative occurring directly after a nominal or a determiner looks odd. An easy way out would be to write a rule that deletes imperatives if the previous word is a nominal or a determiner. It seems, though, that it is possible to write a more general and effective rule that requires certain properties from the immediate left-hand context of an imperative and if these conditions are not satisfied, the imperative is to be rejected. Usually imperatives occur clause-initially. There may be an adverb (including ordinals like "first") between the imperative and the clause-beginning. In addition, coordination is possible, so coordinating conjunctions and commas are legitimate. A clause can also be flanked with various punctuation marks, e.g. quotes; therefore a punctuation mark can occur immediately left of an imperative as well.

```
REMOVE (IMP)
        (NOT -1 >>> OR (ADV) OR (ORD) OR (CC) OR PUNCT);
```

Now that "look" has been disambiguated as a noun, let's re-consider "A". We can assume that a sentence-initial "A" directly followed by an unambiguous singular noun or abbreviation or adjective is a determiner. What else could it be? If it were an abbreviation, the word would contain a full stop "A." or it would be surrounded by parentheses "(A)" which is not the case. So it has to be a determiner.

```
"<a>" SELECT (DET)
        (-1 >>>)
        (1C NSG OR (ABBR SG) OR (A)) ;
```

Consider "changes". The second word to the left is a preposition and there is an intervening adjective. Adjectives can be noun phrase heads, but usually they are modifiers. Let's try a heuristic rule, but let's also minimize the risk of misprediction by making sure the adjective does not have noun-like features like upper case (<>*

signifies upper case) or the feature <Nominal> (that is given to some adjectives often used as NP heads). Furthermore, words analysed by the lexical guesser (<?>) should not be regarded as 'safe' adjectives. It is also reasonable to make sure that there is a noun reading available as an alternative. Even the right-hand context can be used for safeguarding against misanalysis: if the next word looks like the beginning of a NP, the word we are disambiguating could well be a verb, after all. Therefore some typical beginnings of NPs are listed as a further constraint.

```
REMOVE (PRES SG3)
        (-2C (PREP))
        (-1C (A) OR (PCP2) OR (GEN) OR (DET) OR (ADV))
        (NOT -1 (<*>) OR (<Nominal>) OR (<?>))
        (0 (N PL))
        (NOT 1 (DET) OR (ACC) OR (GEN) OR (A) OR (NUM)) ;
```

Consider "in" – a typical ambiguity class, preposition vs. adverb. Adverbs can occur almost anywhere in the sentence, even before noun phrases (just like prepositions), so this looks like a hard case. But it is also known that at least some preposition–adverb homographs are almost always prepositions when they are followed by a noun phrase, so let's capitalize on that. But it's also clear that after some verbs a preposition–adverb homograph, even a prenominal one, is often an adverb. Because those verbs are specific to the particular preposition–adverb homograph, let's write a rule about "in" alone rather than all preposition–adverb homographs collectively: Select the preposition reading if the previous word is not a verb taking "in" as an adverb (for now: "take" and "bring") and the following word indicates the beginning of a noun phrase or a nominal "ing"-clause.

```
"<in>" SELECT (PREP)
        (NOT -1 V-cum-IN/ADV)
        (1C (DET) OR (A) OR (ACC) OR (N) OR (ABBR)
                OR (GEN) OR (NUM) OR (PCP1)) ;
```

The wordform "had" is a finite verb except when preceded by a form of "be" or "have" (functioning as its auxiliaries). If there are no such auxiliaries to the left, the past participle form of "had" is deleted.

```
"<had>" REMOVE (PCP2)
        (NOT *-1 ("be") OR ("have")) ;
```

Consider "practiced". Preceded by a finite form of "be" or "have", an "ed"-form is likely to be a participial form rather than a finite verb. The rule licences any number of adverbs as an intervening category (by disallowing intervening non-adverb categories).

```
REMOVE (PAST)
        (*-1C ("be" VFIN) OR ("have" VFIN) BARRIER NON-ADV)
        (0 (PCP2)) ;
```

Consider "form". A rather obvious rule about the form of the noun phrase would be: readings with the tag V (in effect: infinitive and all finite verb forms, i.e. not participial forms in our lexical description) are illegitimate after a determiner.

However, we can also make a more general statement: infinitives and finite verbs are illegitimate between the beginning of a NP and its head. In practice, we identify some fairly reliable indicators of a beginning NP – DET, GEN, (except INDEPendent genitives like "theirs") and PREP – and use as barrier any category that can occur as a head of a noun phrase, listed in NPHEAD-ALL.

```
REMOVE (V)
        (*-1C (DET) OR (GEN) OR (PREP) BARRIER NPHEAD-ALL
            LINK NOT 0 (INDEP)) ;
```

Coming to the word "law", we see the tag <Rare>, which is given by the lexicon to verb readings of some noun–verb homographs where the noun reading is by far the more frequent one. The following two rules impose rather strict requirements for the verb readings: the contextual licences have to occur more locally than is required for imperatives and infinitives in general. (Note that other constraints could also be imposed on rare readings wrt. the right-hand context.)

In the case of 'rare' infinitives, there has to be an infinitive marker or a coordinating conjunction (in the case of coordinated infinitives) one or two words to the left.

```
REMOVE (<Rare> INF)
        (NOT -1 INFMARKER OR (CC))
        (NOT -2 INFMARKER OR (CC)) ;
```

Rare imperatives have only the most typical distribution of imperatives: close to the sentence-initial position or after a coordinating conjunction (in the case of coordinated clauses).

```
REMOVE (<Rare> IMP)
        (NOT -1 >>> OR (CC))
        (NOT -2 >>> OR (CC)) ;
```

The following two rules capitalize on the assumption that the sequence 'nominal + "that" + "ed"-form' signals a nominal head and the beginning of a relative clause. Hence the following rule selects the relative pronoun reading of "that" if the previous word is in the nominative case and the next word is an unambiguous finite verb.

```
"<that>" SELECT (<Rel>)
        (-1C (NOM))
        (1C (VFIN)) ;
```

The next rule recognizes a nominative and "that" on the left hand side, an ambiguity class (past tense vs. non-finite "ed"-form) on position zero and makes some further tests on the right-hand context to ensure that the assumed finite verb reading is legitimate. (NOT 1 NPHEAD-ALL) rules out possible premodifier uses of an "ed"-form and (NOT 1 (VFIN)) checks that there is not another finite verb too close.

```
REMOVE (PCP2)
        (-2C (NOM))
        (-1 ("that" <Rel>))
        (0 (PAST))
        (NOT 0 (<*>))
        (NOT 1 NPHEAD-ALL OR (VFIN)) ;
```

At "through" we find another preposition–adverb homograph. The solution is similar to the "in" rule except that no verb list was introduced because I couldn't come up with verbs that take pre-NP "through" as an adverb.

```
"<through>" SELECT (PREP)
        (1C (DET) OR (A) OR (ACC) OR (N) OR (ABBR)
                 OR (NUM) OR (GEN) OR (PCP1)) ;

END
```

With these rules we have the beginning of a possible grammar. The next step is to apply the grammar to a manually disambiguated benchmark corpus, in this case of about 122,662 words and containing 80,446 superfluous morphological analyses (i.e. alternative analyses without the tag <Correct!>). The software generates application statistics, so that, about a minute later, the constraints look like this:

```
CONSTRAINTS

REMOVE (SUBJUNCTIVE) # 28     8325     3     0.0004
        (NOT *-2 CS-that) ;

REMOVE (INF) # 31     9103     19     0.0021
        (NOT *-1 INFMARKER) ;
```

Each rule that was applied to the benchmark corpus is followed by a comment starting with the hash sign # and four figures: (i) the line number of the rule in the grammar file, (ii) the number of correct predictions, (iii) number of false predictions and (iv) the ratio (false predictions / all predictions). Note that one rule application may well discard more than one reading.

Overall, 43 correct analyses were discarded, but also the effect of the grammar was quite good: almost 48% of all superfluous morphological analyses were discarded.

Let us examine some cases where certain rules mispredicted. Here is a sample output:

```
"<among>"
        "among" <Correct!> PREP
"<other>"
        "other" PRON NOM SG
        "other" <Correct!> DET POST SG/PL
"<beings>"
        "being" <Correct!> N NOM PL
"<lest>"
        "lest" <Correct!> <**CLB> CS
"<he>"
        "he" <Correct!> PRON PERS MASC NOM SG3 SUBJ
"<be>" D:28!
        "be" <SV> <SVC/N> <SVC/A> V INF
        "be" <SV> <SVC/N> <SVC/A> V IMP VFIN
"<conceived>"
        "conceive" <Correct!> <SVO> <SV> <P/of> PCP2
        "conceive" <SVO> <SV> <P/of> V PAST VFIN
"<as>"
        "as" <Correct!> PREP
        "as" <**CLB> CS
        "as" ADV AD-A>
```

Rule applications are marked with "D:" followed by the rules' line numbers. Here the rule on line 28, namely

```
REMOVE (SUBJUNCTIVE) # 28     8325     3     0.0004
        (NOT *-2 CS-that) ;
```

discarded the correct reading (as indicated by the exclamation mark in "28!"). The reason is obvious: subjunctives are legitimate also in clauses starting with the conjunction "lest". Therefore we add another condition to the rule:

```
REMOVE (SUBJUNCTIVE) # 28     8325     3     0.0004
        (NOT *-2 CS-that OR ("lest" CS)) ;
```

Consider the next error:

```
"<it>"
        "it" <Correct!> <NonMod> PRON NOM SG3 SUBJ
        "it" <NonMod> PRON ACC SG3
"<has>"
        "have" <Correct!> <SVO> <SVOC/A> V PRES SG3 VFIN
"<helped>" D:59
        "help" <Correct!> <SVO> <SV> <InfComp> <P/with> PCP2
"<unify>" D:28, 31!
        "unify" <SVO> <SV> V IMP VFIN
        "unify" <SVO> <SV> V PRES -SG3 VFIN
"<the>"
        "the" <Correct!> <Def> DET CENTRAL ART SG/PL
```

Here the rule

```
REMOVE (INF) # 31      9103      19      0.0021
       (NOT *-1 INFMARKER) ;
```

has discarded a contextually legitimate infinitive. The contextual licence is a verb that acts as an infinitive complementizer and actually this property is indicated with the auxiliary tag <InfComp>. What we need to do is add this tag to the set INFMARKER.

Another error, again due to the infinitive removal rule, is:

```
"<but>"
        "but" PREP
        "but" ADV
        "but" <Correct!> CC
"<do_>"
        "do_" <SVO> <SVOO> <SV> V PRES -SG3 VFIN
        "do_" <Correct!> <SVO> <SVOO> <SV> V IMP VFIN
"<_n't>"
        "not" <Correct!> NEG-PART
"<do>" D:28, 31!
        "do" <SVO> <SVOO> <SV> V PRES -SG3 VFIN
        "do" <SVO> <SVOO> <SV> V IMP VFIN
"<it>"
        "it" <NonMod> PRON NOM SG3 SUBJ
        "it" <Correct!> <NonMod> PRON ACC SG3
"<again>"
        "again" <Correct!> ADV
```

Here the error looks rather trivial: the set INFMARKER contains the base form "do" but not the 'base form' "do_" that is given to enclitic forms split into parts by the tokenizer. In other words, we have to make another addition to the set: "do_". (Actually, these two corrections accounted for all the mispredictions this rule made in the corpus.)

A final example is:

```
"<of>"
        "of" <Correct!> PREP
"<what>"
        "what" <Correct!> <NonMod> <**CLB> PRON WH SG/PL
        "what" <**CLB> DET PRE WH SG/PL
"<can>" D:64!
        "can" N NOM SG
"<be>" D:26
        "be" <Correct!> <SV> <SVC/N> <SVC/A> V INF
        "be" <SV> <SVC/N> <SVC/A> V IMP VFIN
"<regarded>"
        "regard" <Correct!> <as/SVOC/A> <SVO> <SV> PCP2
        "regard" <as/SVOC/A> <SVO> <SV> V PAST VFIN
"<as>"
        "as" <Correct!> PREP
        "as" <**CLB> CS
        "as" ADV AD-A>
```

Here a verb deletion rule was at fault:

```
REMOVE (V) # 64      2925      8      0.0027
        (*-1C (DET) OR (GEN) OR (PREP) BARRIER NPHEAD-ALL
              LINK NOT 0 (INDEP)) ;
```

Pronouns were missing from the set NPHEAD-ALL. The error is corrected by adding the tag PRON to the set NPHEAD-ALL. It is useful to also add several other cautions to this rule, e.g. one to prevent the rule from applying in the case of words with a modal auxiliary analysis as an alternative: (NOT 0 (AUXMOD)). Modals are simply too predominant uses to be risked by a rule like this, so we added the context condition.

Overall, the grammar needed no more than seven corrections to reduce the number of mispredictions on the benchmark corpus to zero. To see how this few hours' worth performs on new text, the corrected grammar was applied to another corpus of about 56,000 words. This corpus is similar in form to the 122,000-word corpus we used in the development of the grammar. However, in the disambiguation of this test corpus, two linguists were used instead of one to reduce the amount of possible human errors and idiosyncrasies to a minimum. The annotation (originally carried out for another evaluation) was carried out as follows: first each annotator disambiguated all words of the corpus alone, using a written annotator's manual. Then the results were compared automatically and the differences were identified (about 0.7% of all words). Then the annotators jointly examined the differences and actually reached an agreement about the correct analysis in virtually all cases (Voutilainen 1995b). As a result, one of the annotated corpora was corrected to reflect the consensus. (For details about a similar annotation experiment, see Voutilainen and Järvinen 1995).

Below you find the grammar that was applied to the 56,000 word test corpus, incorporating the changes made after the first test (described above). The quality of the grammar's predictions on both test corpora are included as (automatically generated) comment lines.

```
# Experimental Constraint Grammar
# Rules for morphological disambiguation

# Atro Voutilainen 5/6/1996

DELIMITERS = "<$.>" "<$!>" "<$?>" "<$HEAD>" "<$2-NL>" ;

SETS

LIST PUNCT = "<$">" "<$.>" "<$!>" "<$?>" "<$HEAD>" "<$2-NL>"
             "<$'>" "<$`>" "<$,>" "<$;>" "<$:>" "<$->" "<$)>"
             "<$(>" "<$/>" "<$[>" "<$]>" "<${>" "<$}>" ;
LIST <<< = <<< ;
LIST >>> = >>> ;
LIST CS-that = ("that" CS) ;
```

```
LIST INFMARKER = AUXMOD "do" "do_" INFMARK> <InfComp> ;
LIST NOM = NOM ;
LIST PLURAL = PL SG/PL PL1 PL2 PL3 ;
LIST NSG = (N SG) ;
LIST NOMINAL = A N PRON NUM PCP1 PCP2 ABBR ;
LIST INDEF = <Indef> ;
LIST V-cum-IN/ADV = "take" "bring" ;
LIST NON-ADV = A N NUM PRON DET ABBR V PCP1 PCP2 CC CS I
               INFMARK> PREP ;
LIST NON-ABBR/NUM/ADV = CC A N PRON DET V PCP1 PCP2 CS I
                        INFMARK> PREP NEG-PART ;
LIST NPHEAD-ALL = ACC NUM NOM INDEP PCP1 A PCP2 PRON ;


# Key:
# - Lines starting with "#" report the performance of the
#   corrected rule on the 122,000-word training corpus
# - Lines starting with "##" report the performance of the
#   rule on the 56,000-word test corpus
# - The first number indicates how often the rule fired.
#   The second number indicates how often the rule removed
#   the correct analysis.

CONSTRAINTS

# "Reliable" rules:
#  - rules that apply at least 250 times to the training corpus
#    and make no mispredictions on it.

# 10855     0 (see Key above)
## 6292     1
REMOVE (IMP)
        (NOT -1 (CC) OR >>> OR PUNCT OR (ADV) OR (ORD)) ;

# 8712     0
## 3482     1
REMOVE (INF)
        (NOT *-1 INFMARKER) ;

# 8330     0
## 4985     1
REMOVE (SUBJUNCTIVE)
        (NOT *-2 CS-that OR ("lest" CS) OR ("whether" CS)) ;

# 2926     0
## 1373     4
REMOVE (V)
        (*-1C (DET) OR (GEN) OR (PREP)
              BARRIER NPHEAD-ALL OR ("there")
              LINK NOT 0 (INDEP))
        (NOT -1 (CC))
        (NOT 0 (AUXMOD)) ;
```

```
# 2604     0
## 870     6
"<in>" SELECT (PREP)
        (NOT -1 V-cum-IN/ADV)
        (1C (DET) OR (A) OR (ACC) OR (N) OR (ABBR)
                 OR (GEN) OR (NUM) OR (PCP1)) ;

# 1494     0
## 559     0
REMOVE (PAST)
        (*-1C ("be" VFIN) OR ("have" VFIN) BARRIER NON-ADV)
        (0 (PCP2)) ;

# 1078     0
## 558     1
REMOVE (<Rare> INF)
        (NOT -1 INFMARKER OR (CC))
        (NOT -2 INFMARKER OR (CC)) ;

# 588     0
## 251     0
REMOVE (<Rare> IMP)
        (NOT -1 >>> OR (CC))
        (NOT -2 >>> OR (CC)) ;

# 265     0
## 92     0
"<had>" REMOVE (PCP2)
        (NOT *-1 ("be") OR ("have")) ;


CONSTRAINTS

# Level 1 heuristics:
# - rules that apply less than 250 times to the
#   training corpus but make no mispredictions on it.

# 35     0
## 30     0
REMOVE (PRES -SG3)
        (-2 >>>)
        (-1 NOM)
        (NOT -1 PLURAL OR (SG1) OR (SG2/PL2)) ;

# 83     0
## 82     0
"<a>" SELECT (DET)
        (-1 >>>)
        (1C NSG OR (ABBR SG) OR (A)) ;
```

```
# 51      0
## 24     0
REMOVE (PRES SG3)
        (-2C (PREP))
        (-1C (A) OR (PCP2) OR (GEN) OR (DET) OR (ADV))
        (NOT -1 (<*>) OR (<Nominal>) OR (<?>))
        (0 (N PL))
        (NOT 1 (DET) OR (ACC) OR (GEN) OR (A) OR (NUM)) ;

# 135     0
## 69     0
"<that>" SELECT (<Rel>)
        (-1C (NOM))
        (1C (VFIN)) ;

# 30      0
## 6      0
REMOVE (PCP2)
        (-2C (NOM))
        (-1 ("that" <Rel>))
        (0 (PAST))
        (NOT 0 (<*>))
        (NOT 1 NPHEAD-ALL OR (VFIN)) ;

# 107     0
## 43     0
"<through>" SELECT (PREP)
        (1C (DET) OR (A) OR (ACC) OR (N) OR (ABBR)
                  OR (NUM) OR (GEN) OR (PCP1)) ;

END
```

The 56,000 word corpus contained 44,445 superfluous morphological analyses, of which 19,528 (43.9%) were discarded. As for errors, the grammar discarded only 14 morphological analyses marked as correct.

Let us have a closer look at the mispredictions. Of the 14 errors, ten were due to two rules. The rule

```
"<in>" SELECT (PREP)
        (NOT -1 V-cum-IN/ADV)
        (1C (DET) OR (A) OR (ACC) OR (N) OR (ABBR)
                  OR (GEN) OR (NUM) OR (PCP1)) ;
```

discarded six correct analyses; all of them were due to the incompleteness of the verb list V-cum-IN/ADV (the verbs “throw”, “lay”, “add”, “hand” and “send” should be added).

Four correct analyses were discarded by the rule

```
REMOVE (V)
        (*-1C (DET) OR (GEN) OR (PREP)
              BARRIER NPHEAD-ALL OR ("there"))
        (NOT *-1 (INDEP))
        (NOT -1 (CC))
        (NOT 0 (AUXMOD)) ;
```

Two of these errors were due to a genitival subject ("Moody's said..", where "said" lost a past tense reading) and the other two were due to a deferred preposition, cf. <make> in the next sample:

```
"<are>"
        "be" <Correct!> <SV> <SVC/N> <SVC/A> V PRES -SG1,3 VFIN
"<now>"
        "now" <Correct!> ADV
"<relied>" D:70
        "rely" <Correct!> <SV> <P/on> PCP2
"<upon>"
        "upon" <Correct!> PREP
"<to>"
        "to" PREP
        "to" <Correct!> INFMARK>
"<make>" D:41, 51, 56!
        "make" <NRare> N NOM SG
"<defeated>"
        "defeat" <Correct!> <SVO> PCP2
        "defeat" <SVO> V PAST VFIN
"<iraq>"
        "iraq" <Correct!> <*> <Proper> N NOM SG
"<behave>" D:41, 51
        "behave" <Correct!> <SV> <SVO> V INF
        "behave" <SV> <SVO> V PRES -SG3 VFIN
```

The remaining four errors were made by four different rules.

14.4.2 Experiences with novices: NorFa'95 CG 'competition'

A possible objection to the apparently promising performance of the above sample grammar is that it was written by a linguist with a considerable experience in writing disambiguation grammars. In order to show how an absolute beginner might do, we now present a report of an experiment with twenty beginners.

In May 1995, there was a Nordic summer school on Natural Language Parsing at the University of Helsinki (NorFa'95). During daytime, the students attended lectures on various parsing approaches. In the evenings, they had a chance to experiment with writing constraint grammars using real data and software. They were provided with:

- a short written introduction to the CG rule formalism and on using the CG parsing software;

Table 14.2 Training and benchmark corpora.

	Train	Benchmark
Words	2338	2419
Errors	0	0
Error-%	0%	0%
Superfl. readings	2085	1927
Readings/word	1.89	1.80

- 2,300 words of morphologically analysed English text where the correct readings were marked;
- a short documentation of the tags and descriptive policies used in the English description.

Their task was to write disambiguation rules, test them on the corpus and improve the rules so that they would also apply with an error margin of less than 1% to other texts. Toward the end of the course, the grammars were submitted to the author, who measured their effect on another benchmark corpus of about 2,400 words (that the students never saw) and reported the results.

The experiment was organized as a kind of competition, with a prize (Karlsson *et al.* (eds.) 1995) for the members of the winning team. The twenty participants were divided into seven teams, so that seven grammars were submitted. None of the participants had any previous experience with writing a Constraint Grammar. According to their own judgements, their competence of English varied from fair to excellent; none claimed a native-level command of the language. The participants were mostly Nordic and Baltic doctoral and masters students of Computational Linguistics.

In tables 14.2 and 14.3 we present detailed information about the corpora and results. The following terms are used in the tables:

- Disambiguation rate (Dis-%): percentage of superfluous readings that were discarded by the grammar.
- Error rate (Error-%): percentage of words that lost the correct reading.
- Hours: hours spent on studying the formalism and writing the grammar.

It is interesting to note that even with the minimal amount of training data and the inexperience of the grammarians, the error rate of even the worst grammar on the benchmark corpus was well under one per cent. Some of the grammars were also

Table 14.3 Results on benchmark corpus (grammars are in a random order).

Grammar	1	2	3	4	5	6	7
Hours	12	?	?	14	?	10	12
Rules	23	19	25	50	57	27	37
Errors	20	4	7	14	7	17	8
Error-%	0.83	0.17	0.29	0.58	0.29	0.70	0.33
Dis-%	24.6	29.6	53.1	63.5	48.6	59.3	45.2
Read/w	1.60	1.56	1.37	1.29	1.41	1.32	1.44

reasonably effective: five of them did about half of the job; the most effective one discarded almost two thirds of the extra readings in the benchmark corpus, with an error rate of 0.58%.

It may also be interesting to know that at least two of the participants have written large-scale lexicons and CGs for two other European languages since then, i.e. at least some of the students seem to have considered the rule-based approach to tagging a worthwhile alternative.

14.5 GENERAL OBSERVATIONS

Some general observations can be made about writing rules.

- Try to make the rules as general and effective as possible. This not only saves you from coming back to the same phenomenon all too often. More importantly, a general constraint will have more applications on your training corpus, i.e. you will get more reliable information about the quality of the rule.

- Heuristics is not a dirty word. Unlike most non-computational linguists, the author of a computational (disambiguation) grammar experiences the consequences of the proposed grammar. If the goal of the computational system is to reduce ambiguity considerably, it is sometimes useful to ignore some grammatical possibilities in the language (i.e. to make generalizations that ignore some possible though rare constructions) to get more disambiguation done.

- Given that the grammarian has sufficiently large tagged corpora available, a useful grammar can be written in a reasonably short time. A month's rule-writing should easily result in a grammar of over 1,000 well-tested constraints.

- The first 200 constraints will probably suffice for resolving well over a half of all initial ambiguities; perhaps even something like three quarters. This means that after the first few days it becomes increasingly difficult to write new constraints that are nearly as effective as the first ones. However, there is a good chance that even the less effective new constraints are reliable in their predictions. This may be due to the increasing tendency of lexicalization in new constraints: once the most common tag-level regularities have been expressed, the grammarian is likely to pay more attention to the combinations of words and tags (rather than of tag sequences alone).

- A realistic disambiguation grammar can contain thousands of constraints. Managing a large grammar – in particular, preventing undesired rule interactions – may seem difficult, but in practice is possible if sufficient attention is paid to making certain that each constraint is sufficiently reliable. One way to increase the reliability of the constraint is to control the (un)ambiguity of the context, e.g. by using the 'careful' mode explained in the above section on the rule formalism. Another desirable facility in the grammar development environment would be a mechanism for identifying pairs of constraints that contradict each other.

- On the basis of new corpus evidence, it usually is quite easy to modify a constraint, mainly by adding new context conditions. A tentative constraint typically expresses, with a couple of context conditions, the kind of construction the grammarian imagined the ambiguity to occur in; further context conditions, added on the basis of observed mispredictions, make this statement more specific.

- Though much ambiguity can be resolved with local rules (whose context conditions can refer, e.g., one or two words to the left or right), it is also obvious that many useful generalizations can (and should) be made about longer tag sequences.

- Writing (and evaluating) a grammar is easier for a tagset that is based on well-defined surface-grammatical distinctions (rather than, e.g., distinctions representing a mixture of grammar, semantics and world knowledge).

14.6 REMAINING AMBIGUITY

A disambiguation grammar that only rarely discards the correct analysis is likely to leave some ambiguities unresolved. There are four types of remaining ambiguity:

- Ambiguities due to an incomplete grammar, i.e. these ambiguities could be (easily) resolved by writing new rules.

- Structurally genuine ambiguities, i.e. morphologically (and syntactically) globally ambiguous sentences like "Time flies like an arrow".

- Ambiguities that cannot be resolved because of some limitations in the parser implementation (for a discussion, see e.g. Voutilainen 1994).
- Ambiguities whose knowledge-based resolution would require a more distinctive grammatical representation, e.g. the use of syntactic (rather than only morphological) descriptors (cf. Voutilainen 1994).

For instance, the 1,150 constraints in the original EngCG grammar leave an average of some 1.05-1.09 readings per word in the output (in the input, there are some 1.7-2.2 readings per word).

Several techniques have been proposed for resolving the remaining ambiguities.

14.6.1 Using statistical models

A combination of rule-based and statistical components is an obvious possibility. A statistical tagger can be trained for the tagset using the hand-annotated (and perhaps also the slightly ambiguous output of the rule-based tagger) as the basis of the statistical language model. If we use the statistical tagger as the final part of the system, it can be applied to those ambiguities left pending by the rule-based system.

One such hybrid is documented by Tapanainen and Voutilainen (1994) where the rule-based EngCG tagger is combined with the Xerox English tagger; in an evaluation, the hybrid reached an accuracy of about 98.5% with fully disambiguated output. As a reservation, one should point out that the ambiguities surviving a well-written grammar are probably very difficult for a statistical component as well. Sometimes hand-written coarse heuristics may turn out to be preferable (Chanod and Tapanainen 1995a).

Another unexplored option might be to use a data-driven rule-learner of the Hindle (1988) or Brill (see the following chapter) type as a first step and then refine the rules manually.

14.6.2 Using collocational information

Many otherwise difficult ambiguities could be resolved by enriching the lexicon with collocational information. For instance, consider the following fragment from a sentence:

```
"<in>"
        "in" PREP
"<the>"
        "the" <Def> DET CENTRAL ART SG/PL
"<piston>"
        "piston" N NOM SG
"<bores>"
        "bore" N NOM PL
        "bore" <SVO> <SV> <P/with> V PRES SG3 VFIN
"<$.>"
```

Here "bores" remains noun–verb ambiguous. However, it seems obvious that it is a noun because it constitutes a modifier–head expression with "piston". If large lists of, e.g., modifier–head expressions were given in the lexicon, many ambiguities like this could be resolved reliably. This kind of useful information can be extracted efficiently from large texts with less-than-perfect taggers, e.g. the modifier–head expression "piston bore" could be identified from some other part of the same text (or from another text) where it occurs in a fully disambiguating context, as shown by Voutilainen (1993).

A likely problem of this approach is that the list of collocations to be generated probably has to be very large – perhaps hundreds of thousands of items – to be practically useful.

14.6.3 Using a syntactic parser

As pointed out above, many morphological ambiguities could be resolved if the linguistic representation were more informative. The idea of using a syntactic representation (i.e. a syntactic parser) in resolving the most difficult morphological ambiguities left pending by the EngCG tagger was investigated in Voutilainen (1995b). In the experiment, the ambiguous output of the EngCG tagger was enriched with syntactic ambiguities which were then resolved using a finite-state syntactic disambiguator. The syntactic disambiguator was allowed to produce only one analysis per sentence. From this unambiguous output, the correctness rate at the morphological level was calculated by comparing it with a double-blind annotated benchmark corpus of over 30,000 words. The correctness rate of the combined EngCG + syntactic disambiguator was slightly above 99% on this text (with no remaining morphological ambiguity), so syntactic information appears to be a worthwhile contributor.

There are obvious drawbacks in this method, too: (i) developing a syntactic parser may require a lot of effort and expertise and (ii) a syntactic parser may be so slow that the resulting tagger, though possibly accurate, might be intolerably slow.

14.6.4 Using observed local regularities

A final possibility is the application of learning methods for improving the tagger's output. The idea is simple: if the predictions of the rule-based tagger are reliable, then regularities could be learned from the fully disambiguated parts of the texts and applied for resolving the remaining ambiguities in the same text. In other words, a learner could be incorporated in the rule-based tagger and the learned generalizations could be applied e.g. when the hand-crafted rules alone predict no more.

Consider a small example discussed in Voutilainen (1995a): in a heterogeneous corpus like Brown or LOB, about 40% of the instances of the wordform "cover" are nouns; the rest are verbs. However, in a single text, this prediction might be poor, e.g. in a car maintenance manual, almost all instances of "cover" might be nouns. How could a rule-based disambiguator indicate this? An early version of EngCG was used for analysing a manual text with 86 instances of "cover": 66 of them were fully disambiguated as nouns; the remaining 20 instances remained ambiguous, i.e. as far as EngCG was concerned, either option was possible in these 20 cases. However, the disambiguating predictions EngCG did make may suggest that "cover" is used solely as a noun in this text, so it might be a good guess (and it was) to disambiguate all remaining instances of "cover" as a noun in this text (and let EngCG carry on the disambiguation of the context of the newly-disambiguated instances of "cover").

15 CORPUS-BASED RULES

Eric Brill

15.1 INTRODUCTION

As we have seen in the previous chapter, we can use rules to select an appropriate tag for each token. We will continue investigating the use of rules in this chapter. However, where in the previous chapter the rules were created manually, based on someone's linguistic knowledge and familiarity with properties of the corpus, we will explore the possibility of learning tagging rules automatically. A potential advantage of automatic rule learning is that such a system could in theory be highly portable, both across domains and across languages. If training material is available, the systems can be retrained with little or no human intervention. A limitation of this approach is that such systems can only learn facts that can be described within the prespecified descriptive language of the learner, which limits the types of rules that can be learned. For example, a person might discover that a word tends to be tagged with one particular tag when it is toward the end of a sentence. If the learner did not have access to the concept of sentence length and position in a sentence, discovering such a heuristic rule would be beyond the capability of the learning algorithm. One thing that differentiates this approach from other machine learning approaches such as training neural networks (cf. Chapter 17) or hidden Markov models (HMMs; cf. Chapter 16) is that the learned information will be in a form suitable for people to understand, edit, improve, etc., just as is the case for manually written rules.

We will first discuss the acquisition of tagging rules using supervised learning, where a manually tagged corpus is available to be used as a 'gold standard' to guide learning.

H. van Halteren (ed.), Syntactic Wordclass Tagging, 247-262.

Tagged corpora are becoming readily available for a number of different languages and genres. Later we will discuss learning without using a manually tagged corpus.

15.2 LEARNING RULES

The idea of using rules for tagging goes back at least to Klein and Simmons (1963), Harris (1962) and Greene and Rubin (1971). There, the system had a lexicon listing all possible tags for each word. Words were initially assigned a set of tags, namely the set of allowable tags for the word as indicated in the lexicon. Then rules were applied to delete tags from the list of possible tags for a word, based on the context that word occurred in. For example, if a word can be a noun or verb and that word appears between a determiner and a verb, the verb possibility should be deleted. For instance:

The/Det race/Noun_or_Verb ended/Verb → The/Det race/Noun ended/Verb

These rules were written manually. As this work predated the existence of large manually tagged corpora, learning these rules automatically was not a possibility at that time. Instead, rule creation was empirically, but manually, driven by writing rules and then examining the quality of the output. If mistakes were made by the system, the rule set was augmented in an attempt to overcome these problems. Not having a gold standard corpus also made it difficult to compare different taggers and to rigorously assess improvements made to a tagger.

With the advent of manually tagged corpora it has become possible to begin exploring methods for automatically learning how to disambiguate tags in context. Initially, virtually all work in the area of automatic tagger creation involved HMM-based taggers (cf. Chapter 16). More recently, people have begun to address the question of whether rule-based methods, such as the taggers originally developed by hand, can learn rules automatically from corpora.[1] Rule-based methods offer a number of potential advantages over HMM-based methods. In the case of sequential rules, the result is the outcome of applying each rule, in order, to the input. In a probabilistic system, all possible outcomes are assigned a probability and the result is found by searching for the tag sequence with highest probability. For tagging, this can be done in linear time using dynamic programming, but there is still search involved in the tagging process to find the most probable tag sequence. Because there is no search involved in applying rule sequences, a rule-based tagger can be made to run very quickly. For instance, in Roche and Schabes (1995) it is shown that a rule sequence for wordclass tagging can be converted into a deterministic finite state transducer, which is optimal in the sense that one state transition per word is all that is needed for tagging, independent of the number of rules. Because for many people a set of simple rules can be easier

[1] Many other learning approaches have also been explored recently for wordclass tagging, including neural networks, decision trees and case-based learning (cf. Chapter 17).

to understand than a large table of probabilities, a rule-based system may be easier to understand and improve. For example, with rule-based systems the lexical information is typically represented simply as a list of allowable tags for a word, possibly with the most likely tag marked. In HMM taggers, lexical information is stored as a set of probabilities $P(word|tag)$. This is the encoding of lexical information about the word "half" used by a rule-based system, extracted from the Penn Treebank tagged corpus:[2].

```
half: CD DT JJ NN PDT RB VB
```

and this is the encoding of lexical information about "half" used in an HMM tagger:[3]

```
P(half | CD)  = 0.000660
P(half | DT)  = 0.000757
P(half | JJ)  = 0.000092
P(half | NN)  = 0.000702
P(half | PDT) = 0.039945
P(half | RB)  = 0.000443
P(half | VB)  = 0.000027
```

Much of the disambiguation information does truly seem more amenable to a rule-based approach than an approach such as an HMM-based tagger. For example, the rule that a word between a determiner and a verb is most certainly a noun can be captured concisely through a single simple rule. In an HMM, this information would be captured much more circuitously, through the relative probabilities of a large set of parameters.

Corpus-based rule-learning systems initially had the disadvantage that unsupervised learning was not possible, compared to well-known techniques for training HMMs without a manually tagged text. In addition, because HMMs assign probabilities to all tag sequences, it is possible to return multiple tags for a word when two tags have very close probabilities. Recently, methods have been developed to do both unsupervised tagging and n-best tagging in the framework of automatic rule acquisition.

15.3 PARSER-BASED WORDCLASS DISAMBIGUATION

One of the early attempts at automatic rule-learning for wordclass tagging is described by Hindle (1989). He had developed a deterministic parser for English and initially had a set of hand-coded rules for tagging within the parser. These rules took two years to develop manually. He was able to devise an algorithm to automatically learn tagging rules and found that the error rate of this automatically trained system was half that of the hand-written rules. Text was parsed using a deterministic parser. Parsing rules

[2] A full listing of the Penn Treebank tagset can be found in Appendix 17.6

[3] One could encode the lexical information in the more intuitive but mathematically less correct form $P(tag|word)$ (cf. 16.3.1). This would result in a less obfuscated lexicon, but also typically results in a degradation in tagging accuracy.

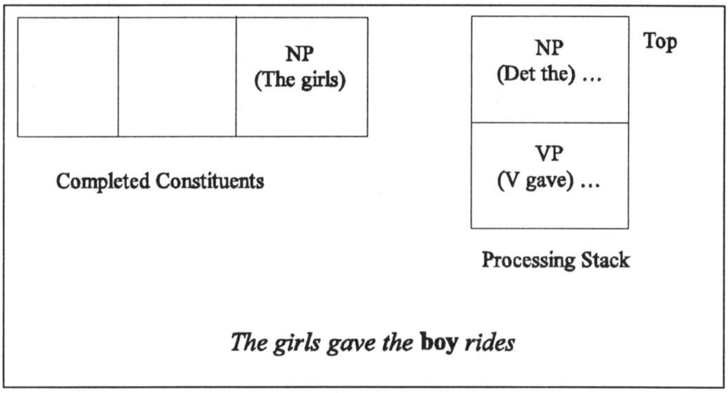

Figure 15.1 Parser state before processing "boy".

in this system are pattern–action rules, where a pattern is a description of the current parser state and an action builds a constituent. A state in parsing is represented by the current incomplete constituent being built and up to three previously completed constituents. For example, in parsing the sentence "The drivers gave the boy rides", when processing the word "boy", the parser might look like figure 15.1.

In this example, the state of the parser is captured by the one completed constituent in the buffer and the incomplete noun phrase at the top of the processing stack, as well as the next word to be processed. Given the state above, the next rule to apply might be to add the word "boy" to the constituent on top of the stack and advance the input by one. Then the parser would complete the top constituent and shift it to the completed constituent buffer. The state of the parser at this point is shown in figure 15.2.

Next, the parser would push a noun phrase onto the stack and pop the completed noun phrase "rides" onto the completed constituent buffer. Given the state of the parser above, it should be clear that the word "rides" is likely to be a noun and not a verb. A tagger looking only two words back would not be able to correctly make this decision. The question is how an effective set of rules can be learned automatically to decide on the wordclass of a word, based on the state of the parser when that word is encountered.

For wordclass tagging, the system initially is provided with general rules which assign words a default tag, ignoring all context. While parsing text for which correct wordclass labels are already available, the system can check whether the parser assigns the correct wordclass to each word, given the current set of disambiguating rules. If the parser assigns the improper tag to a word, then the system attempts to acquire a new rule. In order to assign the proper tag in this context, a new rule is needed which describes the current state of the parser and which is more specific than any other rule which applies to this state. Since the most specific rule is chosen to apply, any new rule will only apply in this context if it is more specific than all other possible rules.

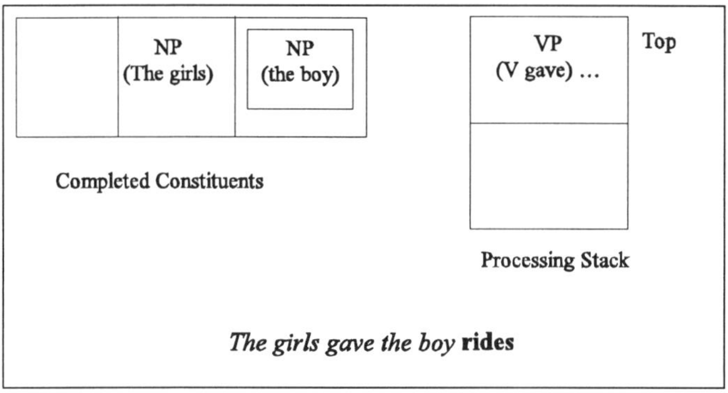

Figure 15.2 Parser state after processing "boy".

In certain cases, it will be impossible to acquire a new rule that will properly tag a word given a particular parser state. This can occur if there already exists a most specific rule for the parser state and that rule assigns the incorrect wordclass. For each rule, the parser keeps track of the number of times it results in the correct tag being assigned, as well as the number of times the incorrect tag is assigned. If a rule ever results in the percentage of correct tags dropping below some threshold, that rule is then deactivated.

When this method was used to learn tagging rules, 35,000 rules were acquired, with a tagging error rate of 3%, which is half the error rate achieved when using hand-coded rules. This was one of the first strong indications that automatic acquisition of linguistic information might result in faster system development and more accurate systems than a system purely built by hand.

15.4 TRANSFORMATION-BASED LEARNING

Another approach to learning rules automatically using supervised learning is called transformation-based learning (Brill 1992, 1995). The rules learned are similar to those created manually in Klein and Simmons (1963), but instead of having rules remove possible tags from a list of tags, in this system each word is assigned exactly one tag and rules change tags based on context. This differs from the earlier manually created rule-based systems in that throughout the tagging process, as rules are applied, every word will have precisely one tag, compared to the variable number of tags in the early rule-based approaches. For instance, a rule might change the tag of a word from Verb to Noun if the word immediately follows a determiner. In this learning paradigm, an ordered set of rules is acquired.

The general architecture of transformation-based learning is illustrated in figure 15.3. Unannotated text is first processed by an initial state annotator. This annotation pro-

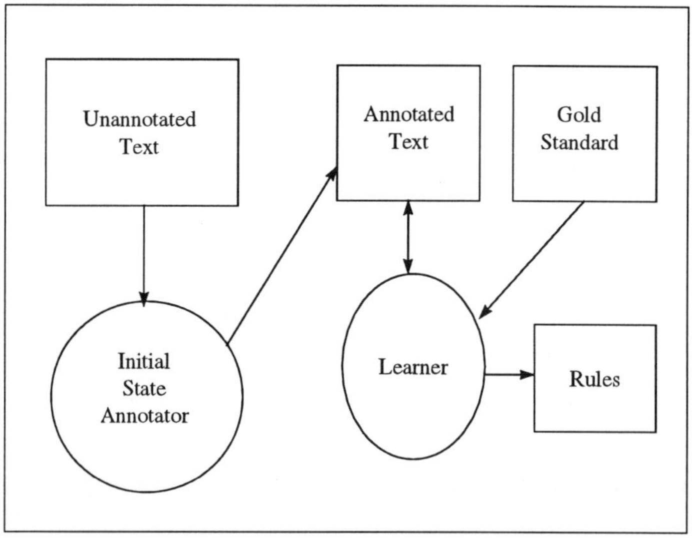

Figure 15.3 Transformation-Based Learning.

gram assigns some initial guess as to the proper annotation for the text. Because the goal of this approach is to minimize the amount of human effort needed to create a tagger, the initial state annotator is typically very simple and quick to create. For supervised learning of rules for tagging, the initial state annotator assigns every word its most likely tag. This can be determined from the training corpus, at least for all words that occur in the training corpus. But this annotator can be any program that takes in unannotated text and outputs annotated text. For instance, learners can be chained together by having the output of one system serve as the initial state annotation for another system. It is possible to have the output of another wordclass tagger serve as the input to a transformation-based tagger. In this case, the transformation-based tagger will learn rules to correct errors made by the initial tagger.

The learning module has a set of rule templates prespecified. Some examples of templates that might be useful for tagging are:

```
Change a tag from X to Y if:
  the previous tag is P
  the next tag is P
  the previous word is W
  the next word is W
  the tag preceding this word by two is P
  the tag following this word by two is P
  the following two words are W and Q
  the preceding two tags are P and R
```

Transformations are restricted to change a tag only to a tag that is an allowed tag for a word, as indicated in the lexicon. It is up to the person developing the system to choose the templates, but not to choose the specific rules that will be used. It is important here that the developer throws a big enough net to catch a large percentage of available cues for deciding on the tag of a word. If the developer only provides the learner with the single template:

```
Change a tag from X to Y if:
  the word three positions to the right has N letters
```

it is unlikely that the learner will be able to learn many effective rules from the set of possible instantiations of this template. However, adding this template to the set of reasonable templates above is unlikely to hurt performance. If there are no good rules to be learned of this type, then no rules of this type will be learned. So other than affecting the efficiency of running the learner, a developer should be quite liberal in deciding what transformation templates to provide to the learner.

In one learning iteration, the learner in essence searches through all possible instantiations of the rule templates it has been provided with, applying each instantiation to the corpus and scoring the resulting tagged corpus. The rule that gives the greatest increase in performance is then appended to the end of the learned rule list. In each learning iteration, a rule is being learned that can be applied to the imperfectly annotated text to bring it closer to the gold standard annotation. It is appended to the end of the learned rule list, applied to the learning corpus and another iteration of rule learning continues on this updated corpus. Learning halts when no further rules can be learned that improve the tagging accuracy of the training corpus or when the improvement a rule makes drops below a threshold. When comparing rule number to training set accuracy, one typically gets a skewed graph, with initial rules reducing the error rate significantly and a large tail of rules at the end which improve accuracy only marginally. The graph in figure 15.4 shows a typical curve of tagging accuracy on a test set as a function of the number of rules learned. As can be seen from this graph, the first few rules account for most of the performance gain and then the tagging accuracy begins to asymptote.

This method is quite robust against overtraining. The rules that are learned early improve performance significantly on the training set. Therefore, the conclusion that these rules will be beneficial is based on a large sample and this conclusion is likely to generalize well to new data. For rules that occur infrequently, the certainty that such a rule will be beneficial on new data is not very great. For instance, a particular rule might result in one tagging error being corrected in the training set. Of these rules with positive scores based on small counts, some might actually result in worse performance on a new data set. However, if the training set is representative of the test set, it is likely that the contribution, either positive or negative, will be minimal. The bias of the set of these rules will be toward making a small positive contribution or no contribution

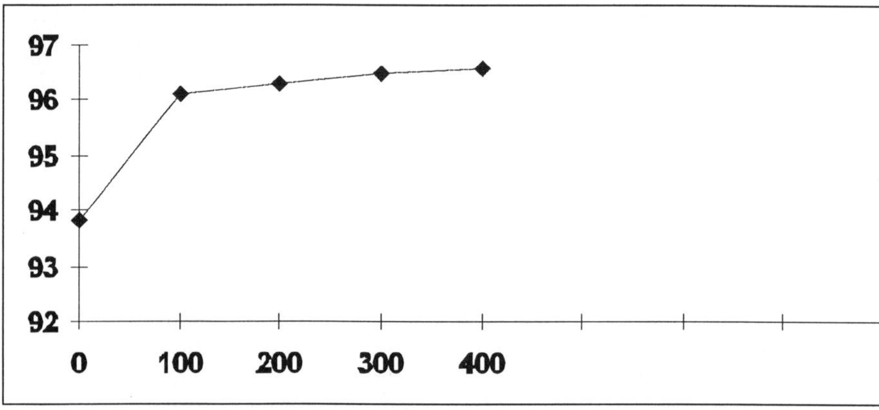

Figure 15.4 Accuracy vs. number of rules learned.

in tagging new data, since it was observed that they made a small positive contribution to the training set.

An actual implementation of the learner needs some thought. The most straightforward implementation would be to generate all possible rules (all instantiations of transformation templates), apply each rule to the corpus, score the result and learn the rule that gives the highest accuracy. The problem with this is that the number of possible rules could be unwieldy. For example, given the templates above, assuming there are 50,000 words and 50 wordclass tags, then there are more than two billion possible rules that would have to be tested by applying them to the entire corpus and scoring the result, which would not be feasible. Fortunately, there is a more efficient method for learning which only requires one pass through the corpus for each rule learned. We will demonstrate the algorithm by example, assuming we only have one template:

```
Change the tag from X to Y if the next word is Z
```

and that we have the following training corpus:

TRAIN:	The/Det	race/Noun	ends/Verb	in/Prep	a/Det	defeat/Noun
GUESS:	The/Det	race/Verb	ends/Noun	in/Prep	a/Det	defeat/Verb

For each iteration of rule learning, one pass is made through the corpus. In this example, the first word is correctly tagged as a determiner. This means that any rule of the form:

```
Change the tag from Det to Y if the following word is ''race''
```

would increase the error rate by one. So we decrement the score of all such rules by one (all rules are implicitly given an initial score of zero). Next, we move to the word "race". In this case, the rule:

```
Change the tag from Verb to Y if the following word is ''ends''
```

would result in a decrease in the error rate by one if Y=Noun and will result in no change in error rate if Y≠Noun. So at this point, we only increment the one good rule by one. Using this data-driven algorithm, we only pass through the corpus once for each iteration of the learner and only need to keep track of a small percentage of the large number of possible rules.

Here are the first five rules learned from one run of this system on English text, in particular the Wall Street Journal corpus, tagged with the Penn Treebank tags:

1	NN	→	VB	Previous tag is TO
2	VBP	→	VB	One of the previous 3 tags is MD
3	NN	→	VB	One of the previous 2 tags is MD
4	VB	→	NN	One of the previous 2 tags is DT
5	VBD	→	VBN	One of the previous 3 tags is VBZ

The first rule states that if a word is tagged as a common noun and the previous tag is TO, then the tag should be changed to a verb. In the Penn Treebank tagset, the word "to" is tagged as TO regardless of whether it is an infinitival "to" or a preposition. But since a singular noun is unlikely to immediately follow a preposition and since a verb is very likely to follow the infinitival "to", this rule will improve the overall tagging accuracy. The second rule captures the fact that if a modal exists, the main verb is untensed. The third rule captures the information that immediately following a modal, a word is much more likely to be a verb than a noun. Note that the learned rules are quite intuitive, which makes the system easy to understand.

Earlier we mentioned that a weakness of the approach of automatically learning rules is that the information sources that can be used by the learner are limited to the descriptive language the learner is provided with. Since the output of this system is something that a person can easily understand, it may be possible for manual and automatic means to be combined. A person can easily add, delete or refine a rule based on linguistic intuition, insight into the problem, or other information sources not available to the automatic learner. This might be faster than writing rules purely manually and might have the advantage that the automatic learner can uncover rules that are clear to a person after they are learned, but which might not have been uncovered manually. Below we list some additional rules that appear later in the learned list, picked randomly from the total set of 284 learned rules:

11	VB	→	VBP	Previous tag is NNS
30	VBD	→	VBN	Next word is "by"
80	RB	→	RP	Current word is "up" and previous tag is VB
139	IN	→	RP	Current word is "out" and next tag is DT
140	DT	→	RB	Next word is "longer"
172	NN	→	VBG	Current word is "operating" and next tag is IN
224	NNS	→	VBZ	Previous tag is WP
265	JJ	→	VB	Next tag is DT

Most of these rules are fairly easy to understand. One thing that makes the rules difficult to understand is that the kth rule applies after the first $k-1$ rules have been applied. A rule is context-dependent, in the sense that it depends on the state of the corpus after applying all of the preceding rules in the list. However, from the point of view of extending the tagger this is actually an advantage. A person can run the tagger and examine the output. If rules can be found manually that would fix errors in the tagger output, these rules can simply be appended to the end of the rule list, without needing to be concerned about how rules interact, since all we need to know for appending a rule to the rule list is the state of the corpus after applying all prior rules.

After learning an ordered list of transformations, new text is tagged by first passing it through the initial state annotator and next applying each of the rules in the list in order. Each rule is applied only once at every site where it can be applied and then the following rule is applied. Given R rules and a corpus of N words, the run-time of the tagger is then $O(R * N)$.[4]

When trained on sufficient data, such an approach typically achieves a tagging accuracy (measured by comparing the output of the tagger to the tags provided by people, for a set of data not used in training) of 96-98% for the Penn Treebank tags, at least when all words to be tagged are in the tagger's lexicon, with accuracies dropping by 0.5-1.0% when there are unknown words.[5] This is comparable to results achieved using other automatic training techniques for inducing a wordclass tagger from data. Various experiments have shown that accuracy seems to asymptote both for rule-based and N-gram taggers at a training corpus size of around 0.5-1.0 $*10^6$ words. In Weischedel *et al.* (1993) and Brill (1995), increasing the training size from 64,000 words to 640,000 words resulted in a 0.4% increase in tagging accuracy (e.g. from 96.3% to 96.7% in one experiment).

15.5 N-BEST WORDCLASS TAGGING

Next, we turn to n-best wordclass tagging. For some applications, it might be desirable for the wordclass tagger not to be forced to make a hard decision in assigning a tag to a word when it does not have adequate information to completely disambiguate between the allowable tags. If the tagger determines that in a particular context a word can be either a noun or a verb, it might be best to ambiguously tag the word as Noun_or_Verb, rather than being forced to commit to one or the other in cases where it knows that the degree of uncertainty is high. For example, we may wish to use tagged text to search

[4] This can actually be improved to $O(N)$, which is optimal in the sense that the run-time is linear with respect to the number of words in the input and is independent of both the number of rules and number of tags (Roche and Schabes 1995).

[5] The transformation-based approach can also be extended to handle words not encountered in the training data. See Chapter 13.

for certain linguistic patterns that can crudely be described by a regular expression on tag sequences. If we want this search to have high recall and are willing to sacrifice precision to achieve this high recall, then it would be advantageous to leave multiple tags on words in situations where a mistagging is likely. If the tagger must commit to a single tag in a context and that tag is incorrect, then this might result in the system being unable to extract this particular instance of the pattern being searched for.

When using a probabilistic tagger, the tagger can assign the probability that a word will be tagged with a particular tag according to the probability model. In this framework, n-best tagging can be done by listing all tags whose probability is sufficiently close to the highest probability tag. The idea is that the tagger is most likely to make mistakes in cases where it has to choose between two tags that are given very similar probabilities by the probabilistic tagging model. This n-best tagger is then parameterized, with the parameter representing the threshold of difference in tag probabilities for the tagger to assign multiple tags. By increasing the threshold (increasing the maximum difference in probability between alternative tags to assign multiple tags for a word), the tagger will assign more tags per word on average, but will increase the probability that the correct tag will be included in the list of tags for a word. This approach is discussed in Weischedel *et al.* (1993).

Rule-based tagging systems that reduce ambiguity, such as the rules used by Klein and Simmons (1963) or those in Constraint Grammar (cf. Chapter 14), also can handle n-best tagging very naturally. In this approach, rules remove tags from the list of possible tags for a word based on the context in which the word appears. If there are not sufficient contextual cues to fully disambiguate a word, the word will end up with more than one tag after all rules have been applied. So only those words ambiguously tagged and in contexts with sufficiently powerful cues to disambiguating will actually be disambiguated.

The transformation-based framework as presented so far does not seem to be suited very well for n-best tagging. However, n-best tagging is easily handled when we apply the framework in a slightly different way. All we have to do is to allow rules which can not only change tags but also add them. Then, after a 1-best tagger has done its best to assign the proper single tag to each word of a text, we can let such rules add tags in those contexts where there is uncertainty. To give an example, in one and the same corpus, the word "yen" can be either a singular or plural noun. In the context "... the yen ...", it is difficult to determine whether "yen" should be singular or plural. The 1-best tagger tags it as a plural noun. So the rule: 'Add the tag Singular Noun if the current word is "yen" and the previous word is "the"' would assign both tags in this ambiguous environment. To learn such rules, we can no longer simply use tagging accuracy as the objective function. If we define accuracy as: 'the word is tagged with only one tag and that tag is the correct tag', then no such rules would ever be learned. If instead we define accuracy as: 'one of the tags a word is tagged with is the correct tag', then the best rule to learn would be: tag every word with every tag regardless of context. Instead, we want to find the rule that increases the accuracy as defined by the correct tag being one of the tags for a word while minimizing the increase in the average number of tags per word. We can accomplish this by scoring each rule by:

$$\frac{Increase\ in\ Tagging\ Accuracy}{Number\ of\ Tags\ Added}$$

i.e. we search for the rule that gives the greatest accuracy improvement per new tag added to the corpus.

Once again, an ordered set of rules is learned. In this case, though, rules all add tags to words in particular contexts. So with the application of each rule, the average number of tags per word increases, as does the number of times one of the tags assigned to a word is the correct tag. This trade-off between average number of tags per word (where fewer is better) and tagging accuracy can be regulated by only applying an initial subset of the learned rules. After learning a set of rules, the n-best tagger is parameterized to only apply the first M rules. Using a small M gives a small average number of tags per word and a small increase in tagging accuracy while a large M will result in a larger number of tags per word and a greater overall tagging accuracy.

15.6 UNSUPERVISED LEARNING

Up to this point, we have only considered the case of learning rules for tagging when the learner is provided with a correctly tagged corpus, in other words when training is supervised in that it has access to a gold standard to use in learning. But what about when we wish to train a wordclass tagger but no tagged text is available, i.e. when we need unsupervised learning? Very little work has been done on completely unsupervised learning, when no human-annotated resources are used at all. In such a system, the training algorithm would take only unannotated text as input. This system would have the huge advantage of being highly portable: there are many more untagged corpora available than tagged corpora. One of the difficulties of such work is coming up with an appropriate measure to evaluate the tagger. If one uses the standard tagging measure, accuracy compared with a manually annotated corpus, such a method is doomed to fail. This is because any manually tagged corpus will be filled with a large number of arbitrary decisions, such as the number of tags, the tag for words in certain ambiguous cases (should "Russian" be an adjective or noun or proper noun in the noun phrase "Russian food"?). There is no way to automatically determine the outcome of these semi-arbitrary decisions, except by observing some manually tagged text. So in terms of tagging for the sake of tagging, it is difficult to evaluate such completely unsupervised systems. If one has a system which takes tagged text as input and then performs some task on this text, then one need not evaluate the precise tags output by the tagger, but rather the usefulness of the automatically tagged text in producing quality output within this system. The only weakness with this is that we then cannot simply compare two tagging approaches, but must compare them in the context of a particular application.

Another approach to unsupervised training of a tagger assumes that there is no tagged text available, but there is a lexicon available listing the allowable wordclasses for known words. This can result in a significant savings in manual effort needed for training in cases where either it is much easier to manually create such a lexicon than to tag sufficient training text, or where an appropriate on-line dictionary is already available which can be used for this purpose. With HMM-based tagging, one can train a statistical model with the same parameters as in the supervised learning case $(P(W_i|T_i), P(T_i|T_{i-1}, ..., T_{i-k}))$ (cf. Chapter 16; see also Merialdo 1994; Jelinek 1990; Kupiec 1992). Here, however, instead of estimating these parameters directly from counts in the training corpus, the Forward-Backward algorithm (Baker 1979) is used to estimate these parameters. An initial guess is made for each parameter and then in each iteration of the algorithm these parameters are adjusted in order to increase the probability the model assigns to the (untagged) training set.

Such unsupervised learning can also be done in the rule-based paradigm. Once again, we will consider a transformation-based learning system, but in this case only a lexicon and an untagged corpus are used for learning. Clearly, the transformation-based learner described above for tagging cannot be used as-is. We cannot initially assign each word its most likely tag, as we would need a tagged text to acquire this information. Also, we cannot compare the efficacy of rules in terms of how much they improve tagging accuracy on a training corpus, as this would require an annotated corpus to measure accuracy. To describe a transformation-based approach to solving this problem, we will once again describe the three components of a transformation-based system: the initial state tagger, the transformation templates and the scoring function.

Initial state tagger. Each word is initially assigned the set of all allowable tags, as indicated in the lexicon. Words not in the lexicon could be assigned a list of likely tags for unknown words (frequent open-class tags). Alternatively, a method such as Mikheev's (cf. Chapter 13) could be used for learning the set of allowable tags for a word not in the lexicon.

Transformation templates. The approach taken for unsupervised learning uses rules of the type introduced in Klein and Simmons (1963), where words are initially assigned the set of allowable tags for that word and then rules are used which reduce the tagging ambiguity based on contextual cues. Some example templates are:

```
Reduce the tagging of a word from X1_X2_..._Xn to Xi if:
  the previous word is W
  the next word is W
  the previous tag is T
  the next tag is T
```

where X1_X2_..._Xn is the set of tags X1 through Xn. An example of a possible transformation is: 'Reduce the tagging of a word from Noun_Verb to Noun if one of the tags of the previous word is Adjective'.

Scoring function. In all of the examples of learning transformation-based rules presented above, the learner had access to a 'gold standard', allowing it to assess the quality of any hypothesized rule. In this case, such a standard is not given and so some quality measure other than tagging accuracy must be used. Rather than present the precise measure, we will give the intuition behind it. Let us examine the case where we are trying to decide whether the word "race" is a noun or a verb in the context "The race ...". All we know is that the word "race" can be either a noun or a verb, information available in the lexicon. We have no disambiguated occurrences of this word to aid us in learning. However, we can look at all words that ever appear in the context "The ___". Of those that only have one permissible tag, we can check whether nouns or verbs occur more frequently in this context. For example, say that when examining the words that appear after "The", those that are exclusively nouns appear in this context 120 times and those that are exclusively verbs appear in this context 10 times. This would be a strong indication that the following rule would be a good one:

```
Change the tag from Noun_Verb to Noun if the previous word is ''The''
```

Therefore, a measure can be created based on this intuition, without requiring any manually annotated text in order for this measure to be computed. This measure will be used to search for the context which gives the *purest* disambiguating information, using words that are unambiguous in the lexicon to determine what tags are and are not allowable in a particular context.

We when use this algorithm on the Penn Treebank Wall Street Journal corpus, the first five rules learned are:

1	NN or VB	→	VB	Previous tag is TO
2	NN or VB or VBP	→	VBP	Previous tag is NNS
3	NN or NNP	→	NNP	Next tag is NNP
4	NN or NNP	→	NNP	Previous tag is NNP
5	NN or VB	→	VB	Previous tag is MD

One problem with both this approach and the HMM-based approach to unsupervised learning using only a lexicon is that they both rely on a clean lexicon that is tuned to the corpus. The optimal set of allowable tags for a word in a lexicon for tagging is not constant across domains within a language. For instance, in the domain of poker, considering "ace" to be only a noun is probably a good idea, whereas in the domain of tennis, it should be considered both a noun and a verb, and in the domain of test taking, it should be considered only a verb. If the lexicon is not well suited to the domain, tagging accuracy drops considerably compared to that which can be achieved with a tuned lexicon. Given this, the question remains whether it is a more efficient use of human resources to spend time making a very clean lexicon or whether it would be better spending this time manually tagging text to be used in supervised training.

15.7 ISSUES OF PORTABILITY

The claim is often made that automatically learning rules is a more efficient process for building a tagger than writing the rules manually. Since the system learns automatically, little or no human intervention is needed, thereby making automated systems more readily portable to new domains and languages. This claim has not been completely substantiated and should be carefully considered. If one wishes to build a tagger for a domain where a significant amount of tagged text or a clean, tuned lexicon already exists, then this claim is most likely true. But what about the case of porting to a domain where no manually prepared resources are available for training? In this case, we need to compare the time it takes to prepare the training material (and possibly tune the learning algorithm or final tagger) to the time it would take to manually develop the tagger. It is likely that the optimal solution is some combination of automatic learning with manual rule writing, having the machine do what machines do best and having people do what they do best.

There are a number of other portability issues that need to be further examined. Given a very small amount of training data, what is the best way to train a tagger? Given a tagger trained on one genre of text, how can it most effectively (both with respect to accuracy and efficiency of creation) be ported to a new genre? We will present rule-based methods for addressing both of these problems.

Because untagged text is readily available but tagged text is more difficult to obtain, there are many situations where one has access to a large untagged text and has access to or the ability to create a relatively small manually tagged text. From what was presented above, one could either train a tagger on the small tagged text using supervised techniques, or train on the large untagged corpus using unsupervised techniques. But is there a way to combine these two information sources? With HMM-based taggers, this can be done (cf. Chapter 16; also Merialdo 1994). In such taggers, training consists of deriving estimates for both lexical probabilities $P(W_i|T_i)$ and contextual probabilities $P(T_i|T_{i-1}, T_{i-2}, ..., T_{i-m})$. Given a tagged corpus, these probabilities can be directly estimated from the corpus. Without a tagged corpus, some initial guess is made for these probabilities and then training consists of iteratively adjusting these parameters in order to increase the overall probability the model assigns to the training set. A large untagged text and small tagged text can be combined by using the small tagged text to compute initial estimates for the probabilities and then running the unsupervised algorithm to adjust the parameters in order to increase the probability the model assigns to the untagged corpus. In practice, this method of combining the two information sources has not proven to be very effective. Merialdo (1994) shows that in such a situation simply training on the small tagged corpus typically gives better performance than can be achieved by combining this with unsupervised learning.

It is possible to utilize both untagged and tagged text in rule-based training as well. Recall that in a transformation-based system, the input is simply an initial guess as to the proper annotation. Therefore, we can first train a transformation-based system using the large untagged corpus as described above. A second set of rules will then be learned using the tagged corpus, where the output of the unsupervised tagger serves as the initial state annotator for the supervised tagger. In a number of experiments it has been shown that combining the two information sources always results in a more accurate tagger than could be obtained solely from training on either the tagged or untagged corpus alone (Brill and Pop, to appear).

16 HIDDEN MARKOV MODELS

Marc El-Beze and Bernard Merialdo

16.1 INTRODUCTION

In the previous chapters we have seen how the context of a word in a sentence is used to help identify the proper tag for that word. It is clear that such consultation of the context is necessary if we want the tagging to reach an acceptable level of correctness. It is not clear, however, that the mechanism used for this consultation should be rule-based.

Generally, most information about the proper tag is contained in the local context. In a rule-based paradigm, this means we can use constraints on sequences of adjacent tags (called *N-grams* for sequences of length N) to define the possible and impossible tags in a given context. Exploitation of the context can also be taken a step further: N-grams are not only possible or impossible, but they also tend to occur with differing frequencies. Several systems (starting with CLAWS; cf. Chapter 2) have exploited statistics on N-gram occurrences to determine the most probable sequence of tags.

The N-gram models used by these systems are reasonably effective, yet they have several shortcomings. First of all, they can only be trained on tagged text. Only if there is a sufficient amount of tagged text available will the model show acceptable performance. Furthermore, even with substantial amounts of tagged text, the statistics on infrequent N-grams will be inexact and some form of *smoothing* is necessary. The second major shortcoming is that the models use some ad-hoc formula to combine the N-gram statistics into a tag or tag-sequence probability. As a result, there is always some doubt whether or not the parameters of the model are optimal.

H. van Halteren (ed.), Syntactic Wordclass Tagging, 263-284.

What is needed, then, is a mathematically more exact technique of modelling the probability distributions that are involved. As it so happens, such a technique already exists and is used widely in other fields, e.g. in speech recognition to represent the acoustic variability of speech. The type of model used with this technique is called a *Hidden Markov Model*, generally abbreviated as *HMM*. The main advantages of HMMs are:

- The proper formulae for using N-gram statistics are mathematically derived in a rigorous way.
- Because we can use the states to represent any kind of information, they provide a flexible framework where N-grams but also other extensions (lemmas, long term context, ...) can be easily added and mixed.
- HMMs can be trained with either tagged text (*supervised training*, by direct observation of probabilities) or untagged text (*unsupervised training*, using the Baum-Welch algorithm, described below).
- There are several efficient algorithms that can be used for the tagging itself.

Before examining how HMMs can be used in the specific application of wordclass tagging, we will first examine them in their more general form.

16.2 HMMS IN GENERAL

16.2.1 Definition and use

Hidden Markov Models are probabilistic finite state automata. They are defined by:

- a set of *states*;
- a set of *transitions*, each going from an initial state to a final state;
- for each state, a probability distribution for all transitions leaving that state;
- a set of *output symbols*, that can be emitted when in a state (or transition);[1]
- for each state (or transition), a probability distribution for all symbols that can be emitted.[2]

---

[1] Some presentations consider that the symbols are emitted from the state, others from the transition. Both types are theoretically equivalent. The difference is mostly a question of personal preference.

[2] It is possible to allow *null transitions*, transitions that do not produce any symbol.

A *path* in an HMM is a sequence of transitions such that the final state of a transition is the initial state of the next transition in the path. The probability of a path is the product of the probabilities of its transitions. Following a path produces a sequence of output symbols. Generally, a given sequence of symbols can be produced by several paths.[3] Therefore, the probability that the model produces this particular sequence of symbols is the sum of the probabilities of these paths.

Mostly HMMs are used in one of two ways. In the first, the probability of a sequence of output symbols is calculated for several competing models, thus deciding which model is most likely to have produced the sequence. An example would be speech recognition, where an audio signal is compared to models for, say, the words "yes" and "no", in order to find out what was (probably) said. In the second way, a model is used to determine which path through the model was most probably used to produce a sequence of output symbols. We will give a more extensive example for this use as this is also the way HMMs are used in wordclass tagging, where we have a sequence of words (output symbols) and want to know the sequence of tags (states).

16.2.2 An example

The basic mechanisms involved when using HMMs are best explained by way of an example.

Suppose we want to use an HMM to model the weather that occurs during a journey along various geographical locations. The states of the model are the locations, say Spain, England and Norway, and the output symbols emitted represent the weather, say Sunny, Cloudy and Rainy. The transition probabilities represent the chances of a random traveller going from one location to another. One possible (and very inexact) way of estimating these would be to count the relative number of plane flights to the other locations. The emission probability for a symbol (weather) at a state (location) is the chance of observing that weather at that location. These can be estimated easily by consulting the weather reports for the last couple of years. We assume the probability of a specific weather type at a location is the number of times that the weather type was observed there divided by the total number of observations for that location.

Now, once we have such a model, what can we do with it? Suppose a friend is travelling through these locations and we wish to know where she went, but she fiendishly only let us know what kind of weather she experienced. This is a chance to use the model: for each path $L_1, L_2, ..., L_n$ we can calculate what the probability of the specific path is (using the transition probabilities $P(L_i|L_{i-1})$ and also what the

[3] That there is an ambiguous relationship between the observed symbols and the path that generates them is the basis of HMMs being called 'hidden'. A model would be non-hidden, and would then be called a *Markov chain*, if the relationship was one-to-one, i.e. if knowing the symbols emitted would unambiguously define the path in the model.

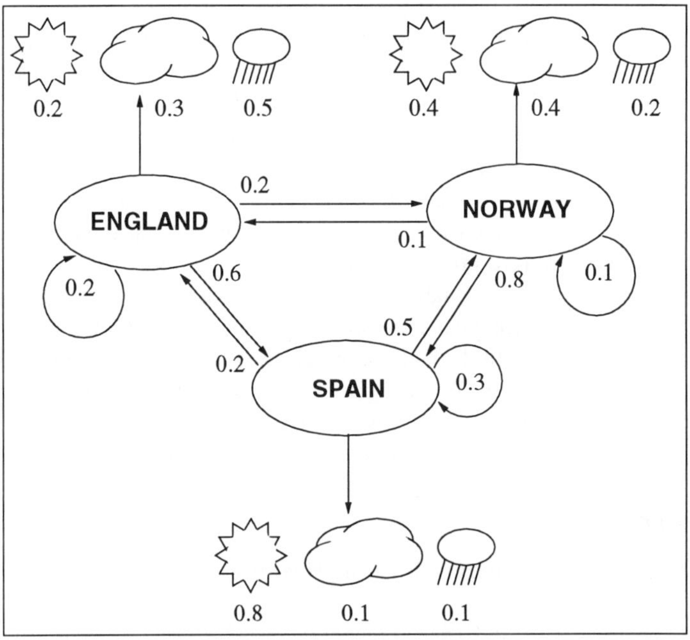

Figure 16.1 Hidden Markov Model of weather on the road.

probability is that the reported weather sequence $W_1, W_2, ..., W_n$ was observed (using the emission probabilities $P(W_i|L_i)$). The path with the highest probability is the best guess for our friend's real trip.

Take the model shown in figure 16.1 and consider the path in figure 16.2. All we would know of this path is the output sequence: Sunny – Cloudy – Rainy – Rainy. Using the probabilities in the model, we can calculate that the most likely path to produce this output is the one in figure 16.3. In this particular case, the difference with the actual path is very small: only one location.

As this exercise shows, considering the weather that was observed during the trip provides a very strong clue to the real path. Obviously we are still guessing, but the procedure described minimizes the chance of errors according to the probabilistic model. How well the most probable path approximates the actual path is determined by the quality of the model, which is not only based on the probability distribution parameters but even more by the fundamental decision which states and output symbols are to be used.

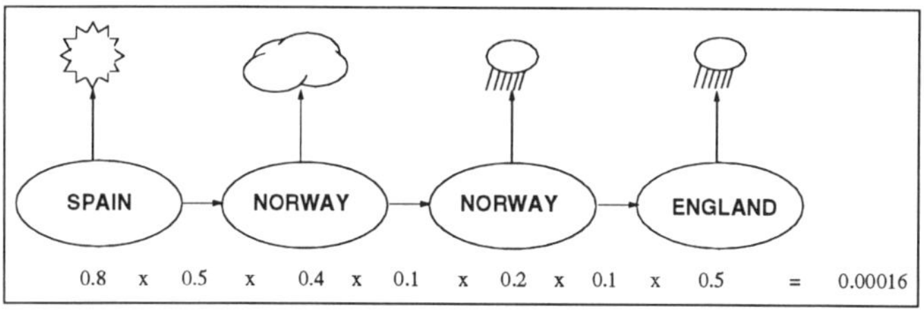

Figure 16.2 A possible path in the weather HMM.

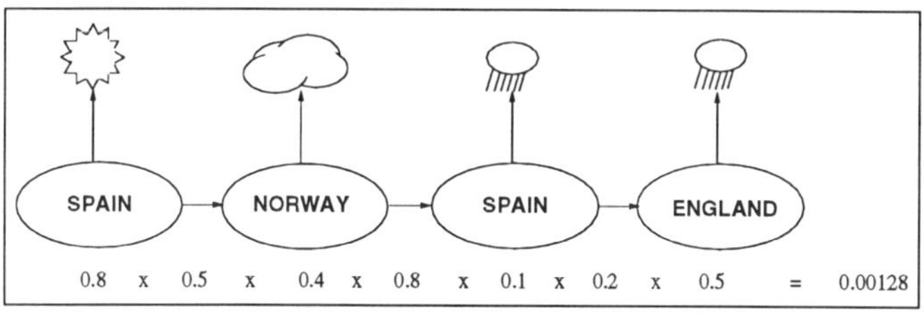

Figure 16.3 A probable path for the given output.

16.2.3 Choosing the underlying topology

After deciding to use an HMM for an application, the first issues are mapping the observations to output symbols and choosing a useful set of states and transitions.

The output symbols are often clearly defined by the application. In the example above, each weather type corresponds to a single output symbol. In the case of tagging, as we will see below, the symbols will be words. We will focus on the case where the output symbols are elements of a finite set, but it should be noted that HMMs may also be used in cases where output symbols belong to an infinite set, e.g. vectors in a N-dimensional space as in speech recognition, where audio signals are represented in the form of multi-dimensional measurements.

How to choose the states of the model and the transitions between them is usually less obvious. For truly hidden models, e.g. in speech recognition, one can simply choose a number of states and use a fully connected network of transitions. However, the efficiency of the model to represent the process of symbol production can be greatly enhanced by encoding knowledge about the real production process into the topology of the network.

As an example, assume that we know that our travelling friend visits specific cities and only some of those cities connect to other countries. Even if the weather in different cities in the same country is the same, it is still useful to introduce the new travel restrictions into the model. As a result, a model in which there is a state for each city will work much better than one in which there are only states for countries.

Since most of the construction of the model is is done by standard algorithms, the choice of the topology is one of the major points where the designer has an influence over the quality of the model.[4] Moreover, this choice is a critical one, for the algorithms for parameter derivation may be very good, but this is meaningless if they are made to derive parameters for an ill-suited transition network.

16.2.4 Training

After the topology of the model has been determined, we have to estimate the transition and emission probabilities. This is usually called the *training* of the model. In our travel example, the training was performed simply by collecting frequencies of observations (of different weather types at each location). This is the easy case where some operations in the model are visible, i.e. we can directly observe both the state and the emitted symbol. Actually, the model is therefore not hidden and training can be done easily by estimating probabilities from relative frequencies.

There are other and more complex cases where the model is really hidden, which means that the training data does not fully specify the path in the model that produced it. Imagine for example that we would have only recordings on the weather that occurred during a number of travels, together with the order in which the locations were visited but not the time that the traveller stayed at each location. We cannot therefore assert precisely where each weather type was observed (except for the first and last ones). Fortunately, there is a mathematical procedure which allows us to compute 'good' estimates of the transition and emission probabilities automatically from a corpus of observations even without a full knowledge of the training situation.[5]

The training algorithm for HMM was originally proposed by Baum (1972) and is known as the *Baum-Welch* (BW; sometimes *Forward-Backward*) algorithm). The mathematical description of the Baum-Welch algorithm and the proof of its correctness are likely to be a bit obscure to the unfamiliar reader. However, the idea that lies behind the BW algorithm is surprisingly simple (cf. Brown *et al.* 1992). If the model is not hidden, i.e. if we can identify the path that was taken inside the model to produce each training sequence, we are able to estimate the probabilities by the relative frequency of taking each transition from each state and of emitting each symbol. If, on the other hand, the model is hidden, there may be several paths that can produce an output sequence. The trick of the Baum-Welch algorithm is the following: if we have initial estimates of the probabilities, we can compute the frequencies of each of the possible paths and then

---

[4] Examples of other points being quantity and quality of the training corpus and the choice of initial values for the parameters.

[5] The probabilities that are estimated in this way are 'good' in the theoretical sense of maximizing the probability of generating the training data, which is not necessarily closely related to the problem tackled by the user, e.g. the accuracy of a tagging procedure.

use the computed frequencies to weight the original probabilities, thus creating new estimates of the probabilities. The Baum-Welch theorem states that these new estimates are better than the initial ones. This means that if we iterate the procedure, we keep improving our estimates until we reach a local maximum. The Baum-Welch algorithm provides an efficient way to compute the new probabilities, by cleverly organizing the computation and avoiding the exponential explosion caused by the enumeration of all paths.

In summary, the Baum-Welch algorithm works as follows:

1. Start with initial estimates for the probabilities. These can be chosen at random or according to some prior knowledge. However, note that in some cases, uniform initial estimates will lead to uniform reestimates, which makes training inefficient.

2. Apply the BW computation on a large training data set and derive new estimates of the probabilities.

3. Iterate step 2 until a convergence criterion is satisfied. The criterion is usually that the probabilities change less than a threshold or the probability of producing the training data increases less than a threshold, but it may also be expressed in terms of computer time used.

16.2.5 Decoding

Once the model is complete, we can use it to either compute the probability of a given sequence of output symbols or find the best path that produces such a sequence. In both cases we are faced mainly with a computational problem: how best to organize a search algorithm in a very large search space, i.e. how to minimize both memory space and response time.

Sometimes, we only need the probability that the model produced a given output sequence. This is generally the case if we have several models that we want to compare, e.g. we have one model generating French text and one generating English text and we want to determine the language of an observed sentence. Under these circumstances we need to determine, for each time t and each state S, what the probability is that the model is in state S at time t, taking into consideration, of course, the corresponding output produced. Since this probability is the sum of the probabilities of all paths leading there, we have to add the probabilities of all incoming paths, each of which can be easily computed from the probabilities of the states at time $t-1$, the transition probabilities and the probabilities of producing the correct output. The probability of the complete output is the probability of the final state when the end of the output is reached.

The computation changes if we also want to find the most probable path which produces a given observation, an activity often called *decoding*. In general, because the model is a hidden model, there are several possible paths that may produce the same output sequence. Therefore, the best choice (according to the probabilistic theory) is to select the path with highest probability. There might be quite a lot of paths, generally an exponential function of the length of the output sequence, so that an exhaustive enumeration could be very time-consuming and often intractable in practice.

Fortunately, we can use the so-called *Viterbi* algorithm (Viterbi 1967; Rabiner and Juang 1986), which takes advantage of a dynamic programming principle to perform the search for the optimal path with much less computation (a linear function of the output length). It is based on the observation that, if you look at the best path that goes through a given state S at a given time t, the path is the concatenation of the best path that goes from the initial state to state S (while emitting symbols from 1 to t) with the best path that goes from state S to the final state (while emitting symbols from $t+1$ to the end). This is because the probability of a path going through state S is simply the product of the probabilities of these two parts (before and after S), so that the maximum probability of the global path is obtained when each part has a maximum probability.

This means that, if we know all optimal paths leading to all possible states at time $t-1$, it is easy to find the optimal paths leading to all possible states at time t. This is just a matter of looking at all the possible extensions of these paths (which produce the right output)[6] and keeping the best one for each state. The number of computations involved is in the order of the number of transitions in the network, while the storage required to store the extensions is in the order of the number of states. It is then possible to construct the extensions sequentially starting from time $t=1$ until the end. The best path that arrives at the final state of the model, while the complete output sequence has been emitted, will be the optimal path that we are looking for.

To recover this best path, we need to keep track of the last transition of the best path that reaches each state at each time t. After arriving at the final state, we can then follow these transitions back recursively and recover, in reverse order, the sequence of states that constitutes the optimal path. The storage required to keep all this information is in the order of the number of states times the length of the output sequence.

There are also variations of the Viterbi algorithm which further reduce the computations and storage required. For example, we can only keep the 'most promising' best paths at each time t, i.e. the best paths with a probability above a predefined threshold or simply the M most likely best paths. Since it is reasonable to expect that sub-paths of the optimal path are 'good' paths, we can expect that a reasonable choice of the threshold will keep them in the search space, while avoiding the computation for low probability paths. Note, however, that by using this intuitively reasonable heuristic, we can no longer guarantee that the true optimal path will be found. As it is often the case, we have to choose a compromise between the cost of computation and the quality

[6] Theoretically, the paths that do not produce the correct output have a probability of zero and therefore will not be among the best ones. This means that, in practice, we can just avoid generating these.

of the result. Such a decision is generally very dependent on the application that is considered.

16.3 HMMS FOR WORDCLASS TAGGING

16.3.1 *The structure of the model*

In wordclass tagging, we are given a sequence of words and are expected to derive a corresponding sequence of tags. The route we will take towards such a derivation is to observe that each sequence of words in a script is in fact a corrupted message which has been transmitted through a noisy channel. The source sent messages made of sequences of words with their wordclass tags W, T, but these messages were subsequently corrupted by an encoder which removed the tags and thus produced word sequences W. Our task is to build a decoder that takes each observed message W and retrieves the original message by restoring the proper tags T. To see how we can accomplish this task we turn to the field which has extensively studied noisy channel transmissions: *Information Theory*. It teaches us that the wordclass assignation problem can be interpreted as the search of an optimal (i.e. most probable) sequence of tags given a sequence of words.

The optimal sequence of tags $\widetilde{T}$ is the one for which $P(T|W)$, the probability of T given the observed sequence W, is the highest. Now, according to Bayes' equation, $P(T|W)$ is equal to $P(T) * P(W|T)/P(W)$ This means that we have to find the sequence $\widetilde{T}$ for which $P(T)*P(W|T)/P(W)$ is maximal. Since the word sequence W is given, this is the same as finding the sequence $\widetilde{T}$ which maximizes $P(T) * P(W|T)$. As a result, to determine the optimal sequence $\widetilde{T}$, we will have to calculate both $P(T)$ and $P(W|T)$ for all T. Unfortunately, neither of these factors can be calculated exactly. Instead we will have to use some form of approximation.

Let us first examine $P(T)$. Using standard probability decomposition formulae, it is possible to show that for a tag sequence T of length L, the probability $P(T)$ is equal to the product of L terms:

$$P(T) = P(t_1) * \prod_{i=2}^{L} P(t_i|t_1, ..., t_{i-1})$$

This decomposition by itself does not help, as the probabilities $P(t_i|t_1, ..., t_{i-1})$ are generally far too numerous. However, we can now approximate each of the terms. The approximation used for N-gram models is to consider equivalent all the contexts $(t_1, ..., t_{i-1})$ that share the same last $N-1$ tags, i.e. we approximate $P(t_i|t_1, ..., t_{i-1})$ by taking $P(t_i|t_{i-N+1}, ..., t_{i-1})$. The most popular cases are $N = 2$, *bigram* models, using $P(t_i|t_{i-1})$, and $N = 3$, *trigram* models, using $P(t_i|t_{i-2}, t_{i-1})$. Higher N are not often used as they tend to lead to rather inefficient use of time and/or space. Given sufficient training data, trigram models are usually superior to bigram ones (Shannon 1951; van Halteren 1996), while still computationally tractable.

The approximation done on the second factor, $P(W|T)$, may seem even stranger but does lead to great simplification, while tagger performance remains at a high level. The probability of the word sequence W given the tag sequence T, both of length L, is also replaced by a product of L terms, working from the (obviously strictly false) assumption that each word depends only on its tag:[7]

$$P(W|T) = \prod_{i=1}^{L} P(w_i|t_i)$$

The two approximations above allow us to use a Hidden Markov Model to model the noisy channel, using the tag histories of length $N-1$ as states and the words as output symbols.[8] The probabilities $P(t_i|t_{i-N+1}, ..., t_{i-1})$ are then the transition probabilities in the model and the $P(w|t)$ the emission probabilities. Usually, the models used in wordclass tagging are triclass models, based on the probabilities $P(t_i|t_{i-2}, t_{i-1})$, which provide a good compromise between accuracy and efficiency.

In order for N-class models to work properly, we need special measures to handle the effects at the start of the sentence, e.g. the probabilities that tags appear as the first tag and at the end, viz. that the model recognizes that the end of the sentence has been reached. Generally, this is done by introducing a dummy symbol O, of which there are $N-1$ prepended to the sentence and one appended. The start state of the model is then $(O, ..., O)$ and the end state the one to which any transition with an output of O leads.

As we will see below, the results obtained with taggers based on N-class HMMs are typically pretty good. When training is done on tagged text, tagger results are comparable to other state-of-the-art data-driven taggers. Furthermore, HMM-based taggers can then be further improved with additional, unsupervised training. Finally, if there is no tagged text available, HMM taggers can even be trained fully on untagged text, although this does lead to somewhat worse tagging results.

[7] In these formulae, and hence in the resulting HMM, the probabilities $P(w|t)$ are used. However, these probabilities are a bit difficult to understand, because it is uncommon to consider all words having a given tag. It would look more natural to use the probabilities $P(t|w)$, because in a dictionary we find for any word the list of its possible tags and the probabilities $P(t|w)$ simply represent the frequency of usage of each tag for this word. Although it is necessary for the mathematical proofs and operations to keep using $P(w|t)$, the two systems are in fact correlated. Using Bayes' formula, we can see that $P(w|t) = P(t|w) * P(w)/P(t)$ which allows us to switch easily from one system to the other.

[8] Such HMM models are often just called N-gram models. The name 'N-gram' literally indicates models where the states are sequences of $N-1$ symbols, whatever these symbols might be. However, as the term 'N-gram models' is also commonly used to designate models where the symbols are words, models using N-grams of wordclasses are more properly named *N-class* (sometimes also *N-POS*) models.

16.3.2 Choice of tagset when using HMMs

The stochastic approach, as used in HMM-based taggers, offers an attractive solution to the tagging problem, because syntactic regularities are automatically captured. However, it would be an error to believe that no expertise at all is required to build a tagger on a new language. One of the things for which expertise is desirable is the choice of the tagset produced by the tagger. Except for special purposes, this tagset should ideally be linguistically motivated. We have already seen (in Chapters 4 and 5) which kind of guidelines to follow when designing such a tagset. In practice, however, tagset design must often also be constrained because of the capabilities of the model used in the tagger.

When using triclass HMMs, for example, syntactic features which are determined by long dependencies cannot be taken into account. Since tenses and modes interact from sentence to sentence it is hopeless to expect that such agreements will be handled within the short size windows used in a triclass model. This does not mean that we cannot include such features in our tagset at all, but if we do, we should expect considerably higher error rates where these features are concerned. Often, they are therefore not included, decreasing the size of the tagset.

The use of an HMM may also lead to an increase of the number of tags. In order for the model to work well, words belonging to a syntactic class should get two different tags if they do not have the same behaviour in an equivalent context, e.g. if the model is expected to handle the constraint between an indefinite article and the following noun, it is necessary that "a" and "an" be assigned two different tags and that nouns have different tags depending on whether they start with a vowel or not. However, such a separation can, if so desired, be kept purely tagger-internal, by mapping the separated tags back to the original one for the final output.

Another point which is related to the efficiency of the model and which hence should be taken into consideration, is that, if ambiguous and non-ambiguous words are mixed within the same class, the training process will converge faster. This is because non-ambiguous words will generate reliable statistics even at the very first stage of the training.

It is also useful to take punctuation into account when tagging with N-class models. Punctuation marks can be considered as symbols produced by the model and assigned tags in the same way as wordforms. While tags associated to punctuation marks can compose a complex system (cf. Nunberg 1990), it is common that punctuation marks are simply gathered in two categories: a) strong punctuation marks, which define the end of a sentence, and b) weak punctuation marks, which can occur in the middle of a sentence.

Apart from by the kind of information represented in the tagset, the difficulty of a tagger's task can be judged by the size of the tagset, or rather by the ambiguity of the tagset (cf. 6.3.2). To give an idea of the difficulty of the task faced in the experiments

Table 16.1 Ambiguity in dictionary and corpus.

Npos	Dictionary	Corpus	Reduced Dictionary
1	84.46	61.93	74.56
2	12.58	14.53	18.78
3	1.72	7.00	3.88
4	1.05	11.07	2.35
5	0.11	2.56	0.27
6	0.05	1.63	0.11
7	0.01	0.24	0.01
8	0.02	0.90	0.03
9	0.00	0.12	0.01

described below, we have examined the ambiguity of the tagset used (BDLEX, consisting of 103 tags) in a large corpus (2 years of the French newspaper Le Monde). The results are shown in table 16.1. The first column shows the number of different BDLEX tags that a word can receive. The other columns list the percentages of words receiving that number of tags. The figures given in the second column show the percentages as calculated on the entries in the underlying French BDLEX dictionary, which contains 201,000 inflected forms. The third column shows the percentages of the tokens observed in our corpus, consisting of about 39 million words. The fourth column, finally, again shows percentages of entries in the dictionary, but ignoring all entries which were not seen in the corpus.

16.4 TRAINING HMMS FOR TAGGING

16.4.1 Training on tagged text

Training of N-class HMMs is generally done using tagged text. The relative frequency counts give first estimates and are subsequently smoothed using some additional sample of text, the so-called *held-out* data.[9] Better estimates of the model parameters may be derived from the data with a few iterations of the training process. However, attempting too many iterations may cause damage to the model (cf. 16.4.3). This problem is denoted as the *overtraining* phenomenon.

The quality of training is obviously dependent on how well the training data covers the N-grams used by the model. In order to get an idea of this coverage, we have

[9] It is important to use held-out data not included in the training, nor in test data. If either were the case, a bias would be introduced in the estimation or in the evaluation of the model.

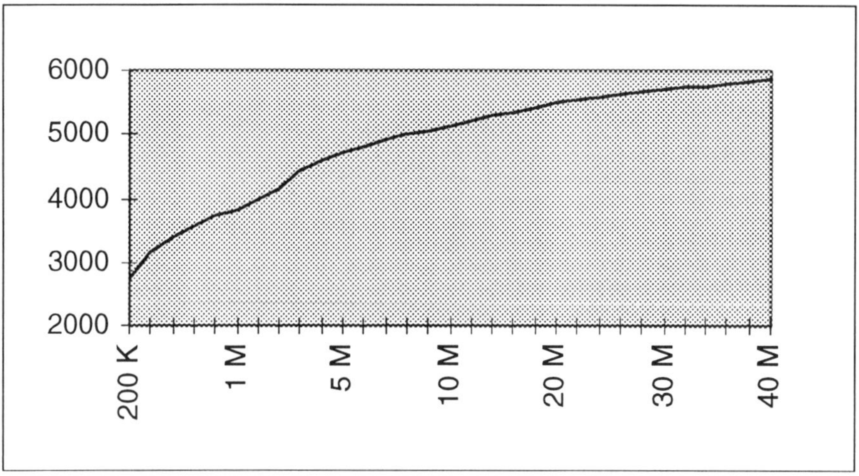

Figure 16.4 Number of biclasses observed in training text.

examined our 39 million word Le Monde corpus, using the abovementioned BDLEX tagset. We first derived a very large size dictionary from the BDLEX vocabulary, adding about 30,000 further entries in order to increase the static coverage of the word list up to 97%, which is not bad for a highly inflected language as French. The number of different biclasses and triclasses versus the size of the training corpus is plotted in the figures 16.4 and 16.5.

Comparing the number of the theoretically possible biclasses (103*103=10,609), with the number of biclasses that really occur in the training corpus (5,877), we note that 45% of the pairs have not been observed even after 39 million words. For the triclasses, the theoretically possible number is 1,092,727, of which only 108,523 (10%) are observed. Now it may well be that unobserved combinations are linguistically impossible, but examination of the graphs indicates that some additional possible combinations can still be expected. Moreover, it is obvious that at lower corpus sizes very many possible combinations are missing.

When we trained an HMM tagger from the full corpus available, its error rate was nearly 6% on a test script not included in the training corpus, which was rather higher than expected. This can be explained for a large part by the fact that there are large amounts of out-of-vocabulary words, mostly proper nouns. This accounts for more or less two thirds of the error rate. The other errors could be split into two equal sets. First, there were two words which were often mistagged and the quality of the tagging has later

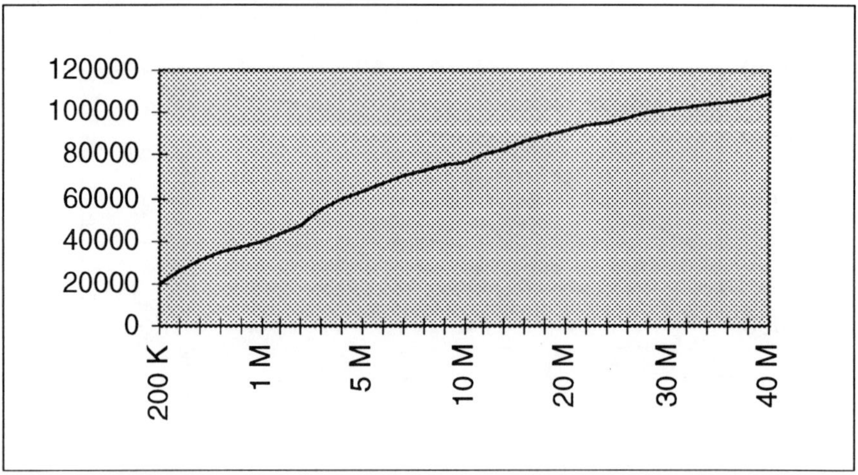

Figure 16.5 Number of triclasses observed in training text.

been improved just by writing specific rules for these two frequent words. Then there were more infrequent errors due to dictionary errors, under-represented structures or long span dependencies.

16.4.2 Smoothing the triclass model

As we saw above, 90% of the triclasses do not occur even within 39 million words of training data. Because the relative frequencies of triclasses based on these counts are hence not reliable enough (they contain a lot of zero counts that correspond to sequences that could exist but have not yet been observed), it is necessary to smooth the triclass model, e.g. with lower order distributions (i.e. biclass, uniclass and so on) in order to give unseen events a chance not to get a null score.

The usual formula for this is

$$
\begin{aligned}
P(t_i|t_{i-2}, t_{i-1}) = \\
&\lambda_3 * P_3(t_i|t_{i-2}, t_{i-1}) + \\
&\lambda_2 * P_2(t_i|t_{i-1}) + \\
&\lambda_1 * P_1(t_i) + \\
&\lambda_0 * P_0
\end{aligned}
$$

where P_3 represents the triclass, P_2 the biclass, P_1 the uniclass and P_0 the uniform distribution. Once these distributions have been estimated, the λ's, which are constrained by $\lambda_0, \lambda_1, \lambda_2, \lambda_3 >= 0$ and $\lambda_0 + \lambda_1 + \lambda_2 + \lambda_3 = 1$, are computed using the Baum-Welch algorithm in such a way that the emission of some held-out data is optimal. This interpolation also contributes to the reduction of the overtraining effects.

Another way to combine triclass and biclass distributions is:

$$\begin{aligned} &P(t_i|t_{i-2}, t_{i-1}) = \\ &\quad \lambda(c(t_{i-2}, t_{i-1})) * P_3(t_i|t_{i-2}, t_{i-1}) + \\ &\quad (1 - \lambda(c(t_{i-2}, t_{i-1}))) * P_2(t_i|t_{i-1}) \end{aligned}$$

where the notation $\lambda(c(t_{i-2}, t_{i-1}))$ means that the λ-coefficients depend on the count of the pair (t_{i-2}, t_{i-1}). If $c(t_{i-2}, t_{i-1})$ is high, the estimate of the P_3 distribution is reliable and $\lambda(c(t_{i-2}, t_{i-1}))$ is expected to be close to 1.

In fact (Derouault and Merialdo 1984), quality of the prediction and corpus representativity are both ensured by the following smoothing:

$$\begin{aligned} &P(t_i|t_{i-2}, t_{i-1}) = \\ &\quad \lambda(t_{i-1}) * P_3(t_i|t_{i-2}, t_{i-1}) + \\ &\quad (1 - \lambda(t_{i-1})) * P_2(t_i|t_{i-1}) \end{aligned}$$

Here, the number of λ-coefficients is equal to the size of the tagset. Furthermore, given the nature of the function that must be maximized for each individual λ, they can be determined much more easily than by using the full Baum-Welch algorithm.

Yet another way to combine several models in order to cope with the sparse training data problem is based on a back-off strategy, introduced by Slava Katz (1987).[10] Assuming that several models of different orders have been trained, the underlying principle of the back-off strategy is that the order of the model to be used will depend on the reliability related to the probability estimated for producing a particular event.

The set of the available distributions may be seen as embedded clusters in a hierarchical classification. Furthermore, the choice of the appropriate distribution is made using a given threshold expressing the concept of reliability. In that way, if a sequence of tags t_1, t_2, t_3 occurs in the test script, three cases have to be distinguished:

[10] See also Jelinek (1990).

Table 16.2 Model quality after unsupervised training.

	Number of tagged sentences used for the initial model						
	0	100	2000	5000	10000	20000	40000
Iter	Correct tags (% words) after BW on 40000 sentences						
0	77.0	90.0	95.4	96.2	96.6	96.9	97.0
1	80.5	92.6	95.8	96.3	96.6	96.7	96.8
2	81.8	93.0	95.7	96.1	96.3	96.4	96.4
3	83.0	93.1	95.4	95.8	96.1	96.2	96.2
4	84.0	93.0	95.2	95.5	95.8	96.0	96.0
5	84.8	92.9	95.1	95.4	95.6	95.8	95.8
6	85.3	92.8	94.9	95.2	95.5	95.6	95.7
7	85.8	92.8	94.7	95.1	95.3	95.5	95.5
8	86.1	92.7	94.6	95.0	95.2	95.4	95.4
9	86.3	92.6	94.5	94.9	95.1	95.3	95.3
10	86.6	92.6	94.4	94.8	95.0	95.2	95.2

1. If the number of times the sequence t_1, t_2, t_3 occurs in the training corpus, say K, is higher than a certain threshold, the probability of the tag t_3 given the history is estimated as $P(t_3|t_1, t_2)$.

2. If K is below the threshold but greater than zero, the probability is estimated as $\alpha{*}P(t_3|t_1, t_2)$, with α between zero and one and dependent on K, which in effect subtracts a small amount of counts from each distribution and dispatches them to the sub-order models (represented by the third case below), thus ensuring the integrity of the model.

3. If K is zero, i.e. the sequence t_1, t_2, t_3 has never been seen in the training corpus, the probability is estimated as $\beta * P(t_3|t_2)$, with the β-coefficients dependent on t_1, t_2 and tuned so that for each distribution (each context) the sum over all the possible events is equal to one.

However, when the number of events is small (as in the case of small tagsets) and enough training data is available, it is not so clear that there is any advantage in using the back-off strategy over a linear combination.

16.4.3 Training HMM taggers with Baum-Welch

It is also possible to train an N-class HMM on untagged text, using the Baum-Welch algorithm. The algorithm will try to find the probabilities of transitions that maximize

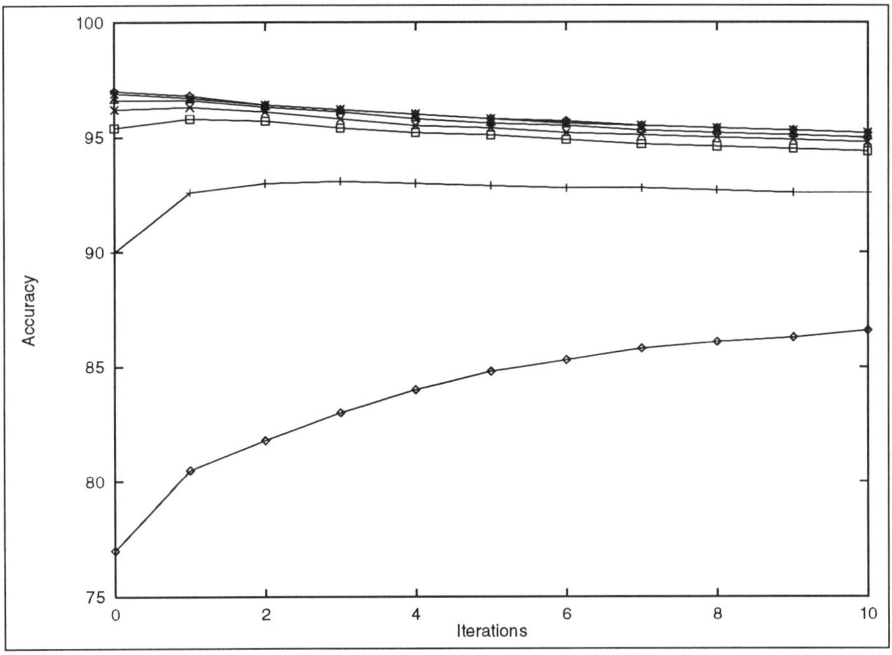

Figure 16.6 Model quality after unsupervised training.

the probability of generating the training data. Since large amount of tagged text are necessary for supervised training and producing such amounts of correctly tagged text is a long and expensive process, it might be interesting to evaluate the performance of an N-class model trained on untagged text. Such experiments were done (Merialdo 1994) and, unfortunately, they show that a model trained in such a way is quite bad at tagging new text when compared with a model trained on tagged text. It is shown that, while training BW can improve tagging performance when the initial model is made from uniform distributions, it might degrade the tagging performance when the initial model is not too bad.

Table 16.2 (represented graphically in figure 16.6) indicates the percentage of correct tags assigned[11] with a model that is initially trained over a number of tagged sentences (from 0 to 40,000), then iteratively trained using the BW algorithm over 40,000 un-

[11]The corpus used in the experiment consists of 42,186 sentences (about one million words) of Associated Press material. The tagging is a 76 tag projection of the set of 159 tags originally used in conjunction with a treebanking effort at Lancaster University. For more details, see Beale (1988).

tagged sentences. When the initial model is 'bad' (in our case, trained with less than 5000 sentences), the BW training induces an improvement in the tagging accuracy. But as soon as the initial model is 'good enough' (in our case, somewhere between 5000 and 10,000 sentences), the BW training only deteriorates the tagging performance.

A possible explanation for this behaviour is that the BW algorithm is guaranteed to improve the ability of the model to predict the training data, which is not the same as the ability to tag new text accurately. While it is reasonable to say that a model which provides good tagging accuracy will probably also predict the training data correctly, the optimum models for these two criteria (training data prediction and tagging accuracy) are probably close but have no reason to be identical. When we start from a bad initial model, the BW algorithm goes in the direction of the prediction optimum, which is also in the direction of the tagging optimum. But when we start from a good initial model, these directions will be sufficiently different that getting closer from the prediction optimum will take the model further from the tagging optimum.

16.5 TAGGING WITH HMMS

16.5.1 Using the Viterbi algorithm

The tagging of text can be seen as a two step process: sentence after sentence, first build a lattice composed of the legal tags associated with each word and then find the most probable path among all the possible paths. It is not always necessary to enumerate all the paths in order to determine which one has the highest probability, namely if we use the Viterbi algorithm.

As we already mentioned above, the states of a triclass HMM correspond to pairs of tags (t_{i-1}, t_i) and the emissions correspond to the words of class t_i, output from all the states $(..., t_i)$. In order to use the Viterbi algorithm, we assume that each word in the output sequence corresponds to one time-slice. In other words, each time-slice moves us to the next tag and produces the next word.[12] The algorithm is then easily implemented, needing only administration of probabilities of each (state, time-slice) pair and the best paths arriving there. In fact, after the move to time-slice i is completed, the probabilities of time-slice $i-1$ are no longer needed, so that at most two time-slice probabilities need to be stored at any time.

[12] The assumption is false if it is possible for a single tag to produce several words, as is the case with multi-token units (MTUs; cf. 4.3.1). As proper handling of MTUs greatly complicates implementation of the Viterbi algorithm, MTU tags are generally broken down into ditto tags, e.g. N:1/2 and N:2/2, which are assumed to be separate tags, so that no special measures are needed. If desired, probabilities around ditto tags can be adjusted to better use the knowledge of their meaning, e.g. the probability that N:1/2 is followed by N:2/2 can be set to one and the probability of N:1/2 following a tag pair can be linked somehow to that of N following the pair.

Normally, the Viterbi algorithm is used to find the optimal path over the whole sentence. However, if two contiguous words are both unambiguous, the time-slice corresponding to those words will contain only a single state with non-zero probability. This state can be seen as a confluent node for all paths, just like the nodes at the start and end of a sentence. It is perfectly possible to apply the Viterbi algorithm between two confluent nodes instead of between sentence separators. Although this does not make a difference in terms of results, it may speed up the tagger and spare memory allocation. Even if confluent nodes are not used explicitly in the tagger, they will still be useful in stopping error propagation.

There are also several technical tricks which can be incorporated in the implementation of the algorithm. The first of these is partly a historical construct, stemming from the time when processors were much slower doing multiplications than additions. Consider the formulation of the Viterbi algorithm (defining the Viterbi probability V of each state/time pair):

$$\begin{aligned} &V(t_{j-1}, t_j) = \\ &\qquad \max_{(t_{j-2}, t_{j-1})} V(t_{j-2}, t_{j-1}) * P(t_j | t_{j-2}, t_{j-1}) * P(w_j | t_j) \end{aligned}$$

This can be computationally simplified by replacing the multiplications with log additions:

$$\begin{aligned} &log(V(t_{j-1}, t_j)) = \\ &\qquad \max_{(t_{j-2}, t_{j-1})} log(V(t_{j-2}, t_{j-1})) + log(P(t_j | t_{j-2}, t_{j-1})) + log(P(w_j | t_j)) \end{aligned}$$

The log values can either be obtained through a table lookup or the whole calculation can be done in log values, with any necessary additions of the original values done by special table lookup (cf. Lee 1989). As current processors are able to perform a stream of multiplications at the same speed as a stream of additions, this reason for using the technique has disappeared. However, working with logarithms does have the advantage that the values do not easily fall below the lowest number which can be represented by the processor. The original values do stand a real chance of this happening as the many multiplications of low values we find in a long path of tags quickly lead to very low numbers.

Another way to avoid the extremely low number problem is to normalize the probabilities of all the partial paths corresponding to a same time-slice. To perform this normalization, one applies to each path a factor which is either the sum of the probabilities of all the partial paths or the maximum of the probabilities. Nowadays, the forward pass is generally implemented using this scheme.

16.5.2 Other forms of decoding

The standard way of using HMMs to tag text is with the Viterbi algorithm. However, there are several possible alternatives of which we will examine two more closely. Others, e.g. the right-to-left or reverse Viterbi model, do not lead to significant enough changes in accuracy or efficiency to be discussed here.

In *Maximum Likelihood* (ML) tagging we do not try to identify the most probable sequence of tags, but rather the most probable tag for each individual word, as determined by summing the probabilities of all paths that give this tag for this word. In this case, the decision of the tag chosen for a word is independent of the choice of the tags of the words next to it. Theoretically, ML tagging should produce better results in the standard tagging measurements as these measurements are based on individual tag scores. However, Viterbi is generally preferred because

- ML tagging is more complicated to implement than Viterbi.
- ML tagging may produce incoherent sequences of tags (e.g. assign tags which cannot follow each other), while Viterbi makes the assignment coherent on the sentence.
- When actually tested (Merialdo 1994), the results are not really different.

ML tagging does have an advantage which may make it more useful under certain circumstances: it can easily produce alternative tags for a given word, so that it can be used in a n-best tagger. This possibility has not yet been explored in practice.

Another possible modification is to use more local optimization. The normal Viterbi algorithm always optimizes over the whole sentence. In the case of a long sentence, this means that processing can be long and memory requirements high. A modified version of Viterbi can make a local decision by tagging words from left to right while looking only at the next k words. It enumerates partial paths over the next k words and selects the tag for the initial word that leads to the most probable paths. In experiments, we found that for $k = 3$ (looking at 3 words at the right of the word being tagged), this procedure gives the same tagging accuracy as complete Viterbi.

16.6 COMBINING DIFFERENT LINGUISTIC LEVELS

16.6.1 Using wordclasses in word models

In many applications, most notably speech recognition, there is need for language models to predict the occurrence of words. A natural choice for such a model is the N-word model. Unfortunately, the N-word model leads to very large numbers of model parameters and the concomitant vulnerability to sparse data. For a vocabulary size V, which for all but restricted domains is typically in the 100,000s, the number of parameters is in the order of V^N.

Table 16.3 Wordclass ambiguity for some French words.

French word	Masculine Noun	Feminine Noun	Verb	Adjective
coche	coach	nick, notch	to nick, notch	
ferme		farm	to close	firm
greffe	clerk's office	graft	to graft	
livre	book, register	pound	to deliver	
manche	handle	sleeve, game, set		
moule	mold, matrix	mussel	to cast, mold	
mousse	ship's boy	moss, lather	to foam, lather	
page	page-boy	page		
passe	master-key	passage, pasado, navigable channel	to pass, go	
pique	spade (cards)	spite, quarrel	to bite, pique	
poste	position, job	post-office	to mail	
ponte	eminence	egg-laying		
tour	turn, trip, wheel	tower, castle (chess)		
voile	veil, cover, fog	sail, canvas	to veil, shade	

One possible solution for this problem is to replace the N-word model by an N-class one. Then, assuming a tagset of size S (at most in the 100s), the order of the number of parameters goes down to only $S^N + S * V$. However, such a large decrease in the number of parameters is generally paid for by a decrease in overall model quality. For example, when the linguistic component of a certain automatic speech recognition system used only a triclass model, 7.2% of the words were misrecognized, while the error rate was only 5% when a triword model was used. If it is technically possible, N-word and N-class models are best used in combination: when the two models in the speech recognition system were combined, the error rate was reduced to 4% (Cerf-Danon and El-Beze 1991).

Another alternative to N-word models is provided by N-lemma models. Trilemma models have the potential advantage of a more accurate prediction than triclass models, while still needing a lower number of parameters than triword models.[13] These characteristics could be important for very large vocabulary systems.

16.6.2 Using lemma models in wordclass tagging

An N-lemma model may also be helpful in order to improve wordclass identification, particularly when semantic features are present in the tagset. For example, the word "sommes" in French is very ambiguous at different levels. It is an auxiliary, like in "nous sommes", which means "we are". As a feminine noun, it means "sums", while, as a masculine noun, it means "naps", and as a verb, like in "tu sommes", it may be understood as "you call upon", "you summon up" or "you sum up". Many similar examples exist, as exemplified in table 16.3.

Using a trilemma model combined with a triclass model, we expect that the context will help identify the tag of such ambiguous words, relying both on the syntactic and the semantic levels, through the class and the lemma sequences. The importance of the trilemma model for the tagging process will be even greater in the case of plural forms, since French determiners like "les" and "des" do not help to choose between a masculine or a feminine lemma.

Creating a trilemma model need not be very difficult if a tagged training text is already available. Each wordform can be associated in a deterministic fashion to a lemma, since, for most inflected languages, practically all pairs (wordform, wordclass) correspond to only one lemma. There are only very few exceptions to this rule, like (for French) "fils", a plural noun coming from both "fil" ("thread") and "fils" ("son"), or "plu", a past participle coming from both "pleuvoir" ("to rain") and "plaire" ("to please").

[13] However, a given lemma may have several meanings, which should ideally be distinguished to refine the prediction. This polysemy disambiguation problem is a difficult one and is not taken into account here.

17 MACHINE LEARNING APPROACHES

Walter Daelemans

17.1 INTRODUCTION

The usefulness and feasibility of *automatically training* a syntactic wordclass tagger instead of hand-crafting it motivated a large body of work on statistical and rule-learning approaches to the problem. Syntactic wordclass taggers trained on corpora are claimed to be equally accurate as, and more robust and more portable than, hand-crafted systems[1]. Moreover, development time is considerably faster. Recently, inductive machine learning approaches such as connectionist learning algorithms, decision tree induction and case-based learning have also been applied to the syntactic wordclass disambiguation problem. In some cases these approaches have interesting properties not present in existing statistical and rule-based approaches.

A succesful inductive machine learning algorithm works by extracting generalizations from a set of examples of a desired input-output mapping. The relations between input and output, implicit in these examples, are discovered by the algorithm and are used to predict the correct output when presented with a new, previously unseen, input

[1] Although we will report published accuracy figures for the taggers discussed, we should be cautious about them. Evaluation of the different methods has not been achieved the same way in each case.

H. van Halteren (ed.), Syntactic Wordclass Tagging, 285-304.

Table 17.1 Tagging as a mapping from sentences to tag strings.

Input	Output
John will join the board	np md vb dt nn

Table 17.2 Tagging as a mapping from focus words with context to tags.

Input					Output
Left context		Focus	Right context		
=	=	John	will	join	np
=	John	will	join	the	md
John	will	join	the	board	vb
will	join	the	board	=	dt
join	the	board	=	=	nn

pattern. In other words, the algorithm *classifies* a new input pattern as belonging to a particular output category.

Many problems in Natural Language Processing (NLP), especially disambiguation problems, can be formulated as classification tasks (Magerman 1994; Daelemans 1995; Cardie 1996). Tagging, e.g., can be seen as a mapping from sentences to strings of tags. In syntactic wordclass tagging, abbreviated tagging from here, a sentence should be mapped into a string of morphosyntactic tags (table 17.1).

By approximating this mapping with a function from a focus word and its context to the disambiguated tag belonging to the focus word in that context (table 17.2), the mapping becomes a *classification* task amenable to symbolic and connectionist Machine Learning (ML) approaches. Context size can vary from one word at each side of the focus word (comparable to trigram models in statistics) to the complete sentence. Some machine learning methods dynamically determine the context size needed to disambiguate a particular focus word. The information provided to the learning algorithm can consist also of 'morphological' features (suffixes, presence or absence of a hyphen, a capital or a digit), syntactic information (features representing the syntactic context in which a word has to be tagged), or any other available linguistic information. In general, there will be as many examples for the learning algorithm as there are words in the training corpus.

In this chapter, we will provide an overview of basic concepts in inductive learning methods and discuss recent research on the application of these methods to tagging.

We will discuss case-based classifiers, decision tree induction methods and neural networks.

Another promising approach is Inductive Logic Programming (ILP; e.g. Muggleton and De Raedt 1994), in which background knowledge and positive and negative examples are used to induce a logic program compatible with the background knowledge and all of the positive examples, but none of the negative examples. The induced logic programs are more expressive than the propositional language of feature vectors, which makes the approach potentially interesting to induce recursive tagging rules. See Cussens (1997) for an early example of this approach.

17.2 INDUCTIVE LEARNING FROM EXAMPLES

17.2.1 Concepts

Machine Learning (ML) is a sub-discipline of *Artificial Intelligence* (AI) which studies algorithms that can learn either from experience or by reorganizing the knowledge they already have. See Langley (1996) and Carbonell (ed.) (1990) for introductory material, Weiss and Kulikowski (1991) for methodological issues and Natarajan (1991) for a formal-theoretical approach.

Conceptually, a learning system consists of a *performance component* which achieves a specific task (given an input, it produces an output) and a *learning component* which modifies the performance component on the basis of its experience in such a way that performance of the system in doing the same or similar tasks improves (figure 17.1). Experience is interpreted rather narrowly here and is represented as a set of examples used to train the system. Examples usually take the form of pairs of inputs with their associated desired output. In tagging, the input is (a description of) a focus word and its context and the desired output is the disambiguated tag to be assigned to the focus word.

To achieve its task, the performance component uses an internal representation. The task of the learning component may therefore be construed as a search in the space of possible representations for a representation that is optimal for performing the mapping. In this chapter, we will consider decision trees, case bases and sets of connection weights as types of languages/formalisms for internal representations for trainable taggers. In most cases, finding *the* optimal representation given a set of examples and a representation language is computationally intractable. Some form of heuristic search is therefore used by all learning systems.

In Machine Learning, the concept of *bias* refers to domain-dependent constraints on the search process: knowledge about the task may be used to make the search simpler. There may also be bias in the way the experience presented to the learning component (the training examples) is preprocessed. The addition of linguistic bias to a learning system is the obvious way to let learning taggers profit from linguistic knowledge about the task.

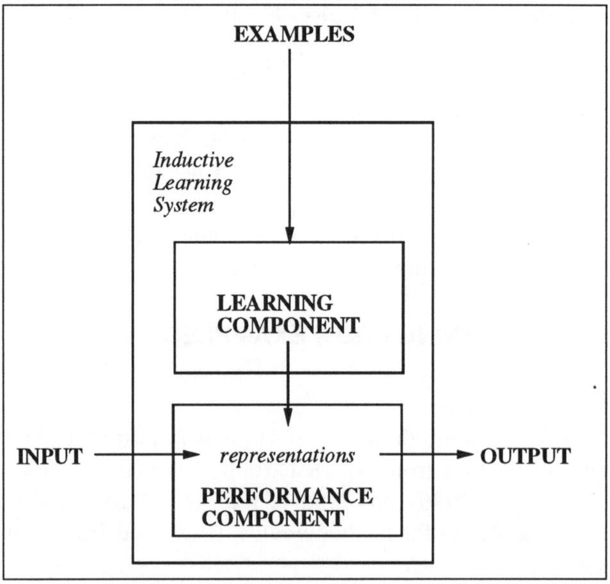

Figure 17.1 General architecture of an inductive learning system.

17.2.2 Classification of learning methods

Given this very general model of inductive learning, a number of dimensions can be distinguished that should be considered in comparing and experimenting with these techniques.

- Amount of *Supervision*. In supervised learning, experience takes the form of *examples* of inputs and the corresponding desired output for each of these inputs. The examples are presented to the system in a training phase. In *unsupervised learning*, examples are presented without information about the desired output. It is up to the system to find similarities in the examples in such a way that they can be exploited by the performance component to solve the task. In *reinforcement learning*, no examples are given; only an indication of the correctness of the output the performance component produces given an input (*feedback*). Most of the research described in this chapter concerns supervised approaches (learning from examples).

- *Input Representation*. Representations used in the ML literature include vectors of bits, vectors of feature-value pairs (numeric or nominal values; compare 'flat' feature structures in linguistics) and complex recursive representations such as

Table 17.3 Comparison of inductive learning methods discussed in this chapter.

	Decision Tree	Case-based	Neural Network
Supervised	YES	YES	YES
Input Rep.	feature value vector	feature value vector	bit string
Output Rep.	symbolic category	symbolic category	bit string
Internal Rep.	tree	cases	connection weights
Incremental	NO/yes[2]	YES	NO
Noise tolerant	YES	YES	YES

semantic nets (compare recursive feature structures in linguistics). For tagging problems, vectors of bits and of nominal feature-value pairs have been proposed.

- *Output Representation.* Output can be a binary category (yes/no) decision, a symbolic category (a finite, discrete set of labels), a continuous category (a real number) or a vector of any of these. In machine learning for tagging, vectors of binary categories and symbolic categories are used (see also Chapter 4).

- *Internal Representation.* The representation used by the performance component, and optimized by the learning component, can be numeric (e.g. connection weights with neural networks) or symbolic (semantic nets, rules, decision trees, examples, ...).

- *Incremental Learning.* A learning system can be incremental. In that case, relevant information in additional examples can be integrated by the learning component into the performance component without re-learning everything from scratch. In non-incremental or *batch learning* systems (such as neural networks), this is not possible. In batch learning, the complete set of examples has to be inspected (sometimes several times) before learning is completed and addition of new examples makes complete relearning necessary.

- *Noise Tolerance.* Different algorithms can be more or less sensitive to noise in the input (wrongly coded examples, missing values, or even ambiguous examples, i.e. examples which have been assigned contradictory outputs in the training set). Algorithms dealing with linguistic data should be noise-resistant.

Table 17.3 gives a characterization of the different inductive learning algorithms discussed in this chapter along these dimensions.

[2] In most versions, the decision tree learning algorithm is batch learning, but incremental versions have been developed (e.g. Utgoff 1989).

17.2.3 Performance evaluations

The success of a learning component in improving performance can be evaluated using a number of different quantitative and qualitative measures:

- *Generalization accuracy.* Performance accuracy of the system on previously unseen inputs (i.e. inputs it was not trained on). This aspect of learning is of course crucial: it gives an indication of the quality of the *inductive leap* made by the algorithm on the basis of the examples. A good generalization accuracy indicates that the learning system has not *overfit* its training examples, as would happen by generalizing on the basis of errors or exceptions present in them. To get a good estimate of the real generalization accuracy, *cross-validation* can be used, e.g. in 10-fold cross-validation an algorithm is tested on ten different partitions (90% training material, 10% testing material) of the full data set available. Each data item occurs once in one of the test sets. The average generalization accuracy on the ten test sets is then a good statistical estimate of the real accuracy (see also Chapter 6).

- *Space and time complexity.* The amount of storage and processing involved in learning (training the system) and performance (producing output given the input).

- *Explanatory Quality.* Usefulness of the representations found by the learning system as an explanation of the way the task is achieved.

17.2.4 Overview of methods

To sum up this introductory section, we will give an intuitive description of how each of the studied algorithms works, using tagging as an example application. We discuss the algorithms in an order of increasing abstraction of the internal representation used by the performance component and created by the learning component. We start from storage and *table-lookup* of the 'raw' examples as a non-learning baseline.

- *Table Look-Up.* Store all examples (patterns of target words with their context and their corresponding disambiguated tag) in a table. When a new input pattern is given to the performance system, look it up in the table and retrieve the output of the stored example. In this approach the system does not actually learn anything and it fails miserably whenever an input pattern is not present in the table (there is no generalization).

- *Case-Based Learning.* Store all examples in a table. When a new input pattern is given to the performance system, look up the most *similar examples* (in terms of number of feature values common to the stored pattern and the new pattern, for example) and extrapolate from the tags assigned to these nearest matches to the new case. Various statistical and information-theoretic techniques can be used to design the *similarity metric*. The similarity metric is also a place where linguistic bias can be introduced in the learning algorithm.

- *Rule and Decision Tree Induction.* Use similarities and differences between examples to construct a decision tree or a rule set (these two are largely equivalent and can be translated to each other) and use this constructed representation to assign a tag to a new input pattern. Forget the individual examples.
- *Connectionism, Neural Networks.* Use the examples to train a network. In back-propagation learning, this training is done by repeatedly iterating over all examples, comparing for each example the output predicted by the network (random at first) to the desired output and changing connection weights between network nodes in such a way that performance increases. Keep the connection weight matrix and forget the examples.

The place of stochastic (statistical) approaches deserves some discussion at this point. In this popular approach to tagging, statistical models (e.g. about the N-grams occurring in a language) are computed on the examples (the corpus) and these are used to extrapolate to the most probable analysis of new input. In terms of abstraction versus data-orientation, stochastic, neural network and rule induction approaches are *greedy learning* techniques. These techniques abstract knowledge from the examples as soon as they are presented. Case-Based Learning is a *lazy learning* technique: generalization only occurs when a new pattern is offered to the performance component and abstraction is therefore implicit in the way the contents of the case base and the similarity metric interact.

One succesful statistical approach, recently applied to the tagging problem, is Ratnaparkhi's use of Maximum Entropy Models (1996). In this classification-based approach, diverse sources of contextual information (comparable to those used by the machine learning approaches discussed below) are expressed as binary features, and are combined in a statistical model that makes no further distributional assumptions on the training data by maximizing the entropy of the distribution subject to the constraints of the training data. The model parameters for the distribution are estimated using an iterative procedure called generalized iterative scaling.

In the remainder of this chapter, we will discuss each of the learning methods and their application to tagging in turn. We conclude with a general discussion and evaluation of the methods described.

17.3 CASE-BASED LEARNING

The case-based learning paradigm is founded on the hypothesis that performance in cognitive tasks (in our case language processing) is based on reasoning on the basis of analogy of new situations to *stored representations of earlier experiences* rather than on the application of *mental rules* abstracted from representations of earlier experiences as in rule induction and rule-based processing.

The concept has appeared in several AI disciplines (from computer vision to robotics) several times, using apart from case-based also labels such as similarity-based, example-based, exemplar-based, analogical, nearest-neighbour, instance-based and memory-based (Stanfill and Waltz 1986; Kolodner 1993; Aha *et al.* 1991; Salzberg 1990). These different names are conveniently captured under the term *lazy learning* (Aha 1997).

17.3.1 Algorithm

Examples are represented as a vector of feature values with an associated category label. Features define a pattern space. During training, a set of examples (the training set) is presented in an incremental fashion to the learning algorithm and added to memory. During processing, a vector of feature values (a previously unseen test pattern) is presented to the system. Its distance to all examples in memory is computed using a *similarity metric* and the category of the most similar instance(s) is used as a basis to predict the category for the test pattern.

In this type of lazy learning, performance crucially depends on the similarity metric used. The most straightforward metric for a problem like tagging with nominal (non-numeric) feature values would be an *overlap metric*: similarity is defined as the number of feature values that are equal in two patterns being compared. In such a distance metric, all features describing an example are interpreted as being equally important in solving the classification problem, but this is not necessarily the case: the category of the word immediately before a word to be tagged is obviously more important than the category of the word three positions earlier in the sentence. We call this problem the *feature relevance problem*. Various feature weighting and selection methods have been proposed to differentiate between the features on the basis of their relevance for solving the task (see Wettscherek *et al.* (1996) for an overview).

Another addition to the basic algorithm that has proved relevant for many natural language processing tasks is a value difference metric (Stanfill and Waltz 1986; Cost and Salzberg 1993). Such a metric assigns different distances to pairs of values for the same feature. In tagging, e.g., such a metric would assign a smaller distance between NP and NN than between NN and VBG. These biases can of course also be added by hand to the learner (e.g. by a domain expert). Several other improvements and modifications to the basic case-based learning scheme have been proposed and should be investigated for linguistic problems. Two promising further extensions are weighting the examples in memory and minimizing storage by keeping only a selection of examples. In example weighting, examples are differentiated according to their quality as predictors for the category of new input patterns. This quality can be based on their typicality or on their actual performance as predictors on a held-out test set. In example selection, memory is pruned by deleting those examples which are bad predictors or which are redundant.

17.3.2 Case-based tagging

KENMORE (Cardie 1994, 1996) is presented as a general framework for knowledge acquisition for NLP using different symbolic machine learning techniques. As an instance of this general methodology, a case-based learning approach is suggested for both morphosyntactic and semantic tagging. The architecture presupposes a corpus, a sentence analyser and a learning algorithm. During knowledge acquisition (training) for a specific disambiguation task (e.g. tagging), a case is created for each instance of the problem in the corpus. Each case is an example of the input-output mapping to be learned; the input part is a context describing the ambiguity and the output part is the solution to the particular ambiguity. The examples may be produced from an annotated version of the corpus or through human interaction. During application, the case-base is used to predict the solution to a new instance of the ambiguity given the input (the context) without intervention.

In a tagging experiment based on 2056 cases from the Tipster JV corpus, a fairly complex case representation based on output from the CIRCUS conceptual sentence analyser is used. Figure 17.2[3] shows the case representation containing context features of the focus word "parts", for which the word sense and the part of speech has to be decided.

Local context features describe the syntactic and semantic information about a five-word window centered on the word to be tagged (the words themselves, their part of speech and their word sense). Global context features provide information about the major constituents parsed already (word sense of the subject and type, concept and semantic feature associated with the last parsed constituent). The class to be predicted is the part of speech and the semantic sense of the middle word. All words (except 129 function words) are initially assumed unknown.

As a solution to the feature relevance problem, Cardie (1993) applies a decision tree learning algorithm (see below) to the dataset and uses only those features which have been selected by the decision tree induction algorithm as being relevant during similarity comparison. This turns out to be an effective way to discard irrelevant features.

MBT (Memory-Based Tagging; Daelemans 1995; Daelemans *et al.* 1996) is a case-based approach in which the feasibility of the approach on a larger scale is investigated, with simpler case representations and with a more elegant solution to the feature rele-

[3] Reprinted from Cardie (1996) with permission.

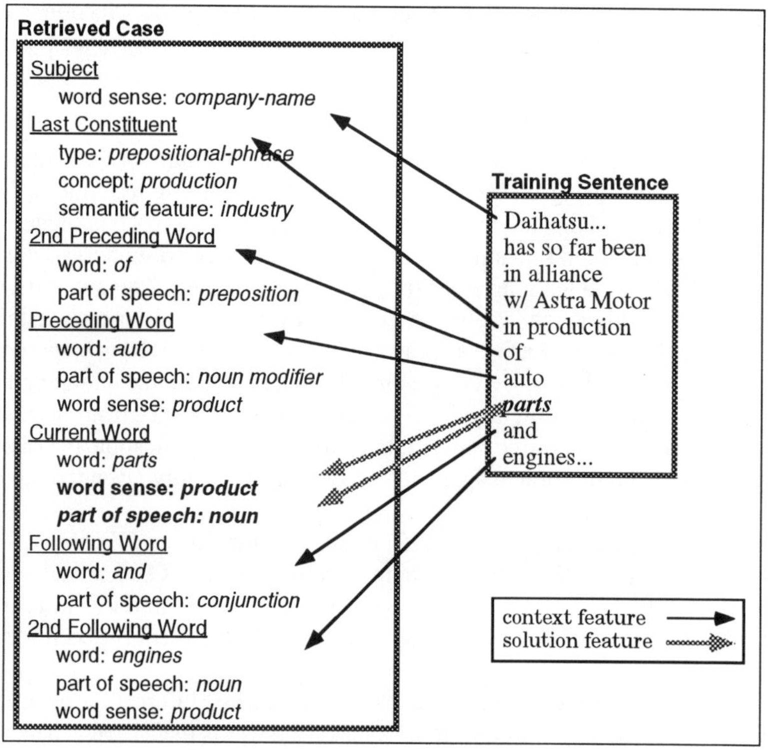

Figure 17.2 Case representation in KENMORE.

vance problem. Experiments were performed on the 3 million word tagged Wall Street Journal corpus[4].

In order to adopt the memory-based approach to the problem of tagging, the following procedure is used:

Lexicon Construction. A lexicon is extracted from the training corpus by computing for each word the number of times it occurs with each category. A new, possibly *ambiguous* tag (e.g. N-V for words which can be both a noun and a verb) is assigned to each word based on this lexical definition.

[4] ACL Data Collection Initiative CD-ROM 1, September 1991; the tagset is the Penn Treebank tagset, consisting of 40 tags. See Appendix 17.6 for a full list.

Case-Base Construction. Two case bases are constructed, one for known words and one for unknown words. The former contains as information the possibly ambiguous category of the word to be tagged (the focus word) and of one context word to the right and the disambiguated category of two words to the left. The latter contains the same context information, but instead of the ambiguous category of the focus word (which is unknown), the first letter and the last three letters of the focus word are added as features. These features provide information about the 'morphology' of the word. The unknown-words case base is constructed on the basis of open class words only.

Tables 17.4 and 17.5 list samples of the known-words and unknown-words case bases for part of the first sentence of the corpus. In the first table, we use the following abbreviations for the known-words case base: f for focus word (the word to be disambiguated, represented by its ambiguous category), d for disambiguated word (a previously contextually disambiguated word to the left of the focus word), a for ambiguous word, a still to be disambiguated word to the right of the focus word, represented by its ambiguous category, and t for the target, disambiguated, category of the focus word. In the unknown-words case base, we find features for left (d) and right (a) context, and instead of the lexical representation of the focus word, we have features representing prefix letters (p) and suffix letters (s) of the focus word.

Tagging. In tagging, new input text is transformed into case representations for case-based reasoning on the basis of either the known- or unknown-words case base.
In MBT, the feature relevance problem is solved by weighting each feature with the average amount of case base information entropy reduction it can provide (i.e. its Information Gain, IG; see Daelemans *et al.* (1996) for more information). The weights for the different features are also listed in tables 17.4 and 17.5. They express the relative relevance of the features and are used as a weight during similarity computation. Advantages of this approach (compared to KENMORE's) are that feature relevance is not interpreted as a yes/no property but as a gradual one and that it does not presuppose using two inductive classification learning algorithms (decision tree induction and case-based learning), one of which is only used for feature selection.

To solve the computational complexity problem inherent in matching all feature values of a new case to the corresponding values of all stored cases, MBT uses IGTREE, a memory- and processing-time-saving heuristic implementation of memory-based reasoning. This formalism (fully described in Daelemans *et al.* 1996, 1997) compresses a memory base into a tree using an information-theoretic heuristic, reducing storage and retrieval complexity considerably without an adverse effect on generalization accuracy. An additional advantage is that this tree structure allows dynamic selection of context width (see also 17.4 below on decision trees).

Table 17.4 Case representation and information gain pattern for known words.

Word	Case Representation				
	d	d	f	a	t
IG	.06	.22	.82	.23	
Pierre	=	=	np	np	np
Vinken	=	np	np	,	np
,	np	np	,	cd	,
61	np	,	cd	nns	cd
years	,	cd	nns	jj-np	nns
old	cd	nns	jj-np	,	jj

Table 17.5 Case representation and information gain pattern for unknown words.

Word	Case Representation						
	d	p	s	s	s	a	t
IG	.21	.21	.15	.20	.32	.14	
Pierre	=	P	r	r	e	np	np
Vinken	np	V	k	e	n	,	np
61	,	6	=	6	1	nns	cd
years	cd	y	a	r	s	jj-np	nns
old	nns	o	o	l	d	,	jj

17.3.3 Evaluation

In the experiment with KENMORE, accuracy of tagging turned out to be 95% overall and 91% on contents words only. In MBT on the Wall Street Journal corpus, overall generalization accuracy was 96.4% (96.7% on known words, 90.6% on unknown words). The level of accuracy attained by probabilistic taggers seems to be well in reach of case-based taggers. Already at small data set sizes, performance is relatively high. We obtained similar scores when adapting the tagger architecture to Dutch by training it on a Dutch tagged corpus.

An important advantage of the case-based approach is the flexibility of case representations: there are several types of information which can be stored in the memory base, ranging from the words themselves to intricate lexical representations. Combined with feature-weighting approaches, this flexibility offers a new approach to information source integration (data fusion) in tagging. The unknown-words case base in MBT, e.g., integrates context information and 'morphological' information (suffix letters) in a smooth way. It would be impossible at present, for reasons of sparseness of data and computational complexity, to estimate probabilities for such intricate contexts in a stochastic approach.

Additional advantages include incremental learning (new cases can be added incrementally to the case bases without need for relearning[5]), explanation capabilities (the best memory matches serve as explanations for the tagging behaviour of the system) and, at least in MBT, fast learning and tagging (more than 1000 words per second).

17.4 DECISION TREE INDUCTION

The *decision tree learning* paradigm is based on the assumption that similarities between examples can be used to automatically extract decision trees and categories with explanatory and generalization power. In other words, the extracted structure can be used to solve new instances of a problem and to explain why a performance system behaves the way it does. In this paradigm, learning is *greedy* and abstraction occurs at learning time. There are systematic ways in which decision trees can be transformed into rule sets (the two representations are equivalent).

Decision tree induction is a well-developed field within AI. See, e.g., Quinlan (1993) for a synthesis of major research findings. More ancient statistical pattern recognition work (such as Hunt *et al.* 1966; Breiman *et al.* 1984) also still makes for useful reading.

17.4.1 Algorithm

A decision tree is a data structure in which nodes represent tests, and arcs between nodes represent possible answers to tests. Leaf nodes represent answers to problems. A problem is solved by following a path from the root node through the decision tree until a leaf node is reached. The path taken depends on the answers that a particular problem provides to the tests at the nodes. Decision tree *learning* works by repeatedly dividing the set of examples into subsets according to whether the examples in a particular subset have a feature-value pair in common, until the subsets are homogeneous, i.e. all examples in the subset have the same category. The algorithm achieves this according to the simplified recursive scheme in Figure 17.3.

To classify new input patterns with a decision tree, start at the root node of the tree and find the value in the input pattern for the corresponding feature. Take the branch corresponding to that value and perform this process recursively until a leaf node is reached. The category corresponding to this leaf node is the output.

[5] Computation of feature weights is not incremental, it presupposes access to a complete batch of training examples, but usually, the weight values become stable after only a few hundred training examples.

Given a set of examples T

If T contains one or more cases all belonging to the same class C_j,
then the decision tree for T is a leaf with category C_j.

If T contains different classes then

- Choose a feature, and partition T into subsets that have the same value for the feature chosen. The decision tree consists of a node containing the feature name and a branch for each value leading to a subset.
- Apply the procedure recursively to subsets created this way.

Figure 17.3 Recursive scheme for constructing decision trees.

Again, we are confronted with a *feature relevance problem* in this approach. In order to obtain a concise tree with good generalization performance (i.e. a tree reflecting the structure of the domain), we have to select at each recursion of the above algorithm a test which is optimal in achieving this goal). The algorithm is non-backtracking and considering all trees consistent with the data is an NP-complete problem, so a reliable heuristic feature selection criterion is essential. Information-theoretic or statistical techniques maximizing homogeneity of subsets by selecting a particular feature are usually applied to this end. Several variants and extensions have been developed to the basic algorithm for pruning (making the tree more compact by cutting off subtrees on the basis of a statistical criterion), grouping similar values of a feature into classes, making tree building incremental, etc.

17.4.2 Decision tree tagging

Work on parsing (including tagging) of text with decision trees was pioneered at IBM (Black *et al.* 1992; Magerman 1994, 1995). SPATTER (Magerman 1995) starts from the premise that a parse tree can be viewed as the result of a series of classification problems (tagging, choosing between constituents, labelling constituents, etc.). The most probable sequence of decisions for a sentence, given a training corpus, is its most probable analysis. In the statistical decision tree technology used (based on Breiman *et al.* 1984), decision trees are constructed for each sub-problem in the parsing task (tagging is one of them). In such a decision tree, leaf nodes contain distributions over categories instead of a single category. E.g., in tagging, the feature associated with the root node of the decision tree might be the word to be tagged. In case its value is "the", the category "article" can be returned with certainty. In case its value is "house", a test at the next level of the tree corresponds to the feature "tag of the previous word". In case its value is "article", the probability distribution returned by the decision tree would be "noun (.8); verb (.2)" (Magerman 1995). In practice, SPATTER uses binary trees, however. Searching for the most probable tag series for a sentence is done by

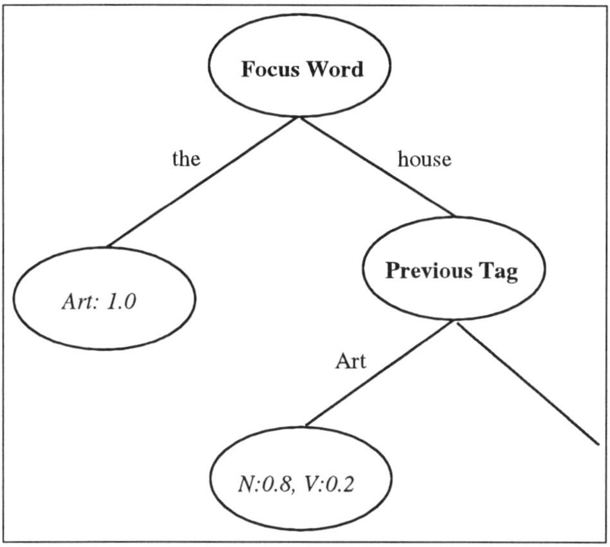

Figure 17.4 A statistical decision tree for tagging.

means of stack decoder search with a breadth-first algorithm and probabilistic pruning. Figure 17.4 shows such a tree, based on Magerman (1995).

Schmid (1994b) describes TREETAGGER, a tagger which takes basically the same approach as SPATTER. Transition probabilities between tags in a tag sequence are estimated using a decision tree induced from a set of N-grams occurring in the Penn Treebank corpus. The features are the tags of the words preceding the word to be tagged. He experimented with one, two and three such context features. The category to be predicted is the tag of the focus word. Using an information-theoretic heuristic feature selection method, the tree is built using the recursive algorithm discussed earlier. As in SPATTER, all tests have binary results. Instead of having a subtree for each possible tag in a context position, corresponding to questions like "what is the tag of the word before the target?", tests are instead individual feature-value combinations with a binary branching, corresponding to questions like "is the tag of the word before the target equal to ADJ?". This results in deeper trees. Again as in SPATTER, at the leaf nodes a probability distribution over the categories is given for those patterns for which the sequence of tests leading to this leaf node are true. The information-theoretic feature selection method ensures that the test chosen maximizes the distinctiveness of the probability distribution of the subtrees. For pruning the resulting decision tree, an information-theoretic heuristic is used as well. Finally, the Viterbi algorithm (Viterbi 1967; see also Chapter 16) is used to find the best tag sequence for a sentence, given the probability distributions obtained by decision tree lookup. The approach is combined with a lexicon system containing a priori tag probabilities for each word and a probabilistic suffix analyser.

17.4.3 Evaluation

Decision tree models are equivalent in expressive power to interpolated N-gram models (Magerman 1995), but whereas in N-gram models the number of parameters to be estimated grows exponentially with N, in decision-tree learning, the size of the model depends on the number of training examples and remains constant with the number of decisions taken into account. Also, the decision tree approach automatically selects relevant context size: uninformative context positions are not used in the tree and because of its computational properties (constant with wider context) larger contexts (corresponding to 4 or 5-grams) can initially be considered. That way, decision tree approaches are potentially more sensitive to context and therefore better equiped to solve long-distance dependencies. Finally, Schmid also reports robustness relative to training set size: TREETAGGER 'degrades gracefully' with smaller training set sizes.

As far as performance is concerned, decision tree methods seem to be comparable to stochastic appoaches: Schmid reports 96.4% generalization on the Penn Treebank using 4-grams (0.3% better than a similar probabilistic trigram tagger, which, however, uses a different lexicon system). Magerman reports 96.5% for sentences up to 40 words in length in the Wall Street Journal corpus. Schmid also reports fast tagging speed performance (10,000 words per second).

In the examples discussed here, decision tree technology does not deliver a solution to the complete tagging problem. The tag probabilities returned by the tree are used by a search mechanism (stack decoder or Viterbi) to find the best series of tags. Also, both versions of the decision tree approach are not completely non-parametric; SPATTER requires smoothing of the decision trees and TREETAGGER requires a pruning threshold. In principle, however, it would be possible to produce a complete tagger on the basis of a learned statistical decision tree. Recently, this approach has indeed been explored (Màrquez & Rodríguez, 1998).

17.5 NEURAL NETWORK METHODS

Multilayer Perceptrons (Rumelhart *et al.* 1986) are the most popular neural network architecture. As an inductive learning technique, supervised neural network learning is a greedy learning approach. During learning, the set of examples is repeatedly inspected to find an optimal set of connection weights between layers of simple units. The training material is thus abstracted into a set of numeric weights which is then used to predict the output of new input patterns. There is an immense literature on neural network algorithms (see Aleksander and Morton (1990), Bishop (1995) and Fausett (1994) for recent introductions). There is also a considerable body of research

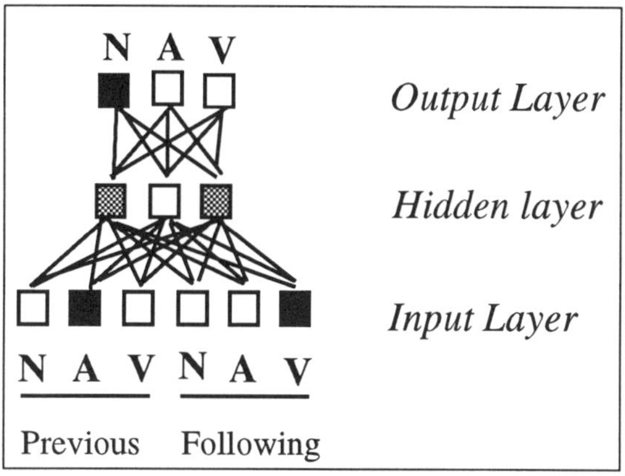

Figure 17.5 A simple neural network architecture for tagging. Activation of units is shown by colouring. In a situation where the previous word is an adjective and the following word a verb, this hypothetical 3-layer network would predict the current word to be a noun.

on applying neural network technology to language processing problems (Reilly and Sharkey (eds.) 1992; Sharkey 1992).

17.5.1 Algorithm

A multilayer perceptron consist of an input and an output layer of simple processing units and one or more intermediate, 'hidden' layers (figure 17.5). The input layer is used to code the input part of an example, the output layer to encode the output part. We will assume a single hidden layer here. All adjacent layers are fully interconnected, i.e. each unit in each layer connected to each unit in the next layer. Each unit has an activation and a threshold. Each connection between two units has a weight. Activation can be expressed as 1 or 0, or as a real number; thresholds and connection weight are usually expressed as real numbers. Activation flows from the input layer to the output layer via the hidden layer.

Two simple rules govern the training and use of a multilayer perceptron. The *activation rule* is a local rule which is used by each unit to compute its activation. Consider for instance the activation flow from input layer to hidden layer. The input activation of a particular unit of the hidden layer (a_j), is equal to $\sum_i a_i \times w_{i,j}$, where the a_i are the activations of the units in the input layer connected to that unit at the hidden layer and the $w_{i,j}$ are the weights of the connections from those input layer units to that hidden layer unit. The threshold value determines whether, given the input activation to a unit, that unit will become active (if the input activation exceeds the threshold the unit becomes active, otherwise it stays 'off'). By making use of such a threshold value, the activation of a unit is a *non-linear* function of activation values of the units in the previous layer connected to it.

The *learning rule* (in this case back-propagation learning) incrementally adapts the connection weights until an acceptable performance is reached by the system (this involves repeatedly cycling through all training examples). This adaptation process works by changing the connection weights depending on the quality of the output of the network (i.e. the activation pattern at the output layer). The output of the network is compared to the desired output. Those connections that contributed to the wrong activation of an output unit are weakened, those that were responsible for not activating an output unit that should have been active are strengthened. These error corrections are also 'back-propagated' to the connections between input and hidden layer.

17.5.2 Neural network tagging

NETGRAM (Nakamura *et al.* 1990) is a multilayer neural network for word category prediction in the context of a speech recognition system. The input consists of the category of two (or more) preceding words, the output is the category of the current word. Words are encoded the following way. Each word is represented by a number of units equal to the number of categories. The unit associated with the category of the word is made active, the other units are inactive. Given two words preceding context and 89 categories, the network has an input layer of 178 units and an output layer of 89 units. Back-propagation is used to train the network. When an input is presented to the network, the two input units corresponding to the categories of the two preceding words are made active and the output unit with the highest activation is taken as the category of the current word.

NET-TAGGER (Schmid 1994a) also used back-propagation learning, but in this case the problem handled is disambiguation (tagging) rather than tag prediction. In the input layer, information about the word to be tagged and one or more preceding and following words is encoded. Again, for each tag and each word pattern position, a unit is created. E.g., supposing 40 tags, 3 words left context, a focus word and 2 words right context, an input layer of 240 (40×6) units is needed. For the left context words, the previous output of the network (the activation levels of the output layer units) is used as input. For the focus word and the right context, the lexical probabilities of the words are used. Adding a hidden layer to a two-layer network did not improve performance. The output layer has one unit for each possible wordclass. The output unit with the highest activation, given an input pattern, is interpreted as the tag of the focus word. In order to attach lexical probabilities to words, a lexicon system based on Cutting *et al.* (1992) is used, combined with an unknown-word guesser using information about suffix letters.

17.5.3 Evaluation

Accuracy of NETGRAM and NET-TAGGER is comparable to stochastic trigram methods, the main advantage being that unseen patterns are interpolated effectively, without requiring expensive special interpolation methods as in stochastic approaches. Connectionist approaches also require the computation of fewer parameters (weights) than statistical models (N-gram probabilities), which becomes especially useful when considering a wider context than trigrams. The number of weight-parameters grows quadratic with the number of input nodes. As in the learning methods discussed earlier, it is possible to output more than one possible tag for a word, e.g. a list of tags ordered according to their likelihood.

17.6 DISCUSSION

With the availability of only a relatively small body of empirical data and theoretical analysis on the applicability of inductive machine learning techniques to tagging, it is too early for strong conclusions. On the empirical side, there is a hard-felt need for a methodologically sound, reliable, empirical comparison of statistical and machine learning approaches to automatic tagging. On the theoretical side, there is a need for more insight into the differences and similarities in how *generalization* is achieved in this area by different statistical and machine learning techniques. In the absence of this knowledge, our discussion will necessarily turn out to be preliminary and superficial. Our discussion will take the form of a number of theses.

- *Learning is preferable to programming.* Compared to hand-crafted rule-based (or constraint-based approaches), an inductive learning approach, such as in the methods discussed here and in stochastic approaches, provides a solution to the knowledge-acquisition and reusability bottlenecks and to robustness and coverage problems. As an example, a fast and accurate tagger for Dutch was learned with the MBT tagger-generator in half a day, with a minimum of linguistic engineering. On the other hand, it should be noted that the accuracy which can be obtained using hand-crafted constraint-based methods (cf. Chapter 14) still seems to be out of reach for automatic learning approaches.

- *Machine Learning is a different type of statistics.* Decision tree induction, case-based learning and neural networks are statistical methods, but they use a different kind of statistics than the well-known maximum-likelihood and Markov model methods, e.g. in case-based learning, no assumptions are made about the distribution of the data whereas most statistical techniques presuppose normal distributions. Different statistical methods have different properties which make them more or less suited for a particular type of application. If only for that reason, the applicability of all types of statistics to the tagging problem should be studied thoroughly. Already from the preliminary empirical data, important advantages of these methods compared to current statistical methods suggest themselves:

1. They require less training data.
2. They require less parameters to be computed and can therefore take into account more context.
3. They provide elegant and computationally attractive solutions to the smoothing problem and to the integration of different information sources.
4. Training is often much faster.

- *Abstraction can be harmful.* In many linguistic tasks, we have found (see Daelemans, Van den Bosch and Zavrel 1999) that an approach keeping complete memory of all training data provides better performance than techniques that abstract from low-frequency and exceptional events, such as rule(learning)-based systems. Neural networks and stochastic approaches are similar to rule- and decision-tree-induction methods in that they abstract from their experience (to a matrix of connection weights in neural networks, to a set of probabilities in stochastic approaches and to a set of rules in rule-induction approaches) and forget about the original data on which these abstractions were based. The effect that full memory of all examples yields better generalization is probably related to the fact that natural language processing tasks such as morphosyntactic disambiguation can be characterised by the interaction of regularities, sub-regularities and pockets of exceptions. Abstracting away from these exceptions causes a performance degradation because new similar exceptions are overgeneralized: being there is better than being probable.

Compared to the well-developed theoretical and empirical foundations of statistical approaches to tagging, the machine learning approach to this problem has only just started. In all methods described, there is still a lot of room for improvement, especially in three areas: exploring variations or extensions of the basic algorithms, adding linguistic bias to the learning algorithms and combining them with other approaches in hybrid architectures. A fourth area where machine learning methods may provide increased tagging accuracy is in the development of machine learning algorithms that take as input the outputs of different taggers (trained, statistical or even hand-crafted) and learn when to trust which tagger. Initial work on this approach is described by van Halteren, Zavrel and Daelemans (1998) and by Brill and Wu (1998).

APPENDIX A: EXAMPLE TAGSETS

In this appendix, we give the full list of tags for three well-known tagsets, viz. those used for the Brown Corpus, for the Penn Treebank and by the EngCG-2 tagger.

There are two reasons to include these full lists. First of all, the three tagsets are used in examples in several chapters of the book and the lists are necessary for a good understanding of these examples. But the tag lists also serve by themselves as an exemplification of complete tagsets, e.g. regarding differences in granularity.

A.1 THE BROWN CORPUS TAGSET

Our first example is the tagset used for the Brown Corpus (Francis and Kučera 1982). It is typical for a whole class of medium granularity tagsets, usually consisting of around a hundred atomic tags.

The list below presents the basic tags. The tagset also includes combination tags. Examples are

- negative forms, e.g. "isn't" is tagged BEZ*
- enclitic forms, e.g. "nobody's" is tagged PN+BEZ
- foreign words, e.g. "esprit" is tagged FW-NN
- cited words, e.g. a citation of the word "book" is tagged NN-NC

H. van Halteren (ed.), Syntactic Wordclass Tagging, 305-310.

- words in headlines, e.g. "book" in a headline is tagged NN-HL
- words in titles, e.g. "book" in a title is tagged NN-TL

Tag	Description	Examples
.	sentence closer	. ; ? !
(	left parenthesis	
)	right parenthesis	
*	"not", "n't"	
–	dash	
,	comma	
:	colon	
ABL	pre-qualifier	quite, rather
ABN	pre-quantifier	half, all
ABX	pre-quantifier	both
AP	post-determiner	many, several, next
AT	article	a, the, no
BE	"be"	
BED	"were"	
BEDZ	"was"	
BEG	"being"	
BEM	"am"	
BEN	"been"	
BER	"are", "art"	
BEZ	"is"	
CC	coordinating conjunction	and, or
CD	cardinal numeral	one, two, 2
CS	subordinating conjunction	if, although
DO	"do"	
DOD	"did"	
DOZ	"does"	
DT	singular determiner	this, that
DTI	singular or plural determiner/quantifier	some, any
DTS	plural determiner	these, those
DTX	determiner/double conjunction	either
EX	existential there	
HV	"have"	
HVD	"had" (past tense)	
HVG	"having"	
HVN	"had" (past participle)	
HVZ	"has"	
IN	preposition	
JJ	adjective	
JJR	comparative adjective	
JJS	semantically superlative adjective	chief, top
JJT	morphologically superlative adjective	biggest
MD	modal auxiliary	can, should, will
NN	singular or mass noun	
NN$	possessive singular noun	
NNS	plural noun	
NNS$	possessive plural noun	

Tag	Description	Examples
NP	proper noun or part of name phrase	
NP$	possessive proper noun	
NPS	plural proper noun	
NPS$	possessive plural proper noun	
NR	adverbial noun	home, today, west
NRS	plural adverbial noun	
OD	ordinal numeral	first, 2nd
PN	nominal pronoun	everybody, nothing
PN$	possessive nominal pronoun	
PP$	possessive personal pronoun	my, our
PP$$	second (nominal) possessive pronoun	mine, ours
PPL	singular reflexive/intensive personal pronoun	myself
PPLS	plural reflexive/intensive personal pronoun	ourselves
PPO	objective personal pronoun	me, him, it, them
PPS	3rd. singular nominative pronoun	he, she, it, one
PPSS	other nominative personal pronoun	I, we, they, you
QL	qualifier	very, fairly
QLP	post-qualifier	enough, indeed
RB	adverb	
RBR	comparative adverb	
RBT	superlative adverb	
RN	nominal adverb	here then, indoors
RP	adverb/particle	about, off, up
TO	infinitive marker to	
UH	interjection, exclamation	
VB	verb, base form	
VBD	verb, past tense	
VBG	verb, present participle/gerund	
VBN	verb, past participle	
VBZ	verb, 3rd singular present	
WDT	*wh*-determiner	what, which
WP$	possessive *wh*-pronoun	whose
WPO	objective *wh*-pronoun	whom, which, that
WPS	nominative *wh*-pronoun	who, which, that
WQL	*wh*-qualifier	how
WRB	*wh*-adverb	how, where, when

A.2 THE PENN TREEBANK TAGSET

Our next example tagset is that designed for the Penn Treebank project (Marcus *et al.* 1993). Because of its projected use, its designers chose a more coarse granularity, leading to a rather small number of tags. For the same reason, the tagset includes a number of compromise tags, such as IN and TO, which serve to avoid 'difficult' choices for the annotators.

Tag	Description	Examples
CC	coordinating conjunction	and, therefore
CD	cardinal number	1987, twenty
DT	determiner	the, any

Tag	Description	Examples
EX	existential there	there
FW	foreign word	je, corporis
IN	preposition or subordinating conjunction	among, on
JJ	adjective	long, third
JJR	adjective, comparative	broader, clearer
JJS	adjective, superlative	closest, darkest
LS	list item marker	C, Third
MD	modal	can, shouldn’t
NN	noun, singular or mass	cabbage, wind
NNS	noun, plural	averages, products
NNP	proper noun, singular	Liverpool, Shannon
NNPS	proper noun, plural	Americans, Andes
PDT	predeterminer	all, such
POS	possessive ending	’, ’s
PRP	personal pronoun	he, myself
PRP$	possessive pronoun	his, your
RB	adverb	fiscally, occasionally
RBR	adverb, comparative	harder, more
RBS	adverb, superlative	earliest, least
RP	particle	along, off
SYM	symbol (mathematical or scientific)	%, >
TO	“to”	to
UH	interjection	uh, man
VB	verb, base form	ask, build
VBD	verb, past tense	registered, wore
VBG	verb, gerund/present participle	focusing, hankerin’
VBN	verb, past participle	chaired, used
VBP	verb, non-3rd ps. sing. present	sue, return
VBZ	verb, 3rd ps. sing. present	bases, pleads
WDT	wh-determiner	what, whichever
WP	wh-pronoun	what, whom
WP$	possessive wh-pronoun	whose
WRB	wh-adverb	how, whereby
#	pound sign	
$	dollar sign	
.	sentence-final punctuation	., !, ?
,	comma	
:	colon, semi-colon	
(	left bracket character	(, [
)	right bracket character	), }
"	straight double quote	
‘	left open single quote	
“	left open double quote	
’	right close single quote	
”	right close double quote	

A.3 THE ENGCG TAGSET

The final example in this appendix is the EngCG-2 tagset, which is featured mostly in chapter 14, where you can also find numerous references to the EngCG system. The information in the table below is current version at the time of writing, as found on the webpage of Conexor (http://www.conexor.fi), which markets the EngCG-2 software. It may differ in places with tags used in the examples in the chapters, e.g. the part-of-speech tags ING and EN used to be PCP1 and PCP2.

The EngCG tagset is different from the other example tagsets in that tokens are not associated with single atomic tags, but rather a sequence of tags, each covering a specific property (see also Chapter 4).

Part of speech	Subfeature	Description
N, ABBR		noun, abbreviation
	NOM	nominative
	GEN	genitive
	SG	singular
	PL	plural
	SG/PL	singular/plural
	<ADV-N>	noun often used adverbially
A		adjective
	ABS	absolutive
	CMP	comparative
	SUP	superlative
NUM		numeral
	CARD	cardinal
	ORD	ordinal
	SG	fraction, singular
	PL	fraction, plural
PRON		pronoun
	NOM	nominative
	GEN	genitive
	ACC	accusative
	SG	singular
	SG1	singular, first person
	SG3	singular, third person
	PL	plural
	PL1	plural, first person
	PL3	plural, third person
	SG/PL	singular/plural
	SG2/PL2	singular/plural, second person
	ABS	absolutive
	CMP	comparative
	SUP	superlative
	PERS	personal
	MASC	masculine
	FEM	feminine

Tag	Description	Examples
PRON		pronoun
	DEM	demonstrative
	RECIPR	reciprocal
	WH	WH-pronoun
	<Interr>	interrogative
	<Refl>	reflexive
	<Rel>	relative
DET		determiner
	GEN	genitive
	SG	singular
	PL	plural
	SG/PL	singular/plural
	ABS	absolutive
	CMP	comparative
	SUP	superlative
	DEM	demonstrative
	WH	WH-determiner
ADV		adverb
	ABS	absolutive
	CMP	comparative
	SUP	superlative
	WH	WH-adverb
ING		ING-form
EN		EN-form
V		verb: finite or infinitive
	INF	infinitive
	IMP	imperative
	PRES	present tense
	SUBJUNCTIVE	subjunctive
	PAST	past tense
	AUXMOD	modal auxiliary
	SG1	singular, first person
	SG3	singular, third person
	-SG1,3	non-singular 1st or 3rd person
	-SG3	non-singular 3rd person
	SG1,3	singular, first or third person
INTERJ		interjection
NEG-PART		"not", "n't"
INFMARK>		to, in+order+to etc.

REFERENCES

Aarts, F. and J. Aarts (1982). *English Syntactic Structures*. Oxford: Pergamon.

Aarts, J., P. de Haan and N. Oostdijk (eds.) (1993). *English Language Corpora: design, analysis and exploitation*. Amsterdam and Atlanta: Rodopi.

Aarts, J. and N. Oostdijk (1997). Handling discourse elements in syntax. In U. Fries, V. Müller and P. Schneider (eds.), *From Ælfric to the New York Times. Studies in English corpus linguistics*. Amsterdam and Atlanta: Rodopi. 107–123.

Aha, D.W. (1997). *Lazy Learning*, Reprinted from: Artificial Intelligence Review, 11. Dordrecht: Kluwer Academic Publishers. 7–10.

Aha, D.W., D. Kibler and M. Albert (1991). Instance-based learning algorithms. *Machine Learning*, 7. 37–66.

Aho, A.V. (1988). *The AWK Programming Language*. Reading, MA: Addison-Wesley.

Aho, A.V., R. Seth and J.D. Ullman (1986). *Compilers: Principles, Techniques and Tools*. Reading, MA: Addison-Wesley.

Aho, A.V. and J.D. Ullman (1992). *Foundations of Computer Science*. W.H. Freeman and Company.

Alam, Y.S. (1983). A two-level morphological analysis of Japanese. *Texas Linguistic Forum*, 22. 229–252.

Aleksander, I. and H. Morton (1990). *An Introduction to Neural Computing*. Chapman and Hall.

Allen, J., M.S. Hunnicutt and D. Klatt (1987). *From Text to Speech: the MITalk*. Cambridge University Press.

Antworth, E.L. (1990). *PC-KIMMO: A two-level processor for morphological analysis*. Dallas, TX: Summer Institute of Linguistics.

Appelt, A.W. and G.J. Jacobson (1988). The world's fastest scrabble program. *Communications of the ACM*, 31:5. 572–578.

Aston, G. and L. Burnard (1998). *The BNC Handbook: Exploring the British National Corpus with SARA*. Edinburgh: Edinburgh University Press.

Åström, M. (1995). A probabilistic tagger for Swedish using the SUC tagset. In *Proceedings of the Conference on Lexicon + Text*. Lexicographica – Series Maior. Tübingen: Niemeyer.

Atwell, E. (1996). Machine learning from corpus resources for speech and handwriting recognition. In J. Thomas and M. Short, *Using Corpora for Linguistic Research*. London: Longman. 151–166.

Baayen, H. and R. Sproat (1996). Estimating lexical priors for low-frequency morphologically ambiguous forms. *Computational Linguistics*, 22:2. 155–166.

Baker, J. (1979). Trainable grammars for speech recognition. In *Speech communication papers presented at the 97th Meeting of the Acoustical Society of America*. 547–550.

Baker, J.P. (1997). Consistency and accuracy in correcting automatically tagged data. In Garside *et al.* (eds.). 241–250.

Bank of English, Collins COBUILD, Birmingham. Information available from: direct@cobuild.collins.co.uk.

Barton, G.E. (1986). Computational complexity in two-level morphology. In *Proceedings of the 24th Annual Meeting of the Association for Computational Linguistics (ACL-86), New York*. 53–59.

Baum, L.E. (1972). An inequality and associated maximization technique in statistical estimation for probabilistic functions of a Markov process. *Inequality*, 3. 1–8.

Beale, A.D. (1988). Lexicon and grammar in probabilistic tagging of written English. In *Proceedings of the 26th Annual Meeting of the Association for Computational Linguistics (ACL-88), Buffalo*. 211–216.

Beesley, K.R. (1996). Arabic finite-state morphological analysis and generation. In *Proceedings of COLING-96, Copenhagen*. 89–94.

Bel N., N. Calzolari and M. Monachini (coords.) (1995). *Common Specifications and Notation For Lexicon Encoding*, MULTEXT D-1.6-B Deliverable. Pisa: ILC.

Berghmans, J. (1994). *WOTAN: WOordklasse TAgger Nederlands*, M.Sc. Thesis, Department of Language and Speech, University of Nijmegen.

Biber, D. (1993). Using Register-Diversified Corpora for General Language Studies. *Computational Linguistics*, 19:2. 219–241.

Bindi R., M. Monachini and P. Orsolini (1991). *Italian Reference Corpus*, NERC Technical Report. Pisa: ILC.

Bishop, C.M. (1995). *Neural Networks for Pattern Recognition*. Oxford: Oxford University Press.

Black, E., R. Garside and G. Leech (eds.) (1993). *Statistically-Driven Computer Grammars of English: The IBM/Lancaster Approach*. Amsterdam and Atlanta: Rodopi.

Black, E., F. Jelinek, J. Lafferty, R. Mercer and S. Roukos (1992). Decision Tree Models Applied to the Labeling of Text with Parts-of-Speech. In *Proceedings of the 1992 DARPA Workshop on Speech and Natural Language Processing*. Morgan Kaufman.

BNC-British National Corpus. Oxford Computing Services, 13 Banbury Road, Oxford.

Bodmer, F. (1994). *WP3 - Converter & Loader D6*, MLAP93-21 MECOLB Final Report WP3. Mannheim: IDS.

Breiman, L., J. Friedman, R. Ohlsen and C. Stone (1984). *Classification and regression trees*. Belmont, CA: Wadsworth International Group.

Brill, E. (1992). A simple rule-based part of speech tagger. In *Proceedings of the Third Conference on Applied Natural Language Processing (ANLP'92), Trento*. 152–155.

Brill, E. (1994). Some advances in transformation-based part-of-speech tagger. In *Proceedings of the Twelfth National Conference on Artificial Intelligence (AAAI'94), Seattle, Washington*. 722–727.

Brill, E. (1995). Transformation-based error-driven learning and Natural Language Processing: a case study in part-of-speech tagging. *Computational Linguistics*, 21:4.

Brill, E. and M. Pop (to appear). Unsupervised learning of disambiguation rules for part of speech tagging. In *Natural Language Processing Using Very Large Corpora*. Dordrecht: Kluwer Academic Publishers.

Brill, E. and Jun Wu (1998). Classifier combination for improved lexical disambiguation. In *Proceedings of COLING-ACL-98, Montreal*. 191–195.

Brown, P., J. Cocke, S. DellaPietra, V. DellaPietra, F. Jelinek, R. Mercer and P. Roossin (1988). A statistical approach to language translation. In *Proceedings of COLING-88, Budapest*. 71–76.

Brown, P.F., V.J. DellaPietra, P.V. DeSouza, J.C. Lai and R.L. Mercer (1992). Class based n-gram models of natural language. *Computational Linguistics*, 18. 467–479.

Calzolari, N. (1994). European efforts towards standardizing language resources. In P. Steffens (ed.), *Machine Translation and the Lexicon*. Berlin: Springer. 121–130.

Calzolari, N., M. Baker and J.G. Kruyt (eds.) (1995). *Towards a network of European reference corpora, Linguistica Computazionale, Vol. XI*. Pisa: Giardini Editori.

Calzolari, N. and J. McNaught (1996). Editor's Introduction. In *EAG-EB-FR1*. Pisa: ILC.

Calzolari, N. and M. Monachini (1994). *Synopsis and Comparison of Morphosyntactic Phenomena Encoded in Lexicons and Corpora. A Common Proposal and Applications to European Languages*. Pisa: ILC.

Calzolari, N. and M. Monachini (1996). EAGLES Proposal for Morphosyntactic Standards: in view of a ready-to-use package. In G. Perissinotto (ed.), *Research in Humanities Computing*, vol. 5. Oxford: OUP. 48–64.

Calzolari, N. and A. Zampolli (1994). Standards to make natural language resources shareable resources. In *Proceedings of the International Workshop on Shareable Natural Language Resources, Nara*. 15–21.

Carbonell, J. (ed.) (1990). *Machine Learning: Paradigms and Methods*. Cambridge, MA: MIT Press.

Cardie, C. (1993). Using decision trees to improve case-based learning. In *Proceedings of the Tenth International Conference on Machine Learning, Amherst, MA*. Morgan Kaufman. 25–32.

Cardie, C. (1994). *Domain-Specific Knowledge Acquisition for Conceptual Sentence Analysis*, Ph.D. Thesis, University of Massachusetts, Amherst, MA.

Cardie, C. (1996). Embedded machine learning systems for Natural Language Processing: a general framework. In S. Wermter, E. Riloff and G. Scheler (eds.), *Connectionist, Statistical and Symbolic Approaches to Learning for Natural Language Processing*, Lecture Notes in Artificial Intelligence. Berlin: Springer. 315–328.

Cerf-Danon, H. and M. El-Beze (1991). Three different probabilistic language models: comparison and combination. In *ICASSP 1991. IEEE International Conference on Acoustics Speech and Signal Processing, Toronto*. 297–300.

Chang, C.-H. and C.-D. Chen (1993). HMM-based part-of-speech tagging for Chinese corpora. In *Proceedings of the Workshop on Very Large Corpora (WVLC), Columbus, Ohio*. 107–120.

Chanod, J.-P. (1994). *Finite-State Composition of French Verb Morphology (MLTT-005)*. Grenoble: Rank Xerox Research Centre.

Chanod, J.-P. and P. Tapanainen (1995a). Tagging French: comparing a statistical and a constraint-based method. In *Proceedings of the Seventh Conference of the European Chapter of the Association for Computational Linguistics (EACL-95), Dublin*. 149–156.

Chanod, J.-P. and P. Tapanainen (1995b). Creating a tagset, lexicon and guesser for a French tagger. In Tzoukermann and Armstrong (eds.). 58–64.

Christ, O. (1994). A modular and flexible architecture for an integrated corpus query system. In *Proceedings of COMPLEX'94: 3rd Conference on Computational Lexicography and Text Research, Budapest*. 23–32.

Church, K. (1988). A stochastic parts program and noun phrase parser for unrestricted text. In *Proceedings of the Second Conference on Applied Natural Language Processing (ANLP'88), Austin, Texas*. 136–143.

Church, K. (1992). Current practice in part of speech tagging and suggestions for the future. In Simmons (ed.), *Sbornik Praci: In Honor of Henry Kučera*. Michigan: Michigan Slavic Studies. 13–48.

Church, K. and P. Hanks (1990). Word association norms, mutual information and lexicography. *Computational Linguistics*, 16:1. 22–29.

Cloeren, J. (1993). Towards a cross-linguistic tagset. In *Proceedings of the Workshop on Very Large Corpora (WVLC), Columbus, Ohio*. 30–39.

Cloeren, J. (1994). *The Minimal Tagset for Morphosyntactic Encoding within MECOLB*, MLAP93-21 MECOLB Final Report WP5. Nijmegen: Department of Language and Speech, University of Nijmegen.

Collins COBUILD English Language Dictionary (1987). London: Harper Collins.

Corazzari O. and M. Monachini (1995). *ELSNET Italian Corpus Sample*, Technical Report. Pisa: ILC.

Cost, S. and S. Salzberg (1993). A weighted nearest neighbour algorithm for learning with symbolic features. *Machine Learning*, 10. 57–78.

Cowie, J., J. Guthrie and L. Guthrie (1992). Lexical disambiguation using simulated annealing. In *Proceedings of COLING-92, Nantes*. 359–365.

Cussens, J. (1997). Part-of-speech tagging using Progol. In N. Lavrac and S. Dzeroski (eds.), *Inductive Logic Programming: Proceedings of the 7th International Workshop (ILP-97)*, Lecture Notes in Artificial Intelligence 1297. Berlin: Springer. 93–108.

Cussens, J., D. Page, S. Muggleton and A. Srinivasan (1997). Using Inductive Logic Programming for Natural Language Processing. In Daelemans *et al.* (eds.). 25–34.

Cutting, D. (1994). Porting a stochastic part-of-speech tagger to Swedish. In R. Eklund (ed.), *Proc. 9:e Nordiska Datalingvistikdagarna, Stockholm 3-5 June 1993*. Department of Linguistics, Computational Linguistics, Stockholm University, Stockholm. 65–70.

Cutting, D., J. Kupiec, J. Pedersen and P. Sibun (1992). A practical part-of-speech tagger. In *Proceedings of the Third Conference on Applied Natural Language Processing (ANLP'92), Trento*. 133-140.

Daelemans, W. (1995). Memory-based lexical acquisition and processing. In P. Steffens (ed.), *Machine Translation and the Lexicon*, Lecture Notes in Artificial Intelligence 898. Berlin: Springer. 85–98.

Daelemans, W., A. Van den Bosch and A. Weijters (1997). IGTree: using trees for compression and classification in lazy learning algorithms. In *Artificial Intelligence Review*, 11, Special Issue on Lazy Learning. 407–423.

Daelemans, W., J. Zavrel, P. Berck and S. Gillis (1996). MBT: a memory-based part of speech tagger-generator. In E. Ejerhed and I. Dagan (eds.), *Proceedings of the Fourth Workshop on Very Large Corpora (WVLC-4), Copenhagen*. 14–27.

Daelemans, W., A. Van den Bosch and A. Weijters (eds.) (1997). *Workshop Notes of the ECML/MLnet Workshop on Empirical Learning of Natural Language Processing Tasks*, Prague.

Daelemans, W., A. Van den Bosch and J. Zavrel (1999). Forgetting exceptions is harmful in language learning. *Machine Learning*, 11, Special Issue on Natural Language Learning. 11–43.

DeHaspe, L. and L. DeRaedt (1997). Mining a natural language corpus for multi-relational association. In Daelemans *et al.* (eds.). 35–48.

DeRose, S.J. (1988). Grammatical category disambiguation by statistical optimization. *Computational Linguistics*, 14:1. 31–39.

Derouault, A.-M. and B. Merialdo (1984). Language modeling at the syntactic level. In *Proceedings of the International Conference on Pattern Recognition, Montreal, Canada*. 1373–1375.

Elworthy, D. (1994). Does Baum-Welch re-estimation help taggers? In *Proceedings of the Fourth Conference on Applied Natural Language Processing (ANLP'94), Stuttgart*. 53–58.

Engel, U. (1988). *Deutsche Grammatik*. Heidelberg: Groos.

Fausett, L.V. (1994). *Fundamentals of Neural Networks: Architectures, Algorithms and Applications*. Prentice Hall.

Feldweg, H. (1995). Implementation and evaluation of a German HMM for POS disambiguation. In Tzoukermann and Armstrong (eds.). 41–46.

Fligelstone, S., M. Pacey and P. Rayson (1997). How to generalize the task of annotation. In Garside *et al.* (eds.). 122–136.

Francis, N.W. and H. Kučera (1982). *Frequency Analysis of English Usage: Lexicon and Grammar*. Boston: Houghton Mifflin.

Fries, C. (1952). *The Structure of English*. New York: Harcourt Brace.

Garside, R. and N. Smith (1997). A hybrid grammatical tagger: CLAWS4. In Garside *et al.* (eds.). 102–121.

Garside, R., G. Leech and A. McEnery (eds.) (1997). *Corpus Annotation*. London and New York: Longman.

Garside, R., G. Leech and G. Sampson (eds.) (1987). *The Computational Analysis of English: A Corpus-Based Approach*. London and New York: Longman.

Gaussier, E. and J.M. Langé (1994). Some methods for the extraction of bilingual termininology. In *Proceedings of the International Conference on New Methods in Language Processing (NeMLaP), Manchester*. 242–247.

GENELEX Consortium (April 1993). *Couche Morphologique*, Version 3.0. ASSTRIL, Gsi-Erli, IBM France, SEMA GROUP.

GENELEX Consortium (September 1993). *Couche Syntaxique, Les Unites Syntaxique Simple*, Tome 1, Version 3.0. ASSTRIL, Gsi-Erli, IBM France, SEMA GROUP.

Greenbaum, S. (1992). *The ICE Tagset Manual*. London: University College London.

Greene, B. and G. Rubin (1971). *Automatic Grammatical Tagging of English*. Providence: Brown University.

Grishman R. and B. Sunheim (1996). Message Understanding Conference - 6: 'A brief history'. In *Proceedings of COLING-96, Copenhagen*. 466–471.

Gros, J., F. Mihelic and N. Pavesic (1994). Sentence hypothesization in a speech recognition and understanding system for the Slovene spoken language. In *Proceedings of the AISB Workshop on Computational Linguistics for Speech and Handwriting Recognition, Leeds*. 91–96.

Gsi-Erli (1993). *Le Dictionnaire AlethDic*. Paris: Gsi-Erli.

Güngördü, Z. and K. Oflazer (1995). Parsing Turkish using the Lexical-Functional Grammar Formalism. *Machine Translation*, 11:4. 293–319.

Hakkani, D.Z and K. Oflazer (1998). Tactical generation in a free constituent order language. In *Natural Language Engineering*, 4. 115–134.

van Halteren, H. (1996). Comparison of tagging strategies, a prelude to democratic tagging. In Hockey and Ide (eds.). 207–215.

van Halteren, H. and N. Oostdijk (1993). Towards a syntactic database: the TOSCA analysis system. In Aarts *et al.* (eds.). 145–161.

van Halteren, H., J. Zavrel and W. Daelemans (1998). Improving data driven wordclass tagging by system combination. In *Proceedings of COLING-ACL-98, Montreal*. 491–497.

Hankamer, J. (1986). Finite state morphology and left to right phonology. In *Proceedings of the Fifth West Coast Conference on Formal Linguistics 5, Stanford University*. 29–34.

Hankamer, J. (1989). Morphological parsing and the lexicon. In W. Marslen-Wilson (ed.), *Lexical Representation and Process*. MIT Press. 392–406.

Hanlon, S. (1994). *A Computational Theory of Contextual Knowledge in Machine Reading*, Ph.D. Thesis, School of Computer Studies, Leeds University.

Harris, Z. (1962). *String Analysis of Language Structure*. The Hague: Mouton and Co.

Hearst, M. (1991). Toward noun homonym disambiguation – using local context in large text corpora. In L. Jones (ed.), *Using Corpora, Proceedings of the Seventh Annual Conference of the UW Centre for the New OED and Text Research*. University of Waterloo and Oxford University Press. 1–22.

Heid U. (1996). About this document. In Teufel (1996a).

Heid U. and J. McNaught (eds.) (1991). Eurotra-7 Study: *Feasibility and Project Definition Study on the Reusability of Lexical and Terminological Resources in Computerised Applications*, Eurotra-7 Final Report. Stuttgart.

van Herweijnen (1993). *The SGML Tutorial, version 1.0*. Dordrecht: Kluwer Academic Publishers.

Hindle, D. (1989). Acquiring disambiguation rules from text. In *Proceedings of the 27th Annual Meeting of the Association for Computational Linguistics (ACL-89), Vancouver*. 268–275.

Hockey, S. and N. Ide (eds) (1996). *Research in Humanities Computing 4. Selected Papers from the ALLC/ACH Conference, Christ Church, Oxford, April 1992*. Oxford: Clarendon Press.

Holstege, M., Y.J. Inn and L. Tokuda (1991). Visual parsing: an aid to text understanding. In *RIAO. Recherche s'Informations Assist'ee par Ordinateur 1991*, Paris: CID. 175–193.

Hopcroft, J.E. and J.D. Ullman (1979). *Introduction to Automata Theory, Languages and Computation*. Reading, MA: Addison-Wesley.

Hughes, J. (1992). Automatic word classification. Paper presented at the ALLC-ACH conference, Christ Church, Oxford, 1992.

Hughes, J. and E. Atwell (1993). Automatically acquiring and evaluating a classification of words. In *Proceedings of the IEE Colloquium on Grammatical Inference: Theory, Applications and Alternatives, University of Essex.*

Hughes, J., C. Souter and E. Atwell (1995). Automatic extraction of tagset mappings from parallel-annotated corpora. In Tzoukermann and Armstrong (eds.). 10–17.

Hunt, E., J. Marin and P. Stone (1966). *Experiments in Induction.* New York: Academic Press.

Janssen, S. (1990). Automatic word sense disambiguation in LDOCE. In J. Aarts and W. Meijs (eds.), *Theory and Practice in Corpus Linguistics.* Amsterdam: Rodopi. 105–135.

Jelinek, F. (1990). Self-organized language modeling for speech recognition. In A. Weibel and K. Lee (eds.), *Readings in Speech Recognition.* Los Altos, CA: Morgan Kaufman. 405–505.

Johansson, S. (1986). *The Tagged LOB Corpus: User's Manual.* Bergen: Norwegian Computing Centre for the Humanities.

Johansson, S. and K. Hofland (1989). *Frequency Analysis of English Vocabulary and Grammar: vol. 2, tag combinations and word combinations.* Oxford: Clarendon Press.

Johns, T. (1994). From printout to handout: grammar and vocabulary teaching in the context of data-driven learning. In T. Odlin (ed.), *Perspectives on Pedagogical Grammar.* Cambridge University Press. 293–317.

Joshi, A. and Hopely, P. (1996). A parser from antiquity. In *Natural Language Engineering,* 2:4. 291–294.

Källgren, G. (1996). Linguistic indeterminacy as a source of errors in tagging. In *Proceedings of COLING-96, Copenhagen.* 676–680.

Kaplan, R. and M. Kay (1994). Regular models of phonological rule systems. *Computational Linguistics,* 20:3. 331–378.

Karlsson, F. (1990). Constraint grammar as a framework for parsing running text. In *Proceedings of COLING-90, Helsinki.* 168–173.

Karlsson, F. (1995). The formalism and environment of Constraint Grammar parsing. In Karlsson *et al.* (eds.). 41–88.

Karlsson, F., A. Voutilainen, J. Heikkilä and A. Anttila (eds.) (1995). *Constraint Grammar. A Language-Independent System for Parsing Unrestricted Text.* Berlin and New York: Mouton de Gruyter.

Karp, D., Y. Schabes, M. Zaidel and D. Egedi (1992). A freely available wide coverage morphological analyser for English. In *Proceedings of COLING-92, Nantes.* 955.

Karttunen, L. (1983). KIMMO: a general morphological processor. *Texas Linguistic Forum,* 22. 163–186.

Karttunen, L. (1993). *Finite-State Lexicon Compiler*. XEROX Palo Alto Research Center.

Karttunen, L. (1994). Constructing lexical transducers. In *Proceedings of COLING-94, Kyoto*. 406–411.

Karttunen, L. and K.R. Beesley (1992). *Two-Level Rule Compiler*. XEROX Palo Alto Research Center.

Karttunen, L., J-P. Chanod, G. Grefenstette and A. Schiller (1996). Regular expressions for language engineering. In *Natural Language Engineering*, Vol. 2, Part 4. Cambridge University Press.

Karttunen, L. and K. Wittenburg (1983). A two-level morphological analysis of English. *Texas Linguistic Forum*, 22. 217–228.

Katz, S. (1987). Estimation of probabilities from sparse data for the language model component of a speech recognizer. *IEEE Transactions on ASSP*, 35:3. 400–401.

Keenan, F. (1993). *Large Vocabulary Syntactic Analysis for Text Recognition*, Ph.D. Thesis, Department of Computing, Nottingham Trent University.

Kempe, A. (1994). *A Probabilistic Tagger and an Analysis of Tagging Errors*. Research Report, Institut für Maschinelle Sprachverarbeitung, Universität Stuttgart.

Khan, R. (1983). A two-level morphological analysis of Rumanian. *Texas Linguistic Forum*, 22. 253–270.

Kirk, J.M. (1994). Taking a byte at corpus linguistics. In L. Flowerdew and A.K.K. Tang (eds.), *Entering Text*. Hong Kong: Language Centre, Hong Kong University of Science and Technology. 18–43.

Klein, S. and R. Simmons (1963). A computational approach to grammatical coding of English words. *JACM*, 10. 334–347.

Kolodner, J. (1992). *Case-Based Reasoning*. San Mateo, CA: Morgan Kaufmann.

Koskenniemi, K. (1983). *Two-level morphology: a general computational model for wordform recognition and production*. Helsinki: Department of General Linguistics, University of Helsinki.

Koskenniemi, K. (1990). Finite-state parsing and disambiguation. In *Proceedings of COLING-90, Helsinki*. 229–232.

Koskenniemi, K. and K. Church (1988). Complexity, two-level morphology and Finnish. In *Proceedings of COLING-88, Budapest*. 335–339.

Koster, C.H.A. (1991). Affix Grammars for Natural Languages. In H. Alblas and B. Melichar (eds.), *Attribute Grammars, Applications and Systems*, Springer Lecture Notes in Computer Science 545. Heidelberg: Springer.

Kučera, H. and W.N. Francis (1967). *Computational Analysis of Present-day American English*. Providence: Brown University Press.

Kupiec, J. (1989). Probabilistic models of short and long distance word dependencies in running text. In *Proceedings of the 1989 DARPA Workshop on Speech and Natural Language Processing, Philadelphia*. Morgan Kaufman. 290–295.

Kupiec, J. (1992). Robust part-of-speech tagging using a hidden Markov model. *Computer Speech and Language*, 6.

Langley, P. (1996). *Elements of Machine Learning*. Los Altos, CA: Morgan Kaufmann.

Lee, K.F. (1989). *Automatic Speech Recognition*. Dordrecht: Kluwer Academic Publishers.

Leech, G. (1993). Corpus annotation schemes. *Literary and Linguistic Computing*, 8:4. 275–281.

Leech, G., R. Garside and M. Bryant (1994). CLAWS4: The tagging of the British National Corpus. In *Proceedings of COLING-94, Kyoto*. 622–624.

Leech, G. and A. Wilson (1993). Invitation Draft, Draft EAGLES Document. Lancaster.

Leech, G. and A. Wilson (1994). *Morphosyntactic Annotation*, EAGLES document EAG-CSG/IR-T3.1. Lancaster: Lancaster University.

Leech, G. and A. Wilson (1996). *Recommendations for the Morphosyntactic Annotation of Corpora*, EAGLES Recommendations. Lancaster.

Longman Dictionary of Contemporary English (1978). Harlow: Longman.

Lun, S. (1983). A two-level morphological analysis of French. *Texas Linguistic Forum*, 22. 271–278.

Magerman, D. (1994). *Natural Language Parsing as Statistical Pattern Recognition*, Ph.D. Thesis, Stanford University.

Magerman, D. (1995). Statistical decision tree models for parsing. In *Proceedings of the 33rd Annual Meeting of the Association for Computational Linguistics (ACL-95), Cambridge, MA*. 276–283.

de Marcken, C. (1990). Parsing the LOB corpus. In *Proceedings of the 28th Annual Meeting of the Association for Computational Linguistics (ACL-90), Newark*. 243-251.

Marcus, M., B. Santorini and M.A. Marcinkiewicz (1993). Building a large annotated corpus of English: the Penn Treebank. *Computational Linguistics*, 19:2. 313–330.

Màrquez, Lluís, and Horacio Rodríguez (1998). Part-of-speech tagging using decision trees. In Claire Nédellec and Céline Rouveirol (eds.), *Machine Learning: ECML-98*, Lecture Notes in Artificial Intelligence 1398. Berlin: Springer. 25–36.

Marshall, I. (1983). Choice of grammatical word-class without global syntactic analysis: tagging words in the LOB Corpus. *Computers in the Humanities*, 17. 139–150.

Marshall, I. (1987). Tag selection using probabilistic methods. In Garside *et al.* (eds.). 42–56.

McEnery, A. and P. Rayson (1997). A corpus/annotation toolbox. In Garside *et al.* (eds.). 194–208.

McEnery, A. and A. Wilson (1994). The role of corpora in computer assisted language learning. *CALL*, 6. 233–248.

Merialdo, B. (1994). Tagging English text with a probabilistic model. *Computational Linguistics*, 20:2. 155–172.

Mikheev, A. (1996). Unsupervised learning of word-category guessing rules. In *Proceedings of the 34th Annual Meeting of the Association for Computational Linguistics (ACL-96), Santa Cruz*. 62–70.

Miller, G.A., R. Beckwith, C. Fellbaum, D. Gross and K. Miller (1993). *Introduction to WordNet: An On-line Lexical Database*. Cognitive Science Laboratory, Princeton University. Available at: http://www.uni-stuttgart.de.

Milne, R. (1986). Resolving lexical ambiguity in a deterministic parser. *Computational Linguistics*, 12:1. 1–12.

Mohri, M. (1997). On the use of sequential transducers in Natural Language Processing. In E. Roche and Y. Schabes (eds.), *Finite-State Language Processing*. MIT Press.

Monachini, M. (1996). *ELM-IT: EAGLES Specifications for Italian Morphosyntax, Lexicon Specifications and Classification Guidelines*, EAGLES Guidelines. Pisa: ILC.

Monachini, M. and N. Calzolari (1994). *Application of EAGLES Proposal for Morphosyntactic Encoding to Italian Lexicon and Corpus*, EAGLES Input Document. Pisa: ILC.

Monachini, M. and N. Calzolari (1996). *Synopsis and Comparison of Morphosyntactic Phenomena Encoded in Lexicons and Corpora. A Common Proposal and Application to European Languages*, EAGLES Recommendations. Pisa: ILC.

Monachini, M. and A. Östling (1992a). *Morphosyntactic Corpus Annotation – A Comparison of Different Schemes*, NERC-WP8-60. Pisa: ILC.

Monachini, M. and A. Östling (1992b). *Towards a Minimal Standard for Morphosyntactic Corpus Annotation*, NERC-WP8-61. Pisa: ILC.

Monachini, M. (coord.) (1995). *Common Specifications and Notation for Lexicon Encoding of Eastern Languages*, Deliverable D1.1 COP Project 106 MULTEXT-East. Pisa: ILC.

Monachini, M. (coord.) (1996). *Lexicon: Morphosyntactic Specifications and Language Specific Instantiations*, MLAP-PAROLE Deliverable WP4.2.2. Pisa: ILC.

Muggleton, S. and L. De Raedt (1994). Inductive Logic Programming: theory and methods. *Journal of Logic Programming*, 19-20. 629–679.

MULTEXT Consortium (1993). *MULTEXT, Technical Annex.*

MULTILEX Consortium (1993). *Standards for Multifunctional Lexicon*. CAP GEMINI, Philips, Univ. of Surrey, Univ. of Bochum, Univ. of Münster.

Nagata, M. (1994). A stochastic Japanese morphological analyser using a Forward-DP Backward-A*N-Best search algorithm. In *Proceedings of COLING-94, Kyoto*. 201–207.

Nakamura, M., K. Maruyama, T. Kawabata and K. Shikano (1990). Neural network approach to word category prediction for English texts. In *Proceedings of COLING-90, Helsinki*. 213–218.

Natarajan, B. (1991). *Machine learning: a theoretical approach*. San Mateo, CA: Morgan Kaufmann.

Nunberg, G. (1990). *The Linguistics of Punctuation*, C.S.L.I. Lecture Notes, Number 19. Stanford, CA: Center for the Study of Language and Information.

Oflazer, K. (1993). Two-level description of Turkish morphology. In *Proceedings of the Sixth Conference of the European Chapter of the Association for Computational Linguistics (EACL-93), Utrecht*. 472.

Oflazer, K. (1994). Two-level description of Turkish morphology. *Literary and Linguistic Computing* 9:2.

Oflazer, K. (1996). Error-tolerant finite-state recognition with apllications to morphological analysis and spelling correction. *Computational Linguistics*, 22:1. 73–90.

Oflazer, K. and I. Kuruöz (1994). Tagging and morphological disambiguation of Turkish text. In *Proceedings of the Fourth Conference on Applied Natural Language Processing (ANLP'94), Stuttgart*. 144–149.

Oflazer, K. and G. Tür (1996). Combining hand-crafted rules and unsupervised learning in constraint-based morphological disambiguation. In *Proceedings of the ACL-SIGDAT Conference on Empirical Methods in Natural Language Processing, Philadelphia, Pennsylvania*. 69–81.

Oostdijk, N. (1991). *Corpus Linguistics and the Automatic Analysis of English*. Amsterdam: Rodopi.

Oostdijk, N. and P. de Haan (1994). Introduction. In Oostdijk and de Haan (eds.). 5–9.

Oostdijk, N. and P. de Haan (eds.) (1994). *Corpus-based Research into Language*. Amsterdam: Rodopi.

PAROLE (1994). *Preparatory Action for Linguistic Resources Organization for Language Engineering, Technical Annex*. Pisa: ILC.

Pereira, Tishby and Lee (1993). Distributional clustering of English words. In *Proceedings of the 31th Annual Meeting of the Association for Computational Linguistics (ACL-93). Columbus, Ohio*. 183–190.

Quinlan, J. (1993). *C4.5: Programs for Machine Learning*. San Mateo, CA: Morgan Kaufmann.

Rabiner, L.R. and B.H. Juang (1986). An introduction to hidden Markov models. *IEEE ASSP magazine*, Januari 1986. 4–16.

Ratnaparkhi, A. (1996). A maximum entropy part-of-speech tagger. In *Proceedings of the ACL-SIGDAT Conference on Empirical Methods in Natural Language Processing, Philadelphia, Pennsylvania*. 17–18.

Reilly, R and N. Sharkey (eds.) (1992). *Connectionist Approaches to Natural Language Processing*. Hove: Erlbaum.

Resnik, P. (1995). Disambiguating noun groupings with respect to WordNet senses. In *Proceedings of the Third Workshop on Very Large Corpora (WVLC-3). Cambridge, MA*. 54–68.

Revuz, D. (1991). *Dictionnaires et Lexiques, Methodes et Algorithmes*, Ph.D. Thesis, Paris: Université Paris.

Ritchie, G.D., G.J. Russell, A.W. Black and S.G. Pulman (1992). *Computational Morphology*. Cambridge, MA: MIT Press.

Roche, E. (1992). Text disambiguation by finite-state automata, an algorithm and experiments on corpora. In *Proceedings of COLING-92, Nantes*. 993–997.

Roche, E. and Y. Schabes (1995). Deterministic part-of-speech tagging with finite-state transducers. *Computational Linguistics*, 21:2.

Rumelhart, D.E., G.E. Hinton and R.J. Williams (1986). Learning internal representations by error propagation. In Rumelhart and McClelland (eds.), *Parallel Distributed Processing, Volume 1*. Cambridge, MA: MIT Press. 318–362.

Salzberg, S. (1990). A nearest hyperrectangle learning method. *Machine Learning*, 6. 251–276.

Samuelsson, C. (1995). A novel framework for reductionistic statistical parsing. In *Proceedings of the 4th International Workshop on Parsing Technologies (IWPT'95), Prague/Karlovy Vary*. 208–215.

Samuelsson, C., P. Tapanainen and A. Voutilainen (1996). Inducing Constraint Grammars. In Miclet and de la Higuera (eds.), *Grammatical Inference: Learning Syntax from Sentences*, Lecture Notes in Artificial Intelligence 1147. Berlin: Springer Verlag. 146–155.

Samuelsson, C. and A. Voutilainen (1997). Comparing a linguistic and a stochastic tagger. In *Proceedings of the 35th Annual Meeting of the Association for Computational Linguistics and the Eighth Conference of the European Chapter of the Association for Computational Linguistics (EACL-ACL-97), Madrid*. 246–253.

Sánchez León, F. (1995). *CRATER-Final Documentation Package*. Madrid.

Santalla, P. and J. Cloeren (1995). *Esquema de Anotación Morfosintáctica para el Corpus de Referencia del Español Actual*, Contribution to Parole-WP4. Madrid: Royal Spanish Academy.

Schabes, Y., M. Roth and R. Osborne (1993). Parsing the Wall Street Journal with the Inside-Outside Algorithm. In *Proceedings of the Sixth Conference of the European Chapter of the Association for Computational Linguistics (EACL-93), Utrecht*. 341–347.

Schachter, P. (1985). Part-of-speech systems. In T. Shopen (ed.), *Language Typology and Syntactic Description. Vol. 1: Clause Structure*. Cambridge University Press.

Shieber, S.M. (1986). *An Introduction to Unification-based Approaches to Grammar*, CSLI Lecture notes. CSLI.

Schmid, H. (1994a). Part-of-speech tagging with neural networks. In *Proceedings of COLING-94, Kyoto*. 172–176.

Schmid, H. (1994b). Probabilistic part-of-speech tagging using decision trees. In *Proceedings of the International Conference on New Methods in Language Processing (NeMLaP), Manchester*. 44–49.

Schütze, H. (1993). Part-of-speech induction from scratch. In *Proceedings of the 31th Annual Meeting of the Association for Computational Linguistics (ACL-93), Columbus, Ohio*. 251–258.

Scott, M. (1996). *Wordsmith Tools*. Oxford: Oxford University Press.

Shannon, C.E. (1951). Prediciton of printed English. *Bell Syst. Techn. Journal*, Januari 1951. 50–64.

Sharkey, N. (1992). *Connectionist Natural Language Processing: Readings from Connection Science*. Dordrecht: Kluwer Academic Publishers.

Silberzstein, M. (1994). INTEX: a corpus processing system. In *Proceedings of COLING-94, Kyoto*. 579–584.

Sinclair, J. (1991). *Corpus, Concordance, Collocation*. Oxford: Oxford University Press

Smadja, F. (1990). Automatically extracting and representing collocations for language generation. In *Proceedings of the 28th Annual Meeting of the Association for Computational Linguistics (ACL-90), Pittsburgh*. 252–259.

Smith, N. (1997). Improving a tagger. In Garside *et al.* (eds.). 137–150.

Sperberg-McQueen, C.M. and L. Burnard (1994). *Guidelines for Electronic Text Encoding and Interchange*, TEI P3.

Sproat, R. (1992). *Morphology and Computation*. Cambridge, MA: MIT Press.

Stanfill, C. and D. Waltz (1986). Toward memory-based reasoning. *Communications of the ACM*, 29. 1212–1228.

Summers, D. (1996). Computer lexicography: the importance of representativeness in relation to frequency. In J. Thomas and M. Short (eds.), *Using Corpora for Language Research*. London: Longman. 260–266.

Svartvik, J. and M. Eeg-Oloffson (1982). Tagging the London-Lund Corpus of Spoken English. In S. Johansson (ed.), *Computer Corpora in English Language Research*. Bergen: Norwegian Computing Centre for the Humanities. 85–109.

Tapanainen, P. (1996). *The Constraint Grammar Parser CG-2*. Department of General Linguistics, University of Helsinki

Tapanainen, P. and A. Voutilainen (1994). Tagging accurately – Don't guess if you know. In *Proceedings of the Fourth Conference on Applied Natural Language Processing (ANLP'94), Stuttgart*. 47–52.

TEI AI 1W2 (1991). *List of Common Morphological Features for Inclusion in TEI Starter Set of Grammatical-Annotation Tags*.

Teufel, S. (1995). *Some Ideas on Meta-Properties of the EAGLES Suggestions*, EAGLES discussion document. Stuttgart.

Teufel, S. (1996a). *ELM-DE: EAGLES Specifications for German Morphosyntax*, EAGLES Guidelines. Stuttgart.

Teufel, S. (1996b). *ELM-EN: EAGLES Specifications for English Morphosyntax*, EAGLES Guidelines. Stuttgart.

Thielen, C. and A. Schiller (1996). Ein kleines und erweitertes Tagset fürs Deutsche. In *Lexikon + Text*, Lexicographica – Series Maior, Bd. 73. Tübingen: Niemeyer.

Tribble, C. and G. Jones (1990). *Concordances in the Classroom*. Harlow: Longman.

Tzoukermann, E. and S. Armstrong (eds.) (1995). *From Texts to Tags: Issues in Multilingual Language Analysis: Proceedings of the ACL SIGDAT Workshop, Dublin*. Geneva: ISSCO.

Tzoukermann, E., D. Radev and W. Gale (1995). Combining linguistic knowledge and statistical learning in French part-of-speech tagging. In Tzoukermann and Armstrong (eds.). 51–57.

Uit den Boogaart, P.C. (ed.) (1975). *Woordfrequenties in geschreven en gesproken Nederlands*. Utrecht: Oosthoek, Scheltema & Hoeksema.

Utgoff, P.E. (1989). Incremental induction of decision trees. *Machine Learning*, 4. 161–186.

Veronis, J. and N. Ide (1990). Word sense disambiguation with very large neural networks extracted from machine readable dictionaries. In *Proceedings of COLING-90, Helsinki, Volume 2*. 389–394.

Veronis, J., L. Khuori and C. Meunier (1994). *Proposal for Morphosyntactic Encoding in MULTEXT*. Aix-en-Provence.

Viterbi, A.J. (1967). Error bounds for convolutional codes and an asymptotically optimum decoding algorithm. *IEEE Transactions on Information Theory*, Vol. IT-13:2. 260–269.

Von Rekowsky, U. (1996). *ELM-FR: EAGLES Specifications for French Morphosyntax*, EAGLES Guidelines. Paris.

Voutilainen, A. (1993). NPtool, a detector of English noun phrases. In *Proceedings of the Workshop on Very Large Corpora (WVLC), Columbus, Ohio*. 42–51.

Voutilainen, A. (1994). *Designing a parsing grammar*, Ph.D. Thesis (Publication No. 22), Department of General Linguistics, University of Helsinki.

Voutilainen, A. (1995a). Experiments with heuristics. In Karlsson *et al.* (eds.). 293–314.

Voutilainen, A. (1995b). A syntax-based part of speech analyser. In *Proceedings of the Seventh Conference of the European Chapter of the Association for Computational Linguistics (EACL-95), Dublin*. 157–164.

Voutilainen, A. and J. Heikkilä (1994). An English Constraint Grammar (ENGCG): a surface-syntactic parser of English. In U. Fries, G. Tottie and P. Schneider (eds.), *Creating and Using English Language Corpora*. Amsterdam and Atlanta: Rodopi. 189–199.

Voutilainen, A., J. Heikkilä and A. Anttila (1992). *Constraint Grammar of English. A Performance-Oriented Introduction*, Publication No. 21, Department of General Linguistics. Helsinki: University of Helsinki.

Voutilainen, A. and T. Järvinen (1995). Specifying a shallow grammatical representation for parsing purposes. In *Proceedings of the Seventh Conference of the European*

Chapter of the Association for Computational Linguistics (EACL-95), Dublin. 210–214.

Voutilainen, A. and P. Tapanainen (1993). Ambiguity resolution in a reductionistic parser. In *Proceedings of the Sixth Conference of the European Chapter of the Association for Computational Linguistics (EACL-93), Utrecht.* 394–403.

al Wadi, D. (1994). *Cosmas-Benutzerhandbuch.* Mannheim: Institut für Deutsche Sprache.

Weischedel, R., M. Meteer, R. Schwartz, L. Ramshaw and J. Palmuzzi (1993). Coping with ambiguity and unknown words through probabilistic models. *Computational Linguistics*, 19:2.

Weiss, S. and C. Kulikowski (1991). *Computer systems that learn.* San Mateo, CA: Morgan Kaufmann.

Wettschereck, D., D.W. Aha and T. Mohri (1996). *A review and comparative valuation of feature weighting methods for lazy learning algorithms*, Technical Report AIC-95-012. Washington, DC: Naval Research Laboratory, Navy Center for Applied Research in Artificial Intelligence.

Wilson, A. and P. Rayson (1993). Automatic Content Analysis of Spoken Discourse: a report on work in progress. In C. Souter and E. Atwell (eds.), *Corpus Based Computational Linguistics.* Amsterdam: Rodopi. 215–226.

Yarowsky, D. (1992). Word-sense disambiguation using statistical models of Roget's categories trained on corpora. In *Proceedings of COLING-92, Nantes.* 454–460.

Zampolli, A. (1995). Introduction. In Calzolari *et al.* (eds.). xi–xxxix.

Zernik, U. and P. Jacobs (1990). Tagging for learning: Collecting thematic relations from corpus. In *Proceedings of COLING-90, Helsinki, Volume 1.* 34–39.

INDEX